A CHEMIST IN THE WHITE HOUSE

FROM THE MANHATTAN PROJECT TO THE END OF THE COLD WAR

A CHEMIST IN THE WHITE HOUSE

FROM THE MANHATTAN PROJECT TO THE END OF THE COLD WAR

Glenn T. Seaborg

American Chemical Society
Washington, DC

Library of Congress Cataloging-in-Publication Data

Seaborg, Glenn Theodore, 1912–
    A chemist in the White House: from the Manhattan Project to the end of the cold war / Glenn T. Seaborg.
        p. cm.
    Includes bibliographical references and index.
    ISBN 0-8412-3347-0
    1. Seaborg, Glenn Theodore, 1912–   . 2. Science and state—United States. 3. Nuclear energy—Government policy—United States. 4. Science consultants—United States—Biography. 5. Chemists—United States—Biography. I. Title.
QD22.S436A3 1997
327.1´74´092—dc20

[B]

2

ADVISORY BOARD

TABLE OF CONTENTS

PREFACE

In recent years I have often augmented my scientific talks and general lectures to students with a brief description of my

national service under the past 10 presidents of the United States. I usually show slides and share a few reminiscences of particular meetings with the presidents and explain how these contacts reflected the differences among the presidents in individual character and style of administration. Such descriptions have proved to be immensely popular, and many people have urged me to record these experiences in written form.

In doing so now, I rely heavily on my daily diary, which I began keeping as a boy of 14 (during the administration of Calvin Coolidge!), and that I continue to keep in a meticulous manner. The reader will note many detailed entries drawn directly from these journals. These entries have been slightly edited, only for the purpose of adding some information about the people mentioned. The journals are in the process of being printed in a report form by the Lawrence Berkeley National Laboratory. A complete set of these will be available in the Library of Congress. Already deposited there (in 1995) are 90 volumes covering the 64 years from 1927 to 1990, with the exception of three years (1977–79) not yet printed. (The entire collection to be deposited there will total about half a million items, including correspondence, administrative files, drafts of publications, photographs, audiotapes, videotapes, films, minutes of meetings, hand-written notebooks, etc.)

Glenn T. Seaborg
University of California, Berkeley

INTRODUCTION

Beginning with Franklin Delano Roosevelt, I have served, in one capacity or another, the last 10 presidents (before our

current president, Bill Clinton, whom I have also met). I knew each president (except Roosevelt) personally, most of them on a first-name basis. My wife Helen met eight of these presidents (all except Roosevelt and Truman). I have also known each of the corresponding 12 vice presidents, beginning with Henry Wallace (who served during the third presidential term of Roosevelt), and all 10 of the first ladies, including Eleanor Roosevelt. Helen also met the last eight first ladies and most of the vice presidents. I might add that I met Herbert Hoover, Roosevelt's predecessor, on a number of occasions during the years following his presidency, including in the 1950s at the Bohemian Grove Encampment (**photo 1**). Thus I have met more than a quarter of all presidents of the United States.

My role in the discovery of plutonium and the development of the processes for its production and extraction for use as an explosive ingredient for the atomic bomb, which brought an end to World War II, is described in the annotated publication of my diary, *The Plutonium Story: The Journals of Professor Glenn T. Seaborg 1939–1946.*

Toward the end of World War II, I headed the chemistry section at the Metallurgical Laboratory at the University of Chicago, which had the responsibility of extracting and purifying plutonium as part of the Manhattan Engineer District efforts. I was an original signatory of the "Franck Report," which was a recommendation by some Manhattan Project scientists to President Truman to demonstrate for the Japanese government the effects of the atomic bomb on an uninhabited island. Our suggestion was either ignored by or did not reach the president.

My introduction to national

Former President Herbert Hoover (left) with General A.C. Wedemeyer at the Caveman Camp of the Bohemian Grove, CA, in the 1950s. Courtesy of A.C. Wedemeyer and the Hoover Institution Archives, Stanford, CA.

public service came when I was appointed by President Truman to serve on the first General Advisory Committee (GAC) to the United States Atomic Energy Commission (AEC) for a term extending from January 1947 to August 1950. This first GAC played an important role in establishing a number of basic AEC policies. The quotes from my diary (Chapter 2) are at times rather technical, but I believe the gist of the message is, in general, understandable by the lay reader.

While chancellor of the University of California, Berkeley, campus, I was appointed by President Eisenhower in 1959 to be a member of the President's Science Advisory Committee (PSAC), on which I served until January 1961, and to the National Science Board of the National Science Foundation (1960–61). Again, some of the quotes here (Chapter 3) are rather technical, but I believe generally comprehensible. While serving on PSAC, I accepted an assignment to chair a special PSAC panel on basic

research and graduate education. From the work of this panel, we issued the report "Scientific Progress, the Universities, and the Federal Government—A Statement on Basic Research and Graduate Education in the Sciences," which called for a partnership between the federal government and the universities to recognize and support the symbiotic relationship necessary between basic research and graduate education. The report, which became known as the "Seaborg Report," gained widespread attention and laid the foundation whereby the federal government is actively engaged in support and funding of scientific inquiry at research universities.

In January 1961 President Kennedy called me to serve as AEC chairman, a position that I held for more than 10 years while on leave of absence from the University of California. This period was longer than those of any previous chairmen and spanned the terms of John F. Kennedy and Lyndon B.

Johnson as well as part of the term of Richard M. Nixon. In this capacity I also served on such bodies as the National Aeronautics and Space Council, the Federal Council for Science and Technology, the Federal Radiation Council, and the President's Council on Marine Resources and Engineering Development. These parts (Chapters 4, 5, and 6) are by far the most comprehensive of the book, which is an understandable result of my full-time service with those three presidents.

One of my proudest moments as AEC chairman came on Aug. 5, 1963, when, as a member of Secretary of State Dean Rusk's delegation, I witnessed the signing of the Limited Test Ban Treaty (LTBT) in Catherine Hall in the Kremlin. To this day, I still support, lobby, write, and speak to urge worldwide adoption of the comprehensive test ban. I have recounted the numerous meetings, negotiations, events, and activities leading to the signing of the LTBT in my book *Kennedy, Khrushchev, and the Test Ban*, written with my friend and AEC colleague Benjamin S. Loeb; we also wrote a follow-up book *Stemming the Tide—Arms Control in the Johnson Years*, which describes the negotiations leading to the Nonproliferation Treaty of 1968. Our third book, *The Atomic Energy Commission Under Nixon: Adjusting to Troubled Times*, is not so devoted to arms control, but it covers my AEC chairmanship under Nixon rather broadly.

I advised Gerald Ford on our national energy problem and Jimmy Carter on arms control issues.

I served as a member of the National Commission on Excellence in Education (NCEE), which worked under the direction of Chairman David P. Gardner from fall 1981 until April 1983 to produce its report, "A Nation at Risk: The Imperative for Educational Reform." We presented this report to President Reagan on April 26, 1983, and it has had a substantial impact on the national reform movement toward improving the status of precollege education in the United States.

My many contacts with George Bush included a special session to warn him about the pitfalls of "cold fusion."

It has been very exciting to play some role in the making of national policy under different administrations and to have the opportunity to learn something of how priorities are established and goals accomplished in the national government. I believe that, through my visits to some 60 countries, I have also made some contributions toward improving international relations between the United States and other countries. I consider myself privileged to have had the opportunity to become acquainted with the men who have done so much to shape history—most of whom, I am pleased to say, I also liked as individuals. I hope that the reader will enjoy this "cruise down memory lane," with its many first-hand accounts of things as they happened, drawn from my diary.

ACKNOWLEDGMENTS

My thanks to a number of people who helped bring this book to publication. ACS senior acquisitions manager Barbara Pralle, copy editor Elizabeth Wood, production specialists Donna Lucas, Amie Jackowski, Kathleen Strum and designer Melinda Grosser expertly guided the book from concept through production; Department of Energy historian F.S. "Skip" Gosling and Ed Westcott of Oak Ridge National Laboratory were exceptionally resourceful in identifying photographs from the Atomic Energy Commission era. I am especially grateful for the efforts of my staff: Sherrill Whyte, Kristin Balder-Froid, David Yan, Carol Harris Earls, Carole Logie, Rachel Starbuck, Perry Hall, and Nancy Lockhart.

chapter 1   FRANKLIN DELANO ROOSEVELT

1933–1945

A RACE AGAINST THE NAZIS FOR THE ATOMIC BOMB

I begin with a description of the discovery of plutonium and

its nuclear properties, which immediately made it a candidate to serve as the explosive ingredient of an atomic bomb. Soon after the United States entered World War II, I moved to Chicago to be in charge of developing the chemical process to be used with the production of this remarkable element. Because the nuclear fission reaction was discovered in Germany and because of the Germans' established prowess, we believed we were losing the race with Adolf Hitler's scientists to be the first to produce this awesome weapon. The winner of this race could be the winner of the war. President Franklin Delano Roosevelt, aware of plutonium and its potential, kept in touch with progress through his scientific adviser, Vannevar Bush. However, he died just a few months before we successfully reached our objective.

The discovery of fission by the two German chemists, Otto Hahn and Fritz Strassmann, in December 1938 soon led to the recognition that uranium might be developed as the explosive ingredient for a

"The search for transuranium elements, a quest born of scientific curiosity, was destined to be the trigger for a series of events that, within a decade, would rock the world and burst upon the consciousness of every literate human being."

bomb of unprecedented explosive capacity. The Hungarian-born American physicist Leo Szilard became very concerned about this possibility and believed that he should give some direct advice to the President to get the United States started on such a project. With the help of Albert Einstein, Szilard's recommendation reached Roosevelt and resulted in the immediate appointment of an Advisory Committee on Uranium (chaired by Lyman J. Briggs, director of the Bureau of Standards) to investigate the problem.

In June 1941 Roosevelt established, by executive order, the Office of Scientific Research and Development (OSRD), with Bush as director, to better coordinate scientific activities related to the war. OSRD, located within the Office for Emergency Management of the Executive Office of the President, oversaw the atomic bomb development and the National Defense Research Committee (NDRC), under the chairmanship of James B. Conant. The Committee on Uranium became the OSRD's Section on Uranium, soon designated cryptically as the S-1 Section.

By the end of summer 1941, Bush, initially somewhat skeptical, was convinced that the possibility of producing an atomic bomb before the end of the war was so strong that every effort must be made as fast as possible. On Oct. 9, 1941, he met with President Roosevelt and Vice President Henry Wallace to seek authority to proceed at a greatly increased level of intensity with commitments to spend millions of dollars. The president, in an historic decision, agreed both immediately and completely. From this point on, the effort proceeded at an accelerated rate.

One approach was to produce enriched uranium-235 (the fissionable isotope of uranium). (My colleague, Ernest O. Lawrence, the inventor of the cyclotron, was a leader in this effort at Berkeley.) The other approach was to produce 94-239, the fissionable isotope of the newly discovered synthetic element with the atomic number 94. This element and its fissionable isotope had been discovered in early 1941 by my co-workers and me at the University of California, Berkeley (and was soon given the name "plutonium").

The story of plutonium is one of the most dramatic in the history of science. For many reasons this unusual element holds a unique position among the chemical elements. It is a synthetic element, the first realization of the alchemist's dream of large-scale transmutation. It was the first synthetic element to be seen by man. One of its isotopes has special nuclear properties that give it overwhelming importance in the affairs of man. It has unusual and very interesting chemical properties. It is alleged to be a dangerous poison. It was discovered, and methods for its production were developed, during the last war under circumstances that make a fascinating story.

The search for transuranium elements, a quest born of scientific curiosity, was destined to be the trigger for a series of events that, within a decade, would rock the world and burst upon the consciousness of every literate human being. These events were the discoveries that led to the exploitation of nuclear energy, particularly as a weapon of mass destruction. Other fundamental scientific discoveries in the past undoubtedly have had an equal, if not greater, effect on man's existence, but no other has exploded in his face as this one did. The announcement to the world of the existence of plutonium was in the form of the nuclear bomb dropped over Nagasaki.

The discovery of plutonium began in the fall of 1940, when I asked a graduate student, Arthur C.

Wahl, to consider studying the tracer chemical properties of element 93 as a thesis problem, a suggestion he was happy to accept. This, and related work on element 94, was carried on in collaboration with Joseph W. Kennedy, who, like myself, was an instructor in the Department of Chemistry at the University of California, Berkeley. After the departure from Berkeley of Edwin M. McMillan (an assistant professor in the Department of Physics) in November 1940, and his gracious assent to our continuation of his search for and possible identification of element 94, our group turned its attention to this problem.

Our first bombardment of uranium oxide with the 16-MeV deuterons from the 60-in. cyclotron was performed on Dec. 14, 1940. Alpha radioactivity was found to grow into the chemically separated element 93 fraction during the following weeks, and this alpha activity was chemically separated from the neighboring elements (especially elements 90 to 93 inclusive) in experiments performed during the next two months. These experiments, which constituted the positive identification of element 94, showed that this element has at least two oxidation states, distinguishable by their precipitation chemistry, and that it requires stronger oxidizing agents to oxidize element 94 to the upper state than is the case for element 93. The first successful oxidation of element 94, which probably represents the key step in its discovery, was effected through the use of peroxydisulfate ion and silver ion catalyst on Feb. 23–24, 1941, in room 307 of Gilman Hall at the University of California, Berkeley. (This room was dedicated as a national historic landmark on Feb. 21, 1966, the 25th anniversary of the discovery.) The particular isotope identified was shown to be of mass number 238.

In view of its apparent importance, the announcement of this discovery was withheld by the discoverers and the editors of *Physical Review* as the result of self-imposed secrecy, even though this work antedated the time of governmental support. It may be of interest to reproduce here extracts from the letters to the editor submitted for publication in *Physical Review* but not published until 1946. The text of a communication dated Jan. 28, 1941, and published under the authorship of Seaborg, McMillan, Kennedy, and Wahl, reads as follows:

We are writing to report some results obtained in the bombardment of uranium with deuterons in the 60-in. cyclotron. . . .

We did observe the growth of alpha-particles in the very carefully purified, as well as in the semi-purified 93 fractions, and the growth curves indicate a half-life of roughly 2 days for the parent of the alpha-emitter. . . .

This alpha-activity is chemically separable from uranium and 93. The chemical experiments so far indicate a similarity to thorium and the activity has not yet been separated from thorium. More chemical experiments definitely must be performed before it can be regarded as proved that the alpha-particles are due to an isotope of element 94.

The report dated March 7, 1941 (by Seaborg, Wahl, and Kennedy) on the oxidation experiment that occurred on Feb. 24, 1941, reads as follows:

We should like to report a few more results which we have found regarding the element 94 alpha-radioactivity formed in the 16-MeV deuteron bombardment of uranium. We sent a first report of this work in a Letter to the Editor of January 28, 1941. . . . With the help of persulfate ion it has been possible to separate quantitatively this radioactivity from thorium, by using the beta-active UX_1 as an indicator for thorium. These experiments make it extremely probable that this alpha-radioactivity is due to an isotope of element 94. The experiments are being continued.

The chemical properties of elements 93 and 94 were studied by the tracer method at the University of California, Berkeley, for the next year and a half. Our group referred to these first two transuranium elements simply as "element 93" and "element 94," or by code names until spring 1942, when the first detailed reports on them were written. The early work, even in those days, was carried out in secrecy—as a matter of fact, I recall that a code name was often used for element 94, even in oral references. Throughout 1941 we referred to it by the code name of "copper," which was all right until we had to introduce the element copper into some of our experiments; we were then faced with the problem of distinguishing between the two. For a while we referred to plutonium as "copper" and to real copper as "honest-to-God copper." This seemed clumsier and clumsier as time went on, and we finally christened the element "plutonium" and began to call it that. To write the original report on the chemical properties, we had to have chemical symbols for the two elements.

This report, by Wahl and myself, dated March 19, 1942, was mailed as a secret report from Berkeley to the uranium committee (the group, headed by L.J. Briggs, that had been coordinating the early U.S. work on possible practical energy from nuclear fission) in Washington, DC, and was issued as Report No. A-135. It was published in its original form in 1948. Below I quote the section in which the name plutonium was first suggested:

> . . . Naming the Elements. Since formulas are confusing when the symbols "93" and "94" are used, we have decided to use symbols of the conventional chemical type to designate these elements. Following McMillan, who has suggested the name "neptunium" (after Neptune, the first planet beyond Uranus) for element 93, we suggest "plutonium" (after Pluto, the second planet beyond Uranus) for element 94. The corresponding chemical symbols would be Np and Pu. The names "eka-rhenium" and "eka-osmium" seem inappropriate in view of the marked dissimilarity of the chemical properties of elements 93 and 94 to those of rhenium and osmium. . . .

Pluto was discovered in 1930 by Clyde Tombaugh, a 24-year-old astronomer. I had the pleasure of meeting Tombaugh during a visit to Albuquerque, NM, in June 1991.

As a result of these tracer investigations during 1941 and early 1942 at the University of California, Berkeley, a great deal was learned about the chemical properties of plutonium. It was established that plutonium had at least two oxidation states, the higher of which was not carried by lanthanum fluoride or cerium fluoride, whereas the lower state was quantitatively coprecipitated with these compounds. It was established that the higher oxidation state could be obtained by treating the lower state with oxidizing agents such as persulfate and argentic ions, dichromate, permanganate, or periodate, and that the upper state could be reduced to a lower (rare earth fluoride-carriable) state by treating it with sulfur dioxide or bromide ion. It was established that plutonium in aqueous solution is not reduced to the metal by zinc and that plutonium does not form a volatile tetroxide. It was shown that a stable lower state of plutonium—probably plutonium(IV)—was carried by $Th(IO_3)_4$. Ether extraction had been used to separate large amounts of uranyl nitrate from plutonium. Methods had also been devised for the separation of plutonium from elements 90, 91, and 93.

Seaborg and astronomer Clyde W. Tombaugh, discoverer of the planet, Pluto, at a press conference at the Sandia National Laboratories on the occasion of a colloquium Seaborg gave about the 50-year history of discovery of the transuranium elements. Seaborg (age 79) and Tombaugh (age 85) spent time reminiscing with the news media about their discoveries. Seaborg is wearing his "Periodic Table" necktie and Tombaugh is wearing his "Walt Disney's Pluto" wristwatch. Albuquerque, NM. June 10, 1991. Courtesy of Sandia National Laboratories, Albuquerque, NM. Photo by Randy Montoya.

On the basis of these facts and others not mentioned here, it was speculated that plutonium in its highest oxidation state is similar to uranium(VI) and in a lower state is similar to thorium(IV) and uranium(IV). It was reasoned that if plutonium existed normally as a stable plutonium(IV) ion, it would probably form insoluble compounds or stable complex ions analogous to those of similar ions, and that it would be desirable (as soon as sufficient plutonium became available) to determine the solubilities of such compounds as the oxalate, phosphate, fluoride, iodate, and peroxide. Such data were needed to confirm deductions based on the tracer experiments.

We conceived the principle of the oxidation–reduction cycle, as applied to the separations processes that became so useful later. This principle applied to any process involving the use of a substance that carried plutonium in one of its oxidation states but not in another. By using this principle, for example, a carrier could be used to carry plutonium in one oxidation state and thus to separate it from uranium and many of the fission products. Then the carrier and the plutonium could be dissolved, the oxidation state of the plutonium changed, and the carrier reprecipitated with the other fission products, leaving the plutonium in solution. The oxidation state of the plutonium could again be changed and the cycles repeated. With this type of procedure, only a contaminating element having a chemistry nearly identical with the plutonium itself would fail to separate if a large number of oxidation–reduction cycles were used. This principle, of course, applies to other processes such as solvent extraction, adsorption, or volatility methods.

The plutonium isotope of major importance is the one with mass number 239. The search for this isotope, as a decay product of ^{239}Np, was being conducted by the same group, with the collaboration of Emilio Segrè, simultaneously with the experiments leading to the discovery of plutonium. The isotope ^{239}Pu was identified, and its possibilities as a nuclear energy source were established during spring 1941, using a sample prepared by the decay of ^{239}Np produced by irradiating uranium with neutrons from the 60-in. cyclotron and later purified by taking advantage of the then-known chemistry of plutonium.

Using neutrons produced by the 37-in. cyclotron in the University of California Old Radiation Laboratory, the group first demonstrated on March 28, 1941, with the sample containing 0.5 µg of ^{239}Pu, that this isotope undergoes slow neutron-induced fission with a cross section even larger than that of ^{235}U. The sample was placed near the screened window of an ionization chamber, which was imbedded in paraffin near the beryllium target of the 37-in. cyclotron. This gave a small but detectable fission rate when a 6-µA beam of deuterons was used. To increase the accuracy of the measurement of the fission cross section, this sample, which had about 5 mg of rare-earth carrier material, was subjected to an oxidation–reduction chemical procedure that reduced the amount of carrier to a few tenths of a milligram. A fission

cross section for ^{239}Pu some 70% greater than that for ^{235}U was found to be in remarkable agreement with the accurate values determined later.

A report of this work was registered on May 29, 1941, for publication in *Physical Review*, but again the information was voluntarily withheld until 1946, when it was published under the authorship of Kennedy, Seaborg, Segrè, and Wahl:

We would like to report that we have observed the fission of 94^{239} with slow neutrons. The cross section for the fission of 94^{239} with slow neutrons is even larger than that of U^{235}. The cross section was determined by comparing the number of fissions obtained with a sample containing 94^{239} with the number obtained with a sample of ordinary uranium under conditions which were as identical as possible in every detail. . . .[The actual value of the cross section was not declassifiable in 1946.] Worth appending here is the information which we have obtained about the alpha-activity of 94^{239}. During the decay of the 125-millicurie sample of 93^{239} the sample was placed near an ionization chamber connected to a linear, pulse amplifier in order to watch for the growth of alpha-particles. A strong magnetic field was used to bend out the beta-particles. An alpha-particle activity was observed to grow with a half-life of about 2.3 days, which is the half-life to be expected for growth from 93^{239}. The alpha-count grew to the value 240 per minute. After the sample was thinned to 0.16 mg per cm^2 the alpha-count became 800 per minute. Correcting for the geometrical factor, as determined with the aid of a known amount of uranium of the same thickness as this sample, the total alpha-emission of the sample amounted to about 60,000 per minute. This corresponds to a half-life of about 3 × 10^4 years. . . . [This value for the half life is in good agreement with the modern value of 24,000 years.]

This demonstration that ^{239}Pu undergoes fission with thermal neutrons with a large probability, showing that all the neutrons emitted in the process are eligible to cause further fissions, established the great value of this isotope. Recognition of this capability led to the wartime Plutonium Project for large-scale production, with an eye toward its possible use in a nuclear weapon. Below is an excerpt from my diary, Saturday, Dec. 20, 1941.

Arthur Compton wrote a memorandum (addressed to Vannevar Bush, James B. Conant, and Lyman J. Briggs) outlining a theoretical and experimental program to be centered at the University of Chicago and carried on with the cooperation of the groups at Columbia, Princeton, and Berkeley for the production of 94^{239} (designated ^{239}94 by modern convention) for use in a nuclear explosive, through the operation of a nuclear chain reaction with natural uranium. Spurred on by Pearl Harbor, he suggested a speeded-up time schedule for obtaining knowledge of the conditions for a chain reaction by June 1, 1942, production of a chain reaction by Oct. 1, 1942, a pilot plant for use of the chain reaction to produce 94 by Oct. 1, 1943, and the production of usable quantities of 94 by Dec. 31, 1944.

Arthur Compton, director of what became known as the Metallurgical Laboratory at the University of Chicago, asked me to join him there and to take charge of developing the chemical process for separating the plutonium to be produced in the chain reaction. I arrived in Chicago on April 19, 1942 (my 30th birthday).

To understand the production problem, let us summarize the reactions of the two isotopes of uranium with neutrons:

^{235}U + n → fission products + neutrons + energy

^{238}U + n → ^{239}U → ^{239}Np → ^{239}Pu

One question that resulted immediately was, Is it possible by use of the mixture of the uranium isotopes as it occurs in natural uranium, consisting of about 0.7% of fissionable ^{235}U and 99.3 percent of ^{238}U by weight, to cause a chain reaction to occur on a very large scale? If so, the extra neutrons produced in the fission of ^{235}U would be absorbed by ^{238}U to form the desired isotope, ^{239}Pu, in large quantity.

The other question that, of necessity, came up for immediate discussion was, Would it be possible to devise, in a reasonable period of time, a chemical means for separating this ^{239}Pu from the uranium and from the tremendous fission product radioactivities attributable to the many fission product elements that would be present with it?

These two staggering problems formed the basis of the Plutonium Project. Their solutions were to a large extent unrelated, and the development program in connection with each problem was, of course, in the hands of research men in different fields: physics and chemistry.

It is beyond the scope of this book to describe the development of the chain-reacting units called "piles" for the production of plutonium, carried out under the direction of the great Italian nuclear physicist Enrico Fermi as well as Walter Zinn, Eugene Wigner, and others.

The problem of separating the new element plutonium from uranium and fission products might not at first seem difficult, for it was primarily a chemical problem. However, it differed in many ways from ordinary chemical problems, and these differences made the solution of the problem as a whole much more difficult, even though some of the differences actually helped the solution. From the beginning, our limited time seemed the most nearly insurmountable difficulty. It was impossible to complete the design and testing of the process before it had to be placed in operation. Even a simple chemical process usually requires a much longer time to place in large-scale operation than did the plutonium separation process, although the latter cannot be regarded as either simple or short.

The problem that had to be solved during fall 1942 was to develop a separation process that would meet the demanding requirements. The process had to accomplish a separation of plutonium in high yield and purity from many tons of uranium in which the plutonium would be present at a maximum concentration of about 250 parts per million. Because of this low concentration, compounds of plutonium could not be precipitated, and any precipitation separation process had to be based on coprecipitation phenomena (i.e., the use of "carriers" for plutonium). At the same time, the radioactive fission products produced along with plutonium in the uranium (as a result of the fission of ^{235}U) had to be separated so that less than one part in 10^7 parts originally present with the plutonium would exist with the final product from the process. This requirement made it safe to handle the plutonium, for without a separation of the fission products, the plutonium from each ton of uranium would have more than 10^5 curies of energetic gamma radiation associated with it. The process of separating fission products was called "decontamination." Thus a unique feature of the process was the necessity of completely separating a wide variety of elements from the final product and doing so by remote control behind large amounts of shielding to protect operating personnel from the radiation. The decontaminated plutonium compound that was to be provided to those responsible for its ultimate use (Los Alamos Laboratory in New Mexico) had to facilitate those final steps; it

had to be a compound or a solution of small bulk that could be shipped without difficulty, and it had to be of a composition that could be easily subjected to further purification. The separation process had to meet the further requirement that a "critical mass" of plutonium, which would lead to a disruptive nuclear chain reaction, should not accumulate at any step of the procedure.

If large amounts of plutonium had been available in fall 1942, and if its chemistry had been as well known as the chemistry of the more familiar elements, the task of developing the chemical process would still have been a formidable undertaking. Essentially all that was known about plutonium was based on secondary evidence from the tracer experiments involving the infinitesimal amounts of the element, which had been produced entirely by cyclotron bombardments. Tracer chemistry itself was a relatively new science; many of its phenomena were not clearly understood; and deductions based on it were often subject to doubt, particularly when applied to a new element. Added to the difficulty of devising the chemical process was the fact that only a few of the fission products had been identified, and many of these proved to be among the least well known of the chemical elements.

Operation of a chemical process by remote control behind massive shielding made it imperative to thoroughly test the process in advance to minimize the possibility of errors in the design of the equipment. Furthermore, operational errors had to be kept to a minimum, and careful chemical control of the operations had to be maintained. In all of these considerations it was obviously desirable to keep the process to be operated as short and simple as possible. Similarly, the process from the standpoint of plant design should consist of steps requiring the same sort of equipment rather than steps that were so fundamentally different as to require many different types of equipment. At the same time, it seemed advisable to design the process and the equipment in such a way as to facilitate changes in case of failure. All of these requirements were met; in fact, the process was operated more successfully than even the most optimistic dared to hope (at the Hanford plant in the state of Washington, built and operated by DuPont) and from the beginning gave high yields and decontamination factors. The plutonium pilot plant, the Clinton Engineer Works, or Clinton Laboratories (code name "Site X"), named for the neighboring village of Clinton, TN, began operating in fall 1943. A nearby village, given the name of Oak Ridge, was constructed to house personnel associated with the operation; this was the responsibility of the University of Chicago, and personnel from DuPont played a key role in the building and operation of the plant.

Although it was felt that such a separation process would depend on the use of the two oxidation states of plutonium, the actual details, such as the best carrier compounds and best oxidizing and reducing agents, had not yet been discovered. Stanley G. Thompson is largely responsible for the conception and early development of the process. The key is the quantitative carrying of plutonium(IV) from acid solution by bismuth phosphate, an unexpected phenomenon discovered in December 1942, and the expected noncarrying of plutonium(VI) by the same carrier material. This method, known as the Bismuth Phosphate Process, operates as follows: Neutron-irradiated uranium is dissolved in nitric acid and, after the addition of sulfuric acid to prevent the precipitation of uranium, plutonium(IV) is

coprecipitated with bismuth phosphate. The precipitate is dissolved in nitric acid, the plutonium(IV) is oxidized to plutonium(VI), and a byproduct precipitate of bismuth phosphate is formed and removed, while the plutonium(VI) remains in solution. After the reduction of plutonium(VI) to plutonium(IV), the latter is again coprecipitated with bismuth phosphate, and the whole "decontamination cycle" is repeated. At this point the carrier is changed to lanthanum fluoride, and a similar "oxidation–reduction cycle" is performed, using this carrier and thereby achieving further decontamination and concentration. The plutonium at this point is sufficiently concentrated that final purification can be accomplished without the use of carrier compounds, and plutonium peroxide is precipitated from acid solution.

Although the outline of a chemical separation process could be obtained by tracer-scale investigations, the process could not be defined with certainty until study of it was made possible at the actual concentrations of plutonium that would exist in the large-scale separation plants. Such a test was particularly necessary in view of the poor understanding of the mechanism by which plutonium(IV) is carried by bismuth phosphate and the skepticism of many researchers that the carrying would be observed at plutonium concentrations that would exist in the Hanford plant. This test had to be carried out as soon as possible after the discovery of the Bismuth Phosphate Process, and it was performed at the Metallurgical Laboratory in early 1943, following the isolation of plutonium in pure form the previous fall.

The question in summer 1942 was, How could any separations process be tested at the concentrations of plutonium that would exist several years later in the production plants when, at this time, there was

"Would it be possible to devise, in a reasonable period of time, a chemical means for separating this 239 PU from the uranium and from the tremendous fission product radioactivities attributable to the many fission product elements that would be present with it?"

not even a microgram of plutonium available? This problem was solved through an unprecedented series of experiments encompassing two major objectives. First, it was decided to attempt the production of an actually weighable amount of plutonium by bombarding large amounts of uranium with the neutrons from cyclotrons. It must be remembered that never before had weighable amounts of transmutation products been produced with any particle acceleration machine. Even extending this possibility to the limit, it was not anticipated that more than a few micrograms of plutonium could be produced. The second aspect of the solution involved the novel idea of attempting to work with only microgram amounts of plutonium but, at the same time, at ordinary concentrations. It was decided to undertake a program of investigation involving volumes of solutions and weightings on a scale of operations much below that of ordinary microchemistry.

By using extremely small volumes, it was possible to arrange matters so that even microgram quantities could give relatively high concentrations, and by developing balances of the required sensitivity, micrograms were sufficient for quantitative gravimetric measurements. The field that embraced the chemical study of material on this minute scale of operation has been given the name "ultramicrochemistry."

From summer 1942 until fall 1943, cyclotron bombardments were the sole source of plutonium, and during this period of time about 2000 μg, or 2 mg, of plutonium was prepared. Ultramicrochemists used this material to maximum advantage to prepare compounds of plutonium and to measure properties such as solubilities and oxidation potentials. In particular, it was possible—and this was of inestimable importance—to

test the Bismuth Phosphate Process, which was being considered for use at Hanford. The various parts of the complicated separation and isolation procedures were tested at the Hanford concentrations of plutonium in the careful and crucial experiments performed by Burris B. Cunningham, Louis B. Werner, Michael Cefola, Daniel R. Miller, Isadore Perlman, and others. Without the possibility of these tests early in 1943, I believe it is fair to say that this process, which went into use at Hanford and turned out exceedingly well, would not have been chosen.

I want to emphasize that the scale-up between the ultramicrochemical experiments to the final Hanford plant amounts to a factor of about 10^9, surely the greatest scale-up factor ever attempted. In spite of these difficulties the chemical separations process at Hanford was successful from the beginning, and its performance exceeded all expectations. High yields and decontamination factors (separation from fission activity) were achieved in the very beginning and continued to improve with time.

Northwest from the village of Pasco in southeastern Washington, the Columbia River makes a 90° bend around a flat, arid expanse of land, which is bounded on its other two sides by high hills. Within this 600-square-mile area were the several plants that made up the plutonium-manufacturing Hanford Engineer Works, known during wartime by the code name "Site W." The site was chosen with especial care for the needs of this unique enterprise. The mild winters and negligible rainfall allowed unimpeded construction year round. Flanking the area at convenient distances were the Grand Coulee and Bonneville power networks.

Not least among the considerations in choice of site was the remoteness of the region from centers of population, a necessary forethought should events have allowed the chain reactions to get out of control or the chemical plants to spew forth their contents, which were almost inconceivably high in radioactivity. It is a tribute to the painstaking care in planning that this massive undertaking offered the safest of industrial jobs to the operators.

The construction camp for the plant was built at the site of the village of Hanford, from which the plant draws its name. At one time more than 50,000 people lived in trailers and barracks at Hanford, but with the completion of construction work it was totally evacuated. At a greater distance from the plant site, in the existing village of Richland, a new village of permanent dwellings was built to house the plant operators.

There were two principal types of plants at the Hanford Engineer Works: those that formed the plutonium within the uranium by a nuclear chain reaction, and those that separated the plutonium from the uranium and the fission products. The former consisted of piles cooled by the circulation of treated Columbia River water. The huge separation units were strictly chemical plants, operated, like the piles, by remote control.

Operation of the first pile began in September 1944, following some initial trouble caused by the absorption of neutrons in the fission product ^{135}Xe, and the first production run of Hanford material in a chemical separations plant began on Dec. 26, 1944. The final operating standards for the Bismuth Phosphate Process, based on recommendations of the Clinton Laboratories, were reviewed at Hanford by a delegation from the Metallurgical Laboratory between Dec. 12 and Dec. 15, 1944. I was part of the delegation and was responsible for final approval of the standards. As had been true in the early operation of the pilot plant at Oak Ridge, TN, a number of minor operational difficulties were encountered, and certain modifications in sampling methods and analytical procedures were required. The yields in the first plant runs, which took place in December 1944 and January 1945, ranged between 60% and 70%, reached 90% early in February 1945, and were up to 93% by early summer and above 95% soon thereafter. From the very beginning, decontamination factors were better than anticipated and reached the overall value of 10^8. The first delivery of plutonium to the Los Alamos Laboratory (code name "Site Y," under the direction of J. Robert Oppenheimer) occurred on Feb. 2, 1945.

The chemical plants were massive structures ingeniously engineered to fit the grave problems inherent in handling the extremely high levels of radioactivity. It is self-evident that no one saw the plutonium as it entered the plant. It is also true that no one saw it until just before it finally emerged as a relatively pure compound.

In the meantime, it had passed through a maze of reaction vessels via thousands of feet of piping with only instruments and an occasional sampling to chart its progress. Probably no process on a grand scale in chemical engineering history received such painful care in its development and engineering designs as that operating in the Hanford plutonium plant—certainly not in such a brief span of time.

Now I turn to a review of some other aspects of the wartime work on atomic energy.

It became apparent early on that adequate supervisory strength for the atomic bomb project was needed. After some attempts at arrangements that I will not describe here, the army was placed in charge. On Sept. 23, 1942, newly promoted Gen. Leslie R. Groves, chosen by Gen. Brehon B. Somer-

"It is a tribute to the painstaking care in planning that this massive undertaking offered the safest of industrial jobs to the operators."

vell and Gen. Wilhelm D. Styer, was assigned to run the project as part of the Manhattan Engineer District (established on Aug. 13), which became known as the Manhattan Project. Lt. Col. Kenneth B. Nichols became his chief aide. Bush and Conant and members of the S-1 Executive Committee continued in positions of responsibility.

On June 29, 1943, President Franklin Roosevelt wrote the following letter to Gen. Groves, which was read to the members of our laboratory:

June 29, 1943
My dear General Groves:

I have recently reviewed with Dr. Bush the highly important and secret program of research, development and manufacture with which you are familiar. I was very glad to hear of the excellent work which is being done in a number of places in this country under your immediate supervision and the general direction of the Committee of which Dr. Bush is chairman. The successful solution of the problem is of the utmost importance to the national safety, and I am confident that the work will be completed in as short a time as possible as the result of the wholehearted cooperation of all concerned.

I am writing to you as the one who has charge of all the development and manufacturing aspects of this work. I know that there are several groups of scientists working under your direction on various phases of the program. The fact that the outcome of their labors is of such great significance to the nation requires that this project be even more drastically guarded than other highly secret war developments. As you know, I have therefore given directions that every precaution be taken to insure the safety of your project. I am sure the scientists are fully aware of the reasons why their endeavors must be circumscribed by very special restrictions. Nevertheless, I wish you would express to them my appreciation of their willingness to undertake the tasks which lie before them in spite of the possible dangers and personal sacrifice involved. In particular, I should be glad to have you communicate the contents of this letter to the leaders of each important group. I

am sure we can rely on the continued wholehearted and unselfish labors of those now engaged. Whatever the enemy may be planning, American Science will be equal to the challenge. With this thought in mind, I send this note of confidence and appreciation.

Very sincerely yours,
Franklin D. Roosevelt

Much of the early incentive for proceeding with the atomic bomb project came from the United Kingdom. As a result, President Roosevelt included discussions of this project in early meetings with U.K. Prime Minister Winston Churchill and Canadian leaders. This cooperation diminished during the later days of the war.

Crucial questions, the answers to which would determine the successful use of plutonium in a weapon, were whether ^{239}Pu emitted neutrons when it underwent fission and the number of such neutrons emitted. These questions had to be answered early, before any ^{239}Pu was available from the reactors. Most of the supply of our chemistry group at the Metallurgical Laboratory was loaned in summer 1943 to Los Alamos for the purpose, and their initial measurement demonstrating the emission of a sufficient number of fission neutrons was made with this sample, which contained a few hundred micrograms. In returning the sample to the Metallurgical Laboratory, advantage was taken of my presence in Santa Fe on a short vacation trip in order to save the time that would have been lost in a cumbersome official transfer of the material. Robert R. Wilson, who had participated in the neutron measurements, met Mrs. Seaborg and me at breakfast at about 5 a.m. in a Santa Fe restaurant, having escorted the sample from Los Alamos with the implied protection of a high-powered rifle; I then transported it by train in my suitcase, without firearms, back to the Chicago laboratory.

As the project progressed, and in view of the potential applications of plutonium as a source of energy for peacetime applications, Bush approached the discoverers of plutonium and its fissionable isotope (Kennedy, Wahl, Segrè, and me) with respect to the assignment of our patent rights to the government (our basic discoveries had been made in the course of our academic research conducted before we had any government contracts). Thus, as recorded in my diary, Kennedy and I met on April 20, 1944, with Bush and Capt. Robert Lavender, overseer of patent matters for OSRD:

Kennedy and I arrived in Washington in the morning and at 10 a.m. went to Capt. Lavender's office on the second floor of the Carnegie Foundation Building at 1530 P St. He took us to see Vannevar Bush in the same building, who spoke to us about President Roosevelt having directed him to obtain all controlling patents in these matters to facilitate international dealings. Bush emphasized they have no intention of coercing us; however, he said, since the University of California is becoming so demanding, he is forced to be legal and tough with UC.

Also on Jan. 8, 1945, Bush wrote us as follows:

... I have discussed with Captain Lavender the purchase by the Government of rights under certain chemical discoveries that have been the subject of negotiations and I have [preliminarily] considered the details of the discoveries in their relation to the work undertaken by the University of California under contracts, and also as to the parts of the discoveries that were made before the contracts were in existence as well as during the periods of the contracts.

It appears to me that the extent of the rights to which you are entitled and the rights to which the Government is entitled are so interwoven that it is very difficult to separate them. As you know a heavy responsibility is placed upon Government officials in authorizing the expenditure of Government funds of the amount that you have mentioned and I feel that the extent of the rights to discoveries made before the contracts and during the contract periods should really be determined by formal Patent Office proceedings.

You, of course, will appreciate the filing of a single application for patents to cover all of the involved subject matter, arising both before and during the contract periods, would secure a much firmer patent position than would result from filing applications piecemeal by you and the Government.

For this reason I have approved in general the proposal made by Captain Lavender to you that a single application be prepared covering all the work done on this particular subject with the understanding, to be confirmed in writing, that the title to the application covering the entire subject will be in you, subject to a non-exclusive license in favor of the Government for all Government purposes.

It is not my desire to influence your decision in this matter as I feel that you should have a free hand in disposing of any rights to your discoveries made before any Government contracts. However, I do wish to express my opinion that both you and the Government would benefit through mutual cooperation in the filing and prosecution of a single application. ...

The complicated relations between the discoverers (inventors), the United States government, and the University of California took years to iron out, and a satisfactory settlement was not reached until May 1955.

The successful test of a plutonium bomb took place at Alamogordo, NM, on July 16, 1945 (known as Trinity); and the use of ^{235}U and ^{239}Pu as the explosive ingredients in bombs detonated over Hiroshima (^{235}U) and Nagasaki (^{239}Pu) occurred on Aug. 6 and Aug. 9, 1945, which led to termination of the war. I was pleased that President Richard Nixon accepted my recommendation to present (on Feb. 27, 1970) special Atomic Pioneer Awards to

"The successful test of a plutonium bomb took place at Alamogordo, NM on July 16, 1945 (known as Trinity): and the use of ^{235}U and ^{239}Pu as the explosive ingredients in bombs detonated over Hiroshima (^{235}U) and Nagasaki (^{239}Pu) occurred on August 6 and Aug. 9 1945 which led to termination of the war."

President Franklin D. Roosevelt gives a campaign speech at Soldier Field in Chicago. Oct. 28, 1944. Courtesy of the Franklin D. Roosevelt Library, Hyde Park, NY.

Gen. Groves, Bush, and Conant (as described later in Chapter 6).

Although I never met Franklin Roosevelt, I did have the opportunity to see him on the occasion of his presidential campaign visit to Chicago on Oct. 28, 1944:

Helen worked at the Met Lab in the afternoon. We had Steve Lawroski over for dinner and then the three of us went to Soldier Field (which was packed full) where we saw President Franklin D. Roosevelt ride around the field in an open car and heard him make a campaign speech.

And then on April 12, 1945:

President Roosevelt died at Hot Springs, Georgia, at 3:35 this afternoon. I first learned of this tragic event from Howard Lange, who came by to tell me after hearing the news on his radio. All of our people are in a state of shock. The major radio stations turned their attention to this news for the remainder of the afternoon and throughout the evening. President Harry S. Truman was sworn into office about six o'clock (EST) this evening.

I first met Henry A. Wallace, who served as Vice President of the

United States during the Franklin Roosevelt's third presidential term (the war years 1941–45), at a conference on atomic energy arranged by University of Chicago Chancellor Robert M. Hutchins. Following is an abridged extract from my diary of Wednesday, Sept. 19, 1945:

At 2:30 p.m. in Room 302 of the Social Science Building I attended the opening session of a three-day conference on Atomic Energy Control, held under the auspices of the University of Chicago. There were 46 invited official participants in the conference: Chester I. Barnard, president, New Jersey Bell Telephone Company; Walter Bartky, professor of applied mathematics, acting dean of Physical Sciences Division, University of Chicago; . . . E.C. Colwell, president, University of Chicago; E.U. Condon, associate director, Westinghouse Research Laboratory, East Pittsburgh, Pennsylvania; . . . Farrington Daniels, professor of Physical Chemistry, University of Wisconsin and director of the Metallurgical Laboratory, University of Chicago; Karl K. Darrow, Bell Telephone Laboratories, New York; . . . James Franck, professor of Physical Chemistry, University of Chicago; . . . Reuben Gustavson, vice president and dean of the faculties, Univer-

sity of Chicago; . . . Selig Hecht, professor of Biophysics, Columbia University; . . . Thorfin R. Hogness, professor of chemistry, University of Chicago; Robert M. Hutchins, chancellor, University of Chicago; Irving C. Langmuir, Research Laboratory, General Electric Company; . . . David Lilienthal, director, Tennessee Valley Authority; . . . Reinhold Niebuhr, professor of applied Christianity, Union Theological Seminary, New York; James J. Nickson, M.D., Metallurgical Laboratory, University of Chicago; Robert Redfield, professor of anthropology, dean of Social Science Division, University of Chicago; . . . Beardsley Ruml, R.H. Macy and Company, chairman Federal Reserve Bank of New York; . . . Eugene Staley, Institute of Pacific Relations, San Francisco; Joyce Stearns, dean of faculties, Washington University, former director Metallurgical Laboratory, University of Chicago; Leo Szilard, Metallurgical Laboratory, University of Chicago; Oswald Veblen, professor of mathematics, Institute of Advanced Study, Princeton, New Jersey; . . . Henry Wallace, secretary of commerce, Washington, D.C.; . . . Harold Urey, professor of physical chemistry, University of Chicago; E.P. Wigner, professor of physics, Institute

for Advanced Study, Princeton, New Jersey.

The scheduled program for the next three days is the following:

I. The Atomic Bomb: General Evaluation
II. Consequences of the Atomic Bomb Under Conditions of National Sovereignty
    A. Influences on Military Strategy and International Relations
    B. Economic Aspects
III. International Control
    A. Control Through Existing International Organizations
    B. Control by Mutual Inspections
    C. Techniques of Moving Toward World Government
IV. If International Control is Unachievable: The Alternatives
    A. Dispersal of Cities
    B. Secrecy in Science
V. Individual Statements on Policy. . .

Then, on Saturday, Sept. 22, 1945, I had lunch with Henry Wallace, secretary of commerce, and others who attended the conference.

Henry Wallace seemed somewhat bitter as a result of being dis-

Left to right: Roosevelt, Vice President-elect Harry S Truman, and Vice President Henry A. Wallace at Union Station. Washington, D.C. Nov. 10, 1944. Courtesy of the Franklin D. Roosevelt Library, Hyde Park, NY.

Former First Lady Eleanor Roosevelt visits the Netherlands Pavilion at the World's Fair. Brussels. Sept. 5, 1958.

placed by Harry Truman as the vice-presidential candidate on Roosevelt's fourth-term campaign ticket. Had Wallace remained as Vice President, he would have ascended to the office of the President of the United States.

Many years later, I met Eleanor Roosevelt during a visit to the World's Fair at Brussels in September 1958.

Franklin Roosevelt obviously had a great appreciation for, and I believe a good understanding of, the work of his scientists and engineers on the Manhattan Project. I had been an ardent admirer ever since he campaigned for the presidency in 1932, when I was too young, by just a few months, to vote

for him. His federal programs for financial aid were "life-savers" for me and my family in the early days of his presidency—the Depression years. My father, who had not been able to find employment as the Depression deepened in the early 30s, found work as a laborer in the federal programs instituted under the New Deal. I also received critically important financial aid during my final year at UCLA. I was inspired by his "fireside chats" on the radio throughout the 1930s. Especially memorable is his radio talk to the nation the day after Pearl Harbor with its inimitable phrase, "Yesterday, December 7, 1941—a date which will live in infamy—the United States. . ."

chapter 2 HARRY S. TRUMAN

1945–1953

HOW BEST TO USE THIS AWESOME POWER

In December 1946 President Harry S. Truman appointed me as a member of the nine-person General Advisory Committee (GAC) of the newly established and appointed Atomic Energy

Commission (AEC). GAC held its first meeting in Washington on Jan. 3, 1947, and on average we met every other month, including meeting with President Truman. I served on GAC until the end of my term on Aug. 1, 1950, attending all but 1 of the 21 meetings. We were very influential in advising AEC on the rehabilitation of the Los Alamos Weapons Laboratory (which had become somewhat disorganized after the end of the war); the operation of AEC facilities for the production of fissionable material; the diminishing role of secrecy in the operation of AEC; the distribution of radioactive isotopes produced in AEC facilities; the instigation of AEC's marvelous program of support of basic research in U.S. universities and colleges; the operation of the national laboratories; the direction of the emerging civilian nuclear power program; the AEC organizational structure; and many other areas where we thought

Missouri senator Harry S Truman accepts the vice-presidential nomination as Franklin D. Roosevelt's running mate, Democratic National Convention, Chicago Stadium's Convention Hall. Chicago, IL. July 21, 1944. Courtesy of The Harry S Truman Library Institute, Independence, MO. Copyright unknown.

our advice, sought or unsought, would be helpful.

An action that gained the most publicity was the recommendation, at a meeting in October 1949 (which I missed because of a visit to Sweden), that AEC not proceed with a high-priority program to develop the hydrogen bomb. I had sent a letter to J. Robert Oppenheimer saying that I had reluctantly come to the conclusion that the United States should proceed with such a program because it was certain that the Soviet Union would do so. GAC members learned on Jan. 31, 1950, of President Truman's decision that the United States should proceed with the development and production of the hydrogen bomb.

I also met Harry Truman on a number of occasions during the years after his presidency.

My wife Helen and I first encountered Harry Truman on July 21, 1944, while we were living in Chicago. We were listening on the radio to the Democratic National Convention; then, on the spur of the moment, we decided to go by streetcar from our apartment on the south side of Chicago to the not-so-distant convention hall in Chicago Stadium, at the Stockyards on the south side of Chicago. We simply walked into the unguarded Convention Hall, where we saw Harry Truman make his acceptance speech for the vice-presidential nomination. We found a place where we stood quite close to Truman as he gave his speech.

The affection that Harry Truman had for his wife Bess and the degree to which he depended on her for support in his arduous duties are so well known that it has become legendary.

Soon after Truman became president, I participated in making an important recommendation to him that was not accepted. In June 1945, six committees were established at the Metallurgical Labora-

A Chemist in the White House: From the Manhattan Project to the End of the Cold War

tory (Met Lab) to make recommendations to the government regarding postwar policy. One was a Committee on Social and Political Implications, headed by German-born James Franck, a venerated Nobel laureate (1925) in physics. I was a member of this group. Other members, all chosen by Franck, were Donald Hughes, James Nickson, Eugene Rabinowitch, Joyce Stearns, and Leo Szilard. The committee's report, shaped mainly by Szilard, with some drafting help from Rabinowitch, was completed on June 11, 1945, and signed by every member of the group. It made, basically, three points: the United States could not avoid a nuclear arms race through a policy of secrecy, the best hope for national and world safety from the consequences of the bomb lay in international control of atomic energy, and the military use of the bomb against Japan was inadvisable because it would "sacrifice public support throughout the world, precipitate the race for armaments and prejudice the possibility of reaching an international agreement on the future control of weapons." We suggested, instead, that the power of the bomb first be demonstrated in an uninhabited desert or barren island. This event would be followed by an ultimatum to Japan to surrender, with the implication that thus many lives would be saved.

The "Franck Report" was delivered to Washington on June 12, 1945, by Franck and Compton. Secretary of War Henry Stimson was not available, so the report was left for review by George L. Harrison (former chairman of the New York Federal Reserve Bank and president of the New York Life Insurance Company). Harrison was serving as deputy chairman (Stimson was chairman) of the so-called interim committee, which was charged with the planning that would be necessary in anticipation of the use of the atomic bomb and the revelation of its existence. Har-

Bess Truman (seated second from left), Truman, and daughter Margaret Truman (seated right) at the Democratic National Convention, Chicago Stadium's Convention Hall. July 21, 1944. Courtesy of The Harry S Truman Library Institute, Independence, MO. Photo by the *Chicago Sun–Times*, Chicago, IL.

WAR DEPARTMENT
P. O. Box 2610
WASHINGTON, D. C.

REFER TO FILE NO. _____

10 October 1945.

Dr. Glenn T. Seaborg
Metallurgical Laboratory
University of Chicago
Chicago 37, Illinois

Noted ___ 10/13/45
Action _____
Recd. OCT 12 1945 GTS
Ans. _____
File ✓ _____

My dear Dr. Seaborg

     The major final factor which determined the surrender of
Japan was the Atomic Bomb. Of course, surrender was an ultimate
certainty in any case; yet the war would have continued for weeks
and perhaps months longer had it not been for the completion and
use of our bomb. That weapon, therefore, helped beyond question
in the saving, from further death, destruction and misery, of friend
and foe alike.

     The contribution which you have made to the development of
the atomic bomb and to the consequent attainment of the historic results for which the whole nation will always be grateful is appreciated
by the War Department and the Manhattan Engineer District, and by me
personally, more than words can convey. The chemical research work and
the fundamental studies of the properties of new materials needed for
the project, which you carried out at the Metallurgical Laboratory
at Chicago, were essential to our success. Your energy and ingenuity,
your scientific skill and judgment, and your self-sacrificing devotion
to our cause are beyond praise. I want you to know how I feel about
this.

     While praying that the forces of nuclear energy, which you
helped so signally to develop for use against the enemy, may be wisely
controlled in the days to come, for the service of a world at peace,
we must realize that no future events can detract from the splendor
of the results attained in the immediate past through our ability to
make military use of these forces. For your indispensable part in
this attainment, in behalf of the War Department as agent for the
American people, I thank you.

                  Sincerely yours,

                  L. R. GROVES,
                  Major General, USA

General Leslie R. Groves's (Head of the Manhattan Engineer District) letter thanking Seaborg for his participation in the United States' atomic bomb efforts during World War II. Oct. 10, 1945.

rison referred the matter to the interim committee's Scientific Advisory Panel consisting of Compton, Enrico Fermi, Ernest Lawrence, and Oppenheimer, who considered it at a meeting on June 16 at Los Alamos and decided that the bomb should be used to help save American lives in the Japanese war; the interim committee, and finally President Truman, came to the same conclusion. In a poll of 150 Met Lab people in July, 60% favored military use of the bomb in Japan. A total of about 25% favored giving a demonstration in the United States with Japanese representatives present, followed by an opportunity to surrender before the bomb would be put to military use in Japan. Also in July, a petition prepared by Szilard with 70 signatures at the Met Lab requested that the bomb not be used without warning. The indications are that neither the "Franck Report" nor the petition reached President Truman (who sailed for Europe on July 6, to confer at Potsdam with Josef Stalin and Winston Churchill), although he may have been briefed on the gist of the recommendations. In any case, the recommendation that there be a demonstration of the atomic bomb was not accepted.

    In a remarkable book (1988), *Danger and Survival—Choices about the Bomb in the First Fifty Years,* the late McGeorge Bundy suggests that influential Japanese observers might have been invited to witness the American bomb test at Alamogordo on July 16. This would have given the Japanese the opportunity to learn of the terrifying impact of the bomb and the option to forestall its use on their cities by agreeing to terms of surrender. This more practical alternative was not suggested by members of the Franck committee, in part because the requirements of secrecy (compartmentalization) prevented them from having any information about plans for the Alamogordo test. However, in the July 1945 poll at

Met Lab, which I mentioned earlier, one of the options was to give an experimental demonstration of the bomb in the United States.

I can understand and do not quarrel with the decision to use the Hiroshima bomb (although I am not convinced that the follow-up Nagasaki bomb was necessary). Accomplishment of the objective of ending the war at the earliest possible date undoubtedly saved hundreds of thousands of lives. An ineffective or failed demonstration of the atomic bomb would probably have prevented so early an end to the war. The use of the bomb dramatized its terrible power and contributed to the subsequent deterrent effect of improved nuclear weapons in forestalling another world war for a record period of more than 50 years and perhaps indefinitely.

In October 1945 Leslie Groves wrote me a letter, commending me for my work at Met Lab.

My first direct contact with President Truman occurred in December 1946, soon after the creation of AEC as the successor to the Manhattan Engineer District. Below is an excerpt from my diary of Dec. 3, 1946:

While attending a meeting this morning, I received a telephone call from AEC Commissioner Robert Bacher (in Washington, DC), inviting me to be a member of the nine-member statutory GAC to AEC. I immediately accepted. Bacher informed me that there will be about six meetings a year; the stipend will be $50 per day. I reported the news of my appointment to Ernest Lawrence.

Soon thereafter, I received a letter from President Truman acknowledging my acceptance of membership on GAC. The initial members of GAC were an impressive group: J. Robert Oppenheimer (who served as chairman and as wartime leader of Los Alamos Laboratory), Enrico Fermi (nuclear physicist, University of Chicago),

THE WHITE HOUSE
WASHINGTON

December 12, 1946

My dear Professor Seaborg:

I am pleased that you find it possible to arrange your schedule of activities so as to accept membership on the General Advisory Committee to the Atomic Energy Commission. This is an opportunity for public service of the highest character and I am particularly appreciative of the spirit of cooperation which prompts you to accept an appointment which gives your government the benefit of your knowledge and experience.

Very sincerely yours,

Harry S Truman

Professor Glenn T. Seaborg,
University of California,
Berkeley,
California.

President Harry S Truman's letter confirming Seaborg's appointment to the General Advisory Committee to the U.S. Atomic Energy Commission. Dec. 12, 1946.

James B. Conant (president, Harvard University), Isidor I. Rabi (nuclear physicist, Columbia University), Lee A. DuBridge (president, Cal Tech), Cyril S. Smith (director, Institute Study Metals, University of Chicago), and industrialists Hood Worthington (DuPont Company) and Hartley Rowe. With such a membership, GAC exerted tremendous influence on the initial commissioners of AEC: David E. Lilienthal (chairman, and ex-chairman of TVA), Lewis L. Strauss (Wall Street financier), Robert F. Bacher (nuclear physicist, Cornell University), Sumner T. Pike (businessman), and William W. Waymack (newspaper editor).

The first three of these meetings set the stage for GAC's actions and, therefore, I quote extensively from my diary:

Friday, Jan. 3, 1947

In Washington. I arrived in Washington at 9 a.m. in ample time for the 10 a.m. GAC meeting in Room 5136, New War Department Building. Present at this meeting were GAC members Conant, Rowe, Fermi, Rabi, Worthington, Smith, and I. Oppenheimer, delayed by travel and weather conditions, arrived at 10:30 a.m. Saturday while DuBridge arrived after adjournment. AEC members present were Lilienthal, Pike, Strauss, and Waymack. Bacher, also delayed by travel and weather conditions, arrived at 4 p.m. Friday. Carroll L. Wilson, general manager of the commission, was also present. Oppenheimer was chosen as chairman (with Rowe as temporary chairman in Oppenheimer's absence). Lilienthal presented each GAC member with an official presidential appointment and then passed out a sheet describing some of the tasks facing the commission.

To the Members of the General Advisory Committee

The members of the Commission have felt that it would be inappropriate to attempt to provide a detailed agenda for the first meeting of the General Advisory Committee. The Commission will look continuously to the General Advisory Committee for advice and assistance "on scientific and technical matters relating to materials, production, and research and development." The methods by which this advice can best be given are in themselves topics for discussion, which it is hoped can be thoroughly explored at this first meeting of the Committee.

The Commission has, however, considered some problems on which the advice of the General Advisory Committee is requested at this time. As a basis for discussion at the meeting, the Commission has concluded that the following general problems should be brought to the attention of the General Advisory Committee.

(1) One of the pressing decisions of the Commission is the selection of a Director of the Division of Research. The Commission's concept of the functions of this division will be outlined to the meeting by its General Manager, Carroll L. Wilson. It is hoped that specific suggestions as to candidates for this position can be made by the General Advisory Committee before the adjournment of its first meeting.

(2) An immediate task facing the Commission is the preparation of a statement of the existing research and development program resulting from the former Manhattan District's research contracts and a review and evaluation of that program. In order to initiate work on this task, the Commission will need the assistance of the principal universities and industrial concerns, which are research contractors. It is hoped that the General Advisory Committee can recommend to the Commission the names of persons from some of the contractors' laboratories who might be brought to Washington for this purpose.

(3) A decision has been made by the Commission to establish a laboratory at Brookhaven and for its construction and operation by Associated Universities, Inc., a

"An immediate task facing the Commission is the preparation of a statement of the existing research and development program resulting from the former Manhattan District's research contracts and a review and evaluation of that program."

nonprofit corporation formed by a group of eastern universities. By transfer, the Commission has acquired a contract providing for the construction of a research laboratory at Schenectady, New York, by the General Electric Company. The Commission would like the General Advisory Committee to consider the need for the establishment of these two laboratories in close proximity and, assuming the need for both laboratories, to consider a reasonable division of research and development programs between them. In particular, the Commission desires an expression of opinion from the General Advisory Committee as to whether in view of the present state of scientific information and technical developments, the research program at the General Electric Laboratory should be one of fundamental investigation or should be more specifically related to industrial and commercial applications of atomic energy.

(4) It is the Commission's hope that affirmative and stimulating relationships can be developed and maintained between the Commission and its research contractors. The General Advisory Committee's suggestions for the establishment of such relationships would be helpful to the Commission. It is suggested that it might be advisable to appoint a special Committee composed of business and scientific representatives of university and other laboratories or foundations to study and report on this problem.

(5) At the time of the appointment of the General Advisory Committee, the Commission informed the president that it would probably appoint other advisory committees in the fields of medicine and biology, geology and mining, and the social sciences. The Commission desires the advice of the General Advisory Committee on the need for and functions of such advisory groups and specific suggestions for appointments to them.

(6) The Commission desires the advice of the General Advisory Committee on the best approach to a study of the extent of contamination and possible corrective measures at the Hanford Engineer Works.

(7) An explanation of certain operating changes at Oak Ridge, adopted by the Manhattan District with the approval of the Commission in December, will be made to the General Advisory Committee. The Commission in approving the changes relied on a report of a Committee of experts to the Manhattan District stating that with suitable safeguards the new operations could be conducted with reasonable safety. In view of the importance of the problem, however, the Commission desires suggestions from the General Advisory Committee for a review of the safety report.

David E. Lilienthal
Chairman

[With respect to item 6, "a study of the extent of contamination and possible corrective measures at the Hanford Engineer Works," some attention was initially given to this, but higher priorities in connection with the soon escalating "cold war" drove this problem into a low level of priority.]

Lilienthal then discussed the functions and role of GAC and described the commission's organization and general nature of its work to date. Carroll Wilson explained the role of the Division of Research as a staff rather than a line operating division. In the ensuing discussion it was agreed that the committee should advise the commission on major policy and program determinations and on the selection of technical consultants. Among the subjects suggested by individual GAC members for future consideration and advice by the committee were the desirability of concentrating on the immediate construction of a reactor for the production of electric power; a re-examination of the raw materials situation, with emphasis on new techniques for the discovery of sources of materials and on the more efficient processing of such materials;

the proper relationship of the development and production of atomic weapons to other research and production programs; the recovery of certain potentially valuable elements now lost in the production of fissionable material (which I suggested) and more effort on studies relating to the use of thorium as fissionable material; the proper balance between freedom in research laboratories to explore the boundaries of scientific knowledge and positive direction by the commission to research work that has reached a concrete stage; the conservation of uranium through re-examination of current means of production of fissionable material at Hanford and Oak Ridge; and the desirability of continuing the wartime emphasis on the engineering aspects of the development of new materials for use in reactors and other processes.

After lunch (12:30 p.m. to 1:30 p.m.), the committee considered candidates for the position of director of research. There was a discussion about the General Electric laboratory at Schenectady (set in an industrial framework) and Brookhaven (academic personnel). Advisory committees were also discussed. I then talked about administrative problems: (1) delay in the publication of the Manhattan Project's technical series, (2) difficulties involved with the employment of students as part-time employees, (3) the unfortunate necessity for all employees to be cleared by the FBI, (4) problems with publications because of policies of the Patent Office of the Manhattan District, and (5) problems with admittance passes. The session adjourned at 5:30 p.m.

Saturday, Jan. 4, 1947

In Washington. GAC meeting reconvened at 9:30 a.m. The following resolution (Conant moved and Fermi seconded) was adopted: The GAC chairman (Oppenheimer), with the help of subcommittees and staff of the commission, will report on existing status of, and future plans for, (a) research and development, (b) materials, and (c) production. There was a general discussion on the conduct and mechanics of the activities of the committee, meeting dates, and so on.

The second meeting of GAC came just one month later:

Sunday, Feb. 2, 1947

In Washington. My train arrived at 9 a.m., and I immediately went to the New War Department Building for the GAC meeting, which began a little before 10 a.m. In addition to the nine GAC members, three commissioners (Lilienthal, Bacher, Waymack), General Manager Wilson, James B. Fisk (director of the Division of Research), Col. James McCormack (director of military application), John H. Manley (GAC secretary), and Military Liaison Committee members (Gen. Lewis H. Brereton, Rear Adm. William S. Parsons), along with the security officer Capt. W.A. Blair and the stenographer Miss E.M. Dashiell, were present.

Oppenheimer explained that he asked Lilienthal to supply the committee with information about the nuclear weapons stockpile and the production rates. Manley, Blair, and Dashiell were not present while Bacher gave us this information. Oppenheimer then led us into a discussion of reports on research and development available to us, especially the "Panel Report" (proposals for research and development in the field of atomic energy, dated Sept. 28, 1945). He stated that this covered the research and development from the Manhattan District at the time the report was written. He summarized the situation reported: No real exploration of new weapons has taken place. Referring to reactors, Oppenheimer said that Walter Zinn's report ("Report on Research and Development," Feb. 1, 1947, Section II-1) states that there has been essentially no new reactor built and no comprehensive reactor program organized. Oppenheimer suggested a recess (10:30 a.m. to 11:30 a.m.) so that we could read the reports.

When we reassembled, Oppenheimer asked for comments about the testing of atomic weapons; Adm. Parsons emphasized the need for testing the presently stockpiled types because they have not been subject to any testing. Everyone agreed that some testing should be made. DuBridge stressed that thorough and adequate preparation is needed because testing is costly in terms of personnel, time, and money, and will involve physical and political dangers. We also discussed some of the details of the weapons and concluded that work toward major increases in energy release was important. We adjourned at 12:30 p.m.

GAC reconvened at 1:30 p.m. in executive session, with Manley and the stenographer present, to discuss reactors. Oppenheimer summarized the need for reactor development as follows: to advance the international aspects of atomic energy through the demonstration of its peaceful utility, to affect public opinion in a similar fashion in this country, and to provide sufficient fissionable material so that questions of allocation become relatively unimportant. The time scale of this effort, he said, should be to obtain some power in the order of a year or two and then an increased power program in five years. Fermi differed from this point of view, saying that nuclear weapons are more important at this time in view of the international situation and that our country should not risk loss of strength in the field of weapon production, development, and research. He stated that it is more important to make Los Alamos healthy again than to develop nuclear reactors. After some discussion, we generally agreed that weapons are a first priority. This then led to an examination of the Los Alamos situation: lack of strength in theoretical matters, the quality of the present direction of the laboratory, the degree of achievement of the past year, the merits of a different location, the community problems existing there, and the possibility of stimulation of Los Alamos by a directive to develop a reactor or to concentrate on thermonuclear explosives. Finally, we talked about the function of GAC, particularly whether members should simply advise the commission using information supplied by the commission or whether it should be more of a working group by obtaining information itself; we tended toward the latter view. [This was an important decision and set the theme for our subsequent activities.]

After a 15-minute recess, we met with McCormack, Parsons, Bacher, and Fisk to summarize our discussions. GAC stated that we view the weapon situation as vital and an important and priority job of the commission. Additional emphasis on thermonuclear weapons over that suggested in the "Panel Report" should be now given, we said; this may serve to strengthen Los Alamos and to improve its performance with regard to present weapons. We also stated that more work is required on the initiator problem, and we stressed that we agreed with the "Panel Report" that weapons tests should be carried out. We agreed to assist in the recruitment of personnel for the AEC's program. The session was adjourned at 5:30 p.m.

Monday, Feb. 3, 1947

In Washington. GAC convened at 9:15 a.m. with all members present, plus Manley and Louise Johnson (stenographer). Oppenheimer summarized the previous day's conclusions, saying that it seemed to him that the heart of the problem was reached with surprising speed: The making of atomic weapons is something to which we are now committed, much as we dislike it. He reiterated our comments that GAC should help as much as we can, and there may be a time when a group of us should spend time (as long as six weeks) at Los Alamos to help. Oppenheimer stated that we probably do not want to recommend the removal of Los Alamos to another site but we should point out to the commission: (1) a group of us might be willing to go to Los Alamos to live and work for a period of time, (2) the position of technical director may be easier to fill than that of the theoretical position, and (3) the "super" may not be the best way to revitalize Los Alamos; a reactor program might provide the proper incentive. We discussed the summary, and Rabi pointed out that he believes the various commission laboratories are not pulling together.

We then talked about reactors, and Oppenheimer pointed out four things to keep in mind: (1) we need more plutonium and more plant than we have; (2) we need a research reactor with high flux to permit entrance into fields now only marginally accessible; (3) we must look into the long-range aspects of power breeders, probably those using plutonium; and (4) real atomic power should be produced as quickly as possi-

"Oppenheimer summarized the previous day's conclusion, saying that it seemed to him that the heart of the problem was reached with surprising speed: The making of atomic weapons is something to which we are committed, much as we dislike it."

ble. Fermi suggested the following order: (1) an improved version of Hanford, (2) a breeder power unit of the Zinn type should be built as soon as possible, and (3) a high-flux reactor should be built as soon as possible. Oppenheimer commented that he would like an even higher priority given to a reactor of the 10,000-kW useful power range. We also talked about the Clinton situation and the present state of the Daniels' high temperature pile, both of which are unsatisfactory. The session adjourned at 12:15 p.m.

Our afternoon meeting opened as an executive session at 2 p.m. with Manley and Lucille Ross (stenographer) present. We talked about how to transmit the results of our meetings and agreed that Oppenheimer will transmit a letter to Lilienthal, replying to the specific questions raised by Lilienthal in regard to the military applications of atomic energy. Oppenheimer then appointed the following subcommittees: Weapons—Conant (chairman), Rowe, Rabi, Fermi; Reactors—Smith (chairman), Fermi, Worthington; and Research—DuBridge (chairman), Rabi, Conant, Seaborg. Oppenheimer will be an ex-officio member of each committee. We also decided to question the policy of holding one Hanford production plant in stand-by condition to guarantee production of adequate amounts of polonium for use as initiators. This executive session was adjourned at 4:30 p.m. At 4:45 p.m. Lilienthal, Bacher, Pike, Waymack, Strauss, Fisk, Wilson, and McCormack joined us to hear the results of our executive sessions. I left shortly thereafter to take a cab to the train station to catch my train, the Capitol Limited, at 5:30 p.m., for Chicago.

By the time of GAC's third meeting, AEC had moved to a new headquarters:

Friday, March 28, 1947

In Washington. I went by cab from the Hotel Statler to the new Washington headquarters of AEC (the newly vacated Public Health Building) at 1901 Constitution Ave., N.W. Here GAC's third meeting started at 9:45 a.m. in Room 213. All members were present. Manley (executive secretary), Commissioner Bacher, AEC General Manager Wilson,

AEC Director of Research Fisk, Gen. McCormack (director of military application), and Mr. Shivers (a security officer) also attended.

Wilson then summarized the actions and considerations of the commission since the last GAC meeting. He described how the Monsanto plant at Oak Ridge has increased the rate of production of initiators for atomic weapons and their plans for further increase of production. The commission, he said, has decided to keep the weapons laboratory at Los Alamos and has extended the contract with the University of California for its operation, and he reported that a small group is responsible for preparing a paper about the need for tests of nuclear weapons. Provision has been made, he said, for production of Zinn's fast reactor at the site of Argonne Laboratory, and arrangements have been made with Stewart Oxygen Company for the distribution of deuterium. Wilson began a discussion of the three commission memoranda by describing the problems the commission faced when it took over operations from the Manhattan District: budget, scope of activities, and so on. [The first GAC memorandum (Report of Subcommittee on Reactors) follows.]

The Subcommittee on Reactors has carried out a general survey of proposals for pile development. In the course of this survey it became plain that the construction of additional and replacement piles at Hanford is in a category by itself, and with the associated development and installation of the Redox process must have the highest priority. The design should not be substantially changed; thus, no change in the coolant should be considered, but changes that will not sensibly delay construction should be incorporated, such as modification of the graphite structure. No experimental facilities for power generation or for research other than simple irradiation should be included. This does not preclude the possibility of incorporating the results of experience as the program of construction advances. The possibility of using portions of the present water treatment plants for the new units should be considered.

The main objective of the Hanford work should be steady production of plutonium until the operation of breeders is able to meet all demands for plutonium. There is no point in planning a large expansion of production without the installation of new methods of conserving raw material. This involves chemical recovery, as by the Redox process; high plutonium concentration (which forces a reconsideration of the effect of Pu^{240} on the weapon); and the use of isotope separation to make possible the reuse of depleted material by enrichment.

We list below four other reactors in the estimated order of decreasing importance: the high-flux reactor, the Argonne fast reactor, the G.E. intermediate reactor, and the Clinton gas-cooled pile. We recommend that the detailed engineering design, including provision for recovery of fissionable materials, of the first two of these be authorized immediately, that construction be authorized as soon as possible, and that the necessary fissionable material be allocated unless unanticipated difficulties appear. . .

A serious study of the use of plutonium as a reactor fuel should be started, and facilities for its metallurgical manipulation should be established at one or two more sites. . . .

There was a discussion of the time scale for developing nuclear power for civilian use; there was agreement that this is probably years off in view of the military problem, the raw material situation, and the long time scale for breeder operation. Fermi estimated that it would be 50 years before any nuclear electric power production at the magnitude of the electric power now consumed will be available. He suggested that breeding must be the ultimate basic solution for the shortage of uranium raw materials problem. We adjourned at 12:35 p.m.

*Conant and Fermi then gave the report of the Weapons Subcommittee. Conant described the need for tests for nuclear weapons, and Fermi urged that realistic theoretical studies of thermonu-*clear designs be made. I reported on the *Chemical Problems Attendant with Immediate Increased Production of Plutonium,* based on the March 17, 1947, memorandum I sent to the commission, which stated that a substantial increase in plutonium production would depend more on additional reactors at Hanford than on breeders. I urged the early development of a plan for handling the irradiated uranium wastes at Hanford, explaining that the Redox Process, which will use solvent extraction techniques to recover both uranium and plutonium, will help in the waste storage problem. [This was a recommendation that was not adopted as the exigencies of the cold war took over.] Successful development of breeding, I said, might well depend on the development of a process such as the Redox Process. We adjourned at 5:15 p.m.*

Saturday, March 29, 1947

In Washington. Attending this morning's 9:30 a.m. meeting were all nine members plus Manley, Fisk, Roger S. Warner (director of engineering), and Shivers (security officer). Fisk introduced a fourth commission memorandum about Los Alamos, which we agreed that the Weapons Subcommittee will investigate during its visit there next week. We then went back to discussing the first commission memorandum, the report of the Research Subcommittee, and consideration of a central laboratory to carry out primary research and development responsibilities of the commission. I again brought up the possible delay in the Redox program, pending unification of research in such a central laboratory.

GAC focused on the concept of a central laboratory for power reactor development and agreed that a high-flux reactor is the backbone of a long-range reactor program. The most attractive site for a central research laboratory, we believe, is the new site of Argonne Laboratory in DuPage County near Chicago. We adjourned at 12:30 p.m.

During this noon's intermission, which lasted until 3:40 p.m., I talked with Jim Fisk about Spofford English, who is being considered for an important position in Fisk's office. I recommended Spof highly.

"Fermi estimated that it would be 50 years before any nuclear electric power production at the magnitude of the electric power now consumed will be available."

Sunday, March 30, 1947

In Washington. I again took a taxi to the AEC building to attend a 10 a.m. GAC meeting with all the members except Rowe. Shivers was present as a security officer. Oppenheimer brought up the material flow problem and appointed a new subcommittee—Fermi (chairman), Worthington, and me to gather information so that GAC can make recommendations on this problem. We are directed to report as early as possible on an investigation of ^{240}Pu tolerances in weapons as they affect Hanford operating procedure and on the question of increasing the reactivity of spent uranium fuel recovered from Hanford operations by adding enriched ^{235}U from the K-25 plant to it or recycling some of the spent Hanford uranium fuel through a specially designed diffusion plant. We adjourned at 12:05 p.m.

The afternoon session began at 1:10 p.m. with the same group. Bacher, Waymack, Wilson, Fisk, and McCormack joined us at 2:30 p.m. Oppenheimer, in his usual masterful way, summarized our conclusions on the commission's memoranda 1, 2, and 3, said that the Subcommittee on Weapons will report back to GAC on memorandum 4 after the group visits Los Alamos, and told of the creation of a new subcommittee to examine the efficient utilization of raw materials. Oppenheimer also said that GAC believes a theoretical physicist is required on the matter of the super. Wilson asked if it would be possible to proceed with pile construction on the assumption that the Redox Process would be successful—I replied that I had no doubt as to the success, but only as to the possibility of meeting the time schedule for construction and operation; however, I said that I do not feel parallel work by other companies will cut the schedule because of competition for the same personnel. The meeting adjourned at 3:50 p.m.

At our July 1947 meeting we considered, among other topics, the future of atomic energy with a somewhat pessimistic forecast with respect to time scale.

Monday, July 28, 1947

We returned to executive session at 2:55 p.m. and continued talking about the draft of our statement on the feasibility of atomic power. After a number of suggestions, Oppenheimer agreed to redraft the statement for presentation to the commission at tomorrow morning's sessions.

Then, for the next hour and a half (until 6:30 p.m.), we worked on Oppenheimer's preliminary draft of our statement on atomic power. After dinner, we reconvened in executive session at 8:20 p.m. and again worked on our statement on atomic power. We also talked about our recommendations concerning classified areas and declassified subjects, plus the problem of Clinton Laboratories. We concluded that we shall express no opinion on the matter of the takeover of the management of Clinton Laboratories by the University of Chicago. This session concluded at 10:20 p.m.

Tuesday, July 29, 1947

Oppenheimer brought up the question of our draft statement on atomic power, and we agreed to transmit it to the commissioners for their comments at our joint meeting.

In the course of our duties as advisors to the Atomic Energy Commission we have re-examined the situation as it appears today with regard to the generation of atomic power for industrial uses. It may be that this situation is not fully understood by our scientific colleagues and the public; and it therefore seems desirable to give a brief account of it. We are aware that in this field as in other new fields discoveries and inventions may make our present views obsolete—indeed, this is likely. Nevertheless, it is desirable that what is foreseeable on the basis of present knowledge be distinguished from hopes which are necessarily vague as to future changes in that picture.

It does not appear hopeful to use natural uranium directly as an adequate source of fuel for atomic power. The reactivity of systems based on natural uranium is low; even the fraction of U^{238} which can be consumed without replenishing the fuel is small. Because of this the raw material requirements, if such reactors are to play an important role in power economy, are economically prohibitive. It is true that one could re-enrich this natural fuel, but the power expenditure involved in such isotope separation by any present methods makes this likewise prohibitive.

For these reasons the hopeful approach for the future lies in the development of reactors which (a) generate power at high temperature, and (b) create more fuel from U^{238} or thorium than they consume. Although both of these requirements appear to be capable of being met in principle, neither has so far been realized in fact. An intensive development program will be required to achieve these ends. No one can estimate how long this program will take, but it appears to be reasonable to us to anticipate a period of about 10 years before the series of very difficult metallurgical, engineering, and chemical problems can be solved.

Even when reactors can be designed and built along these lines it must be anticipated that the fractional increase of nuclear fuel per year will be quite small. It can be increased by increasing the specific power of the reactors (that is the rate of energy generation per unit of nuclear fuel invested), by maintaining a good neutron economy, and by a good system of chemical techniques to prevent the loss of material when it is removed and reprocessed. This chemical requirement again demands both a long time between reprocessings, and high standards of chemical operation to reduce losses in an operation which will of necessity involve much recycling of material. These many problems, all of which involve their own difficulties, must be solved together if a reasonable economy of nuclear fuel is to be achieved. To date none of them has been solved at all.

When and if these problems are solved, the generation of industrial power may indeed become profitable, but it will still be on a small scale. Even according to the most optimistic estimates, the rate of accumulation of nuclear fuel by regeneration will be slow. Thus even if the years of development are also used to accumulate a substantial initial stockpile of nuclear fuel, and even if favorable technical answers appear for the designs of these reactors, decades will elapse before stocks of nuclear fuel can have accumulated which will supplement in a significant way the present power resources of the industrial nations of the world. . . .

In retrospect, our forecast was flawed because of the assumption

that nuclear fuel would be in short supply and, hence, it would be necessary to develop breeder reactors.

At 3:30 p.m. we were joined by Lilienthal, Waymack, Strauss, Pike, Wilson, Fisk, McCormack, Walter Williams (director of production), and Belsley (a security officer), as well as Gen. Brereton and Adm. Parsons. Oppenheimer summarized our opinions about nuclear weapons tests, the problems with reactor development, the civil service question, the pattern of research distribution, and our statement about atomic power. The following discussion was concerned primarily with our statement on atomic power, which Strauss felt was very strong and so pessimistic that it will adversely affect the ability of the commission to operate, for example, in regard to the use of public funds. Waymack said the statement will be meaningless to the general public and will not result in a gain of education toward the understanding of atomic energy problems. Pike remarked that this means the commission is on trial, and to say now that atomic power is very far distant means that we now have, in atomic energy, only a military tool. Lilienthal said the last paragraph holds no hope for peaceful uses of atomic energy. Strauss suggested that it would be prudent to consider the statement in the interval between now and the next GAC meeting; we agreed to this suggestion. Finally, Fisk said that he thinks the real shock comes from the implied situation about the nonexistence of adequate raw materials for real commercial exploitation of atomic energy unless fuel can be obtained through breeding and utilization of fuels other than ^{235}U. The joint meeting ended at 5 p.m.; we went into executive session at 5:10 p.m. for a further discussion of our statement, considering the implications and the factual basis of the draft. We agreed that Oppenheimer will prepare a new draft incorporating some of the suggestions and changes. The meeting was adjourned at 6:20 p.m., and I went to the airport to catch a flight home.

I had some interesting meetings to discuss a number of issues with the five AEC commissioners, AEC laboratory directors, and oth-

"In retrospect, our forecast was flawed because of the assumption that nuclear fuel would be in short supply and, hence, it would be necessary to develop breeder reactors."

ers at the Bohemian Grove in August 1947:

Monday, Aug. 18, 1947

    I picked up Eugene Wigner (director, Research and Development, Clinton Laboratories) and Frank Spedding (director, Ames Laboratory) at the Durant Hotel at 9 a.m. and drove to San Francisco and the Fairmont Hotel, where I picked up Guy Suits (director of research, General Electric); the four of us then proceeded to the Bohemian Grove on the Russian River, arriving in time for lunch. Among those in attendance at the Grove were Lilienthal, Ernest O. Lawrence (director, Radiation Laboratory), Donald Cooksey, Fisk, Walter Zinn (director, Argonne National Laboratory), Philip M. Morse (director, Brookhaven Laboratory), Norris Bradbury (director, Los Alamos Scientific Laboratory), Strauss, Waymack, Bacher, Pike, Oppenheimer, and McCormack.

    The informal discussions today centered primarily on the role AEC should play in the support of basic research.

Tuesday, Aug. 19, 1947

    In Bohemian Grove. The group met to discuss a number of issues, including the time scale for the development of nuclear power for the production of electricity. The commissioners met privately to debate the question of the foreign distribution of radioisotopes. Lewis Strauss is adamant in his view that radioisotopes should not be allowed to go to European countries, because they can use them to further their military research. GAC members and most of the other commissioners strongly disagree with him about this.

Wednesday, Aug. 20, 1947

    In Bohemian Grove. I took part in some continuing discussions and later drove back to Berkeley.

Our meetings in October, November, and December 1947 focused on, among other topics, basic research:

Thursday, Oct. 2, 1947

    In Washington. I took a taxi from the Hotel Statler to the AEC building

and attended a 2 p.m. meeting of GAC's Subcommittee on Research (DuBridge, Seaborg, Rabi, and Conant). We decided to propose that the full GAC recommend to AEC members that AEC support basic research rather broadly in universities in order to fill the time gap pending the establishment of a National Science Foundation.*

Friday, Oct. 3, 1947

    In Washington. I again took a taxi from the Statler to the AEC building to attend the 9:10 a.m. session of GAC and AEC. All members were present, in addition to Manley, our support secretary; Anthony Tomei, our assistant support secretary; and Palazzalo, a security officer.

    Lilienthal told us that authorization has been made for distribution of isotopes to foreign countries and that the commission has also asked for exemption for its employees from civil service regulations for a three-year trial period. Bacher, in commenting on the statement on atomic power, said AEC is now grateful for its preparation; he feels it should be rewritten in a style that would make it more understandable to the general public, after which it could be submitted for declassification. Worthington said he believes the statement should include a discussion of the economic aspects of atomic power. This session was over at 12:55 p.m.

Saturday, Oct. 4, 1947

    In Washington. I again went to the AEC building for the 9 a.m. GAC session. Most of the morning was taken up with a discussion of our attempts to revise our statement on atomic power; we finally decided that Fermi and Smith will collaborate on one form of revision and that Oppenheimer and Rabi will do so on another. These will be prepared for comparison at the next GAC meeting. We also approved the revised form of the report from our Subcommittee on Reactors, which incorporated the points we brought out yesterday. This session adjourned at 1 p.m. for lunch.

Sunday, Oct. 5, 1947

    In Washington. At 10:45 a.m. our committee met at the AEC headquarters with everyone except Conant present. We again went over the revised drafts of the reactor report and the research report; these were approved for transmission to the commission. We also talked about the problem of declassification and agreed that as much secret information as possible should be declassified in order to make maximum progress in the research programs. The session adjourned at 12:50 p.m.

After lunch GAC had a joint session with AEC Commissioners Waymack, Pike, and Bacher; Military Liaison Committee members Rear Adm. William S. Parsons, Col. John H. Hinds, Adm. Ralph A. Ofstie, Rear Adm. Thorvald Solberg, and Gen. Lewis H. Brereton (chairman); and AEC staff members Wilson, Fisk, Williams, and McCormack. Oppenheimer then made his usual masterful summary of GAC's deliberations over the last two and one-half days. He called attention to the problem of detection of any nuclear or atomic bomb tests by a rival country and to the question of radiological warfare, in which one must recognize that defense is not separable from offense. With respect to declassification, Oppenheimer reported our conclusion that it is not sound policy to retain classification on something that may be of use only in the order of 10 years, that declassification policy should be enlightened by information supplied by intelligence reports, and that declassification may be warranted to remove pernicious public misconceptions (such as those current on the power aspects of atomic energy and super bombs).

Sunday, Nov. 23, 1947

In Washington. I again attended the GAC session at 9:30 a.m. in the AEC building. We worked and adopted the draft on the future of atomic power that Fermi, Smith, Conant, and DuBridge produced last night and discussed the urgency of expansion and improvement of the nuclear weapons production facilities at Sandia. Our five-page statement on nuclear power described the technical problems and complex economic factors to be overcome in building a nuclear power system; we conclude with the summary that we "do not see how it would be possible under the most favorable circumstances to have any considerable portion of the present power supply of the world replaced by nuclear fuel

before the expiration of twenty years." We discussed our first annual report to President Truman (which Oppenheimer will draft), the need for declassification (we will stress with the commissioners the nuclear power and superweapon misunderstanding by the public), and the need for a policy on the support of pure research (e.g., the Bevatron).

At 2 p.m. we held our windup joint session with the AEC commissioners and staff and with members of the Military Liaison Committee. Oppenheimer summarized our recommendations on Sandia, Fisk's three questions (Nuclear Energy for the Propulsion of Aircraft, Naval propulsion, Bevatron), nuclear weapons tests (there will be a GAC meeting on Dec. 30 in Chicago to discuss this), the Military Liaison Committee's view on keeping the gt (grams per ton of uranium) level of plutonium and amount of tritium classified as top secret, our views on the status of the nuclear weapons tests of our rivals, and the need for a public statement on big weapons.

Monday, Dec. 29, 1947

In Chicago. I took a cab to the New Chem Building on the University of Chicago campus for a special meeting of GAC at 9:55 a.m., attended by all of GAC members except Rowe; that is, Oppenheimer, Conant, DuBridge, Fermi, Smith, Worthington, Rabi, and I, in addition to Manley. Oppenheimer suggested five items for the agenda: (1) a trip through Argonne National Laboratory, which we tentatively decided to forego because of time considerations; (2) the problem of Clinton Laboratories; (3) the Los Alamos weapons test program; (4) a report from the Subcommittee on Material Flow; and (5) the report of GAC through the chairman to President Truman. Item (3) we decided will be considered in executive session from 8:30 to 10 a.m. tomorrow, to be followed by a discussion on the same topic with members of the Military Liaison Committee and AEC. Worthington said that his Subcommittee on Material Flow (agenda 4) has not found anything urgent to report, so the matter can be held over.

At 10:05 a.m. GAC was joined by Commissioner Bacher and Director of Research Fisk. Bacher told us that, since

We did "not see how it would be possible under the most favorable circumstances to have any considerable portion of the present power supply of the world replaced by nuclear fuel before the expiration of twenty years."

a number of candidates have refused to take the position of director of Clinton Laboratories, the commission has decided to remove reactor development from there. DuBridge asked what would be left at Clinton, and Bacher replied, both stable and radioactive isotope production, the Oak Ridge Institute, a strong chemical development based on facilities that exist there and nowhere else, and biological research. Fermi expressed some concern about this move, indicating that some of the people engaged in reactor work at Clinton Laboratories might not want to move to the Argonne laboratory. We agreed that, among the key people, it would be questionable as to whether Alvin Weinberg (physicist) and Gale Young (physicist) would be willing to transfer. The consensus was that there was a lack of unity of direction of effort at Clinton and that, in spite of AEC's urgings, the laboratory has actually avoided taking responsibility for a vigorous program of reactor development. We concluded that, even though the proposal to centralize at Argonne may lead to loss of men, it appears that we may be left with an inadequate reactor program if the transfer to Argonne is not made.

[Actually, such a transfer of reactor work from Clinton Laboratories to the Argonne Laboratory never took place.]

We then approved Oppenheimer's draft of the report on atomic energy activities for 1947 to President Truman with the addition of a couple of editorial changes and adjourned at 1:15 p.m.

[Our report was sent to President Truman on Dec. 31, 1947.]

December 31, 1947
The President
The White House
Washington, D.C.
My dear Mr. President:

A year has passed since you named us to the General Advisory Committee, to advise the Atomic Energy Commission on scientific and technical matters relating to materials production and research and development. We were prepared to understand the importance of the work of the Commission for the welfare of the United States. Thus we have taken our duties as advisors very earnestly, have devoted to them much time and study, and for at least fifteen days, at intervals throughout the year,

have held meetings at which all of us were present. We have had frequent and candid discussions with the Atomic Energy Commission, with its staff, and with the Military Liaison Committee, and have reported such recommendations as we were able to make to the Commission in seven detailed reports, which necessarily have a very high classification.

Our activity during this year reflects not only the sense of great importance which we attach to successful development of this field; it also reflects the difficulties with which the Commission was faced in assuming its responsibilities, and the unsatisfactory state of its inheritance. We very soon learned that in none of the technical areas vital to the common defense and security, nor in those looking toward the beneficial applications of atomic energy, was the state of development adequate. Important questions of technical policy were undecided, and in many cases unformulated. Giant installations and laboratories were operating with confused purposes and with inadequate understanding of the importance and relevance of the technical problems before them. Our atomic armament was inadequate, both quantitatively and qualitatively, and the tempo of progress was throughout dangerously slow. This state of affairs can in large measure be attributed to the long delays in setting up an atomic energy authority, and to the inevitable confusions of policy and of purpose which followed the termination of the war. The difficulties were increased by the fact that the war-time installations and laboratories, which served so well their primary function of developing atomic weapons for early military use, were in most cases not suited to continue the work as the nature of the technical problems altered, and as the transition from war-time to peace-time operation changed the conditions under which rapid progress might be possible.

It has thus been our function to assist the Commission in formulating technical programs, both for the short and for the somewhat longer term. These programs are aimed in the main at three objectives;

(1) The development, improvement and increase of atomic armament.

(2) The development of reactors for a variety of purposes.

(3) The support of the physical and biological sciences which in one way or another touch on the field of atomic energy.

As to the improvement of our situation with regard to atomic weapons, we are glad to report that the year has seen great progress, and that we anticipate further progress in the near future. From the beginning, we shared with the Commission an understanding of how dangerous complacency could be with regard to our work in this field. We have been much gratified at the establishment of Pacific proving grounds, where the performance of altered and improved weapons can be put to the test of actual proof and measurement. While much yet remains to be done, and while the long term program of atomic armament is only in its earliest beginnings, we nevertheless believe that steps already taken to improve our situation, and others which will follow as time makes them appropriate, have gone very far toward establishing this activity on a sound basis.

Atomic reactors have many purposes. They can produce the fissionable materials which can be used in atomic weapons or as fuel for other reactors. They can be useful instruments of research in the physical and biological sciences and in technology. They may, within a decade, be developed to provide sources of power for specialized application, for instance, for the propulsion of a limited number of naval craft. They may, within a time which will probably not be short, and which is difficult to estimate reliably, be developed to provide general industrial power, and so make important contributions to our whole technological and economic life. This variety of purpose, the novelty of the field, and the relatively small number of men trained to work in it, makes substantial progress in the development of atomic reactors difficult to realize. Many steps have been taken by the Commission during the past year to encourage work in this field, to invite the participation of industry, to promote the completion and construction of promising specific designs, and to enlist the participation of qualified experts. Yet, it is the opinion of the Advisory Committee that much yet remains to be done, that new personnel and new talent must come to contribute, and that many years will elapse before our work in this field has the robustness and vigor which its importance justifies. As an aid to the Commission, we have attempted to formulate the prospects, and to give some estimates of the nature of the effort required, for attainment of the various objectives. We believe that a more widespread understanding of the nature of the problems, and of the contributions which engineers can and must make, and of the way in which industry can helpfully participate, are essential for the health of these efforts.

In the support of basic science, we have welcomed the broad interpretation of its responsibilities which the Atomic Energy Commission has maintained. We studied in detail the proposals recently adopted for making certain radioactive isotopes available, primarily for biological and medical research, not only within this country, but abroad. We see in this a prudent but inspiriting example of the extension to others of the benefits resulting from the release of atomic energy, an extension sure to enrich our knowledge and our control over the forces of nature.

During the last year we have frequently come upon a problem, the further consideration of which seems to us essential. We have been forced to recognize, in studying the possible implementation of technical policy, how adverse the effects of secrecy, and of the inevitable misunderstanding and error which accompany it, have been on progress, and thus on the common defense and security. We believe that in the field of basic science, the Commission has inherited from the Manhattan District, and has maintained, an essentially enlightened policy. Even in the fields of technology, in industrial applications, in military problems, the fruits of secrecy are misapprehension, ignorance and apathy. It will be a continuing problem for the Government of the United States to re-evaluate the risks of unwise disclosure, and weigh them against the undoubted dangers of maintaining secrecy at the cost of error and stagnation. Only by such re-evaluation can the development of atomic energy make its maximum contribution to the securing of the peace, and to the per-

"This variety of purpose, the novelty of the field, and the relatively small number of men trained to work in it, makes substantial progress in the development of atomic reactors difficult to realize."

petuation and growth of the values of our civilization.

We are, my dear Mr. President

Very sincerely yours,
James B. Conant
Lee A. DuBridge
Enrico Fermi
I.I. Rabi
Hartley Rowe
Glenn T. Seaborg
Cyril S. Smith
Hood Worthington
J.R. Oppenheimer, Chairman

"At our meetings in June 1948, we considered the issues of radiological warfare, custody of atomic weapons, material balance in the production of fissionable material, and the urgent need for the reorganization of the commission's method of operation."

Tuesday, Dec. 30, 1947

In Chicago. I again took a taxi to the University of Chicago campus and met at 8:50 a.m. with the same GAC members as yesterday, in addition to Dr. Norris E. Bradbury (director of Los Alamos Laboratory), who was there to present the plan of shots for the test of nuclear weapons, the Sandstone series, scheduled for the Eniwetok Atoll in the Pacific next spring. At the beginning of the discussion, I raised the additional question about continuous monitoring of the accumulation of atmospheric radioactivity, particularly radiokrypton gas, resulting from pile operation, as a method of measuring the level of plutonium production; we agreed to make this an item for discussion at our next meeting. After Bradbury's presentation, DuBridge moved that we support the Los Alamos plan for nuclear weapons testing; the motion was seconded by Conant and carried unanimously. This motion signifies our approval of the types of weapons and order of their testing but leaves open some details of the design and the precise criteria for determining whether a weapon is good or bad. This executive session adjourned at 10:05 a.m.

We were joined at 10:15 a.m. by Gen. Lewis H. Brereton (chairman of the Military Liaison Committee), Rear Adm. William S. Parsons, Rear Adm. Thorvald Solberg, Commissioner Bacher, Gen. McCormack (director of military application), and Capt. James S. Russell (McCormack's deputy). Oppenheimer summarized the previous discussion and said that GAC approved the Los Alamos plan for nuclear weapons tests. There was then a discussion of the laboratory techniques used for

investigation of the assembly behavior of the nuclear weapons and of the measurements to be made in the tests. We talked about the possibility of subsequent operations at Eniwetok Atoll and whether there should be two or three tests in the Sandstone series, on the basis that GAC had presented evidence of the need for three shots. Finally, it was agreed that the test plan would be considered for final action by the research and development board, the Military Liaison Committee, and AEC after it has been carefully outlined for consideration by these three bodies. We adjourned at 11:30 a.m.

At our meetings in June 1948, we considered the issues of radiological warfare, custody of atomic weapons, material balance in the production of fissionable material, and the urgent need for the reorganization of the commission's method of operation:

Friday, June 4, 1948

In Washington. I went by taxi to the AEC building, where I attended the 10:10 a.m. opening session of the tenth GAC meeting. All members except DuBridge were present (Conant, Fermi, Oppenheimer, Rabi, Rowe, Smith, Worthington and I) plus GAC Secretary Manley, Assistant Secretary Tomei, and Mr. DeMatteis (security officer). We discussed a number of general items, such as engineering training (GAC believes that more should be done than merely increasing the number of conventionally trained engineers, that the engineers should be trained to face new situations). Other topics brought up by memoranda and staff papers were support of ONR, material balance in the production of ^{235}U and ^{239}Pu, GAC–AEC relations, and so on.

At 11:35 a.m. Gen. McCormack, Col. Kenneth E. Fields (Military Application), and David B. Langmuir (executive secretary, Program Council) joined us, and the discussion turned to weapons. We were told about the results of the Sandstone tests and talked about the interim stockpile and the Los Alamos program (McCormack suggested that serious consideration be given to the Booster-type weapon). He then brought up the subject of radiological warfare, saying that the panel is seriously considering the production of radioactive materials by the neutron irradiation of special materials rather than the use of separated fission products. Oppenheimer pointed out that problems of dispersal and military use had not been considered, and McCormack said that the Armed Forces Special Weapons Project will study the military problems. The discussion continued until 12:30 p.m.

After lunch (1:40 p.m.) we met in executive session and continued talking about radiological warfare, particularly the matter of dispersal. Fermi said he felt we should study critically the issue of irradiated substances vs. fission products. Oppenheimer, too, noted that he would like to see tests of effectiveness before any actual production is initiated. We agreed to re-emphasize to the commission the importance of the problems associated with dispersal and military use.

[The notion of radiological warfare was soon abandoned.]

At 2:05 p.m. Robert Bacher came into the meeting to inform us about custody of atomic weapons. (Deputy General Counsel Joseph A. Volpe, Waymack, and DuBridge also came in during the discussion.) Bacher went over the background of the subject; we discussed three points: (1) the policy question as to whether the National Military Establishment should have custody; (2) the technical question of whether appropriate arrangements can or cannot be made for satisfactory surveillance and refabrication; and (3) the effect of the complexity of these arrangements on the time at which transfer from AEC might be considered. The chairman ruled that the first point was not a matter for GAC action. This evoked a long, serious discussion on the various points, such as: the danger that an international emergency might suddenly explode with the use—militarily perhaps sound but politically disastrous—of atomic weapons in the absence of extraordinary control over this type of military power. It was pointed out that the danger of being too late in the use of atomic weapons is much less than the danger of being too early. Eventually, the chairman summarized with: (1) there seem to be some technical arguments for the military to have possession of some atomic weap-

ons, but no arguments can be made for a wholesale transfer; (2) the problems associated with surveillance and reworking of the stockpile would require establishing such complex arrangements—arrangements that would introduce potential friction—that one could question the worth of the transfer; and (3) the draft resolution presented for GAC's information does not appear to express a satisfactory basis for the transfer of custody.

[In actuality, a gradual transfer of custody to the military took place during the ensuing years.]

The session adjourned at 6:20 p.m.

Saturday, June 5, 1948

In Washington. GAC convened at 9:10 a.m. with all members present, although Fermi was absent from 10 a.m. until 12:15 p.m. (in order to consult an eye doctor). Brosnan replaced DeMatteis as security officer. Since we covered so much material yesterday, the committee decided to cover such items, although not strictly within the duties of GAC, as necessary for carrying out a successful technical job (e.g., the organization of AEC). We talked about this for a while, and the chairman asked Conant to choose two people (DuBridge and Rowe) to prepare a statement about this matter.

At 1:30 p.m., with Conant and his committee absent to work on their statement, we again met and approved the report of the Subcommittee on Material Balance (Fermi [chairman], Oppenheimer, Worthington, and I), which attempted to show how to obtain the optimum stockpile of ^{235}U and ^{239}Pu through coordinated operation of Oak Ridge and Hanford, subject to limitations of raw materials supply and cost. Then, at 2:10 p.m., Conant's committee on the organization of AEC presented its draft. Some suggestions were made; the chairman will make an oral introduction and presentation to the commissioners this afternoon, although GAC has not yet accepted the paper.

We had a short recess and then talked about thermonuclear weapons. Oppenheimer said that the reason for the urgency in discussing this is that it will be necessary to increase the production of tritium if the work on thermonuclear weapons is to be encouraged. He feels the Booster bomb design *should be encouraged. Conant expressed concern about the possible development of more devastating weapons; however, Fermi suggested that it is not wise to remain ignorant. Fermi then went on to explain the mechanisms of the Booster bomb and the differences between it and the Alarm Clock and the Super Bomb.*

At 4:30 p.m., we met, along with Manley, in the office of the AEC chairman with Lilienthal, Strauss, Bacher, and Waymack. Oppenheimer said that GAC had thought about the situation with respect to atomic energy and GAC's role, especially with the present uncertainties for the future of AEC and of some GAC members. GAC, Oppenheimer said, is concerned with the lack of progress in the question of understanding security, the failure to achieve adequate methods for public accountability, and so on. He then read the draft prepared by Conant, DuBridge, and Rowe.

The General Advisory Committee is venturing to make a report to the Commission on a subject not within their jurisdiction. We shall speak frankly because we feel the situation is extremely serious. It is possible to give technical and scientific advice effectively only if one has confidence in the soundness of the organization which is in a position to benefit from the counsel. We are sorry to report that in our collective judgment this condition is lacking. We do not want this to be a destructive criticism, for we know all the difficulties which have confronted the Commission. Furthermore, we congratulate the Commission on success of the weapon development program. We, therefore, are going to make a specific proposal as to what we think should be done.

We feel that the original decision of the Commission to decentralize their operations on a geographic basis was a mistake. The fear that was expressed by the chairman of the Commission at our first meeting, namely, that a functional organization run from Washington would be over-centralized, is in our opinion unjustified. Watching the operations of the organization for one and a half years has convinced us that satisfactory results can only be obtained by a drastic reorganization. Fundamentally, this would mean a reversal of the initial decision and the establishment of a line rather than a staff organization. This would be a functional rather than a geographic arrangement. We envisage five key men, in addition to the General Manager, of relevant experience and high competence, and to whom are delegated large responsibilities. These would be the Director of Research, Director of Weapons, Director of Reactors, Director of Production, and an overall administrative officer. We are aware that the law does not set up these Divisions but we feel it would be possible to put this scheme into effect within the framework of the law. Otherwise the law should be changed. For example, the Division of Engineering might be the Division of Reactors without doing violence to the English language. A second difficulty that we recognize is that the work is actually carried out by contractors. We believe, however, that each installation can be made the responsibility of one of the four operating Divisions listed above. It should not be too hard to work out a flexible arrangement which in fact gives the technical director of the contractor direct access to the Commission's representative and vice versa.

We venture to point out that the operation of the Commission which has been successful, namely, improvement of new weapons, has been the one in which the administrative setup is similar to that outlined above. The geographic area of Los Alamos coincides with a functional division and the contractor being a university has never been called upon to exercise a technical judgment. We believe if five top men such as envisaged above could be assembled and given responsibility they, together with the General Manager, should form an Executive Committee. The Commission should delegate to this Committee operating responsibility for all technical matters and should regularly obtain through this Committee a working knowledge of the technical problems. In its main function of formulating broad national policies the Commission will surely need staff advisors. Some of us feel that the Commission has been so concerned with the details of the operation that they have been unable to formulate general policy.

The GAC in presenting this report does not endorse the details given above which are the product of a subcommittee's thinking. They do believe, however, that a drastic reorganization is important and that something along the line suggested can and should be undertaken. We are afraid we can be of little use to the Commission under the present organization. We despair of progress in the reactor program and see further difficulties even in the areas of weapons and production unless a reorganization takes place.

There was a long discussion after Oppenheimer finished reading the draft, with some disagreement by the commissioners. [Actually, the commissioners, especially Chairman Lilienthal, were dismayed by the report. However, a reorganization, much along the lines suggested, actually took place in the following months.]

Sunday, June 6, 1948
In Washington. I again attended the GAC meeting, which began at 9:40 a.m. in the AEC building. Everyone except Conant was present. The session began with some rather routine items: minutes of the last meeting and a discussion of the continuation or replacement of GAC members whose terms expire Aug. 1, 1948. We then went on to the question of tritium production and the Booster bomb. The conclusion was reached that Los Alamos and Hanford people should be consulted about the amount of time that could be devoted to the tritium production, and that Los Alamos should proceed with design considerations and experimental investigations of the possibility of a Booster bomb. I then raised the question of personnel security, which has become an issue between the scientists and the commission. Oppenheimer agreed, saying that the investigative agency should go after the real difficulties—those individuals who are poten-

tial or actual traitors—but should take a reasonable risk in favor of not hurting individuals. Fermi also agreed and said the fetish for security with its ridiculous aspects should be removed. GAC will include the following in its report:

> *The GAC has not understood the basic policy at the root of the rules at present in force with regard to security and secrecy of information. We most strongly recommend that the Commission, if necessary through the work of an ad hoc panel, make a fundamental study of the issues involved, particularly with regard to the use of secrecy as an instrument for maintaining security. . . .*

On July 24, 1948, I made the following note in my journal:

A noon press release presented a statement from President Truman, announcing the release of the Fourth Semi-annual Report of the Atomic Energy Commission. President Truman said:

> As President of the United States, I regard the continued control of all aspects of the atomic energy program, including research, development, and the custody of atomic weapons, as the proper functions of the civil authorities. Congress has recognized that the existence of this new weapon places a grave responsibility on the President as to its use in the event of a national emergency. There must, of course, be very close cooperation between the civilian Commission and the Military Establishment. Both the military authorities and the civilian Commission deserve high commendation for the joint efforts which they are putting forward to maintain our nation's leadership in this vital work. . . .

With regard to secrecy, he added:

> Secrecy is always distasteful to a free people. In scientific research, it is a handicap to productivity. But our need for security in an insecure world compels us, at the present time, to maintain a high order of secrecy in many of our atomic energy undertakings. When the nations of the world are prepared to join with us in the international control of atomic energy, this requirement of secrecy will disappear. Our Government has sought, through its representatives on the United Nations Atomic Energy Commission, to find a common basis for understanding with the other member nations. However, the uncompromising refusal of the Soviet Union to

"We are afraid we can be of little use to the Commission under the present organization. We despair of progress in the reactor program and see further difficulties even in the areas of weapons and production unless a reorganization takes place."

> "Secrecy was an issue that improved with time but has not yet been satisfactorily resolved, especially with respect to the cumbersome and difficult declassification of so-called secret material."

participate in a workable control system has thus far obstructed progress. . . .

I am pleased to see that President Truman has adopted a view on AEC civilian control of atomic weapons in line with GAC's strongly held views.

Secrecy was an issue that improved with time but has not yet been satisfactorily resolved, especially with respect to the cumbersome and difficult declassification of so-called secret material. On April 6, 1949, GAC members had an interesting meeting with President Truman:

The nine GAC committee members (Oppenheimer, Oliver E. Buckley [president, Bell Telephone Laboratories], Conant, DuBridge, Fermi, Rabi, Rowe, Smith, and I) met with President Truman in the Oval Office. The meeting was very interesting and successful. The president expressed great optimism about the international situation. To our surprise, he said he felt that in two years there would be general settlement with Russia that would include the outlawing of atomic weapons and said he has concluded this on the basis of the information available to him. He said he appreciated very much the helpful services of GAC members in advancing the atomic weapons program. Continuing in his optimistic mood, he said that indications are that we won't have to work at advancing the atomic weapons program ever again.

Oppenheimer emphasized to the president that GAC has been urging the commission to release more and more information to the public, which will establish a good foundation for the commission's program and be consistent with democratic processes in general. The president expressed approval of such a program of release of information.

The president also discussed the matter of custody and production of atomic weapons. He said he had heard from Senator Millard E. Tydings that the custody and production of atomic weapons should be the responsibility of the military. The president said that he disagreed with this and that he believed firmly in the civilian control of atomic energy.

All in all, it was a very satisfactory meeting.

In June 1949, some of the problems were the same, some different. Oliver E. Buckley had replaced Worthington on GAC; Henry D. Smyth (Princeton University physicist) and Gordon E. Dean (a lawyer) had replaced Bacher and Waymack on the commission; and Kenneth S. Pitzer (University of California, Berkeley, chemist) had replaced James Fisk as director of research:

Thursday, June 2, 1949

In Washington. I took a cab to the AEC building in time to attend the 9 a.m. session of GAC. All members were present (J.R. Oppenheimer, O.E. Buckley, J.B. Conant, L.A. DuBridge, E. Fermi, I.I. Rabi, H. Rowe, C.S. Smith and I). We went over the agenda of this meeting—a few items were added. We decided to plan for an evening meeting tonight in order to finish by 4 p.m. on Saturday. I then gave a brief report of the meeting yesterday on the chemical problems with plutonium and uranium recovery. We decided to meet in Berkeley on July 14 and 15.

Commissioner Strauss and new AEC commissioners H.D. Smyth and G.E. Dean (replacing Robert Bacher and William Waymack), along with K.S. Pitzer and Shields Warren (director of biology and medicine) came in at 9:30 a.m. Strauss expressed concern about the foreign distribution of isotopes, stating there would be indirect benefit to the military potential of any nation as the result of their receipt of isotopes; he gave the example of the shipment to the Norwegian Defense Research Establishment. Several members took issue with Strauss's position. At this point Strauss and Dean left the meeting, and Pitzer and others continued the discussion about the foreign distribution of isotopes and the fellowship program. Pitzer asked for input from GAC on these matters, as well as on the magnitude of the support of science appropriate for AEC to assume.

We reconvened at 2:05 p.m. with L.R. Hafstad (director of reactors) and U.S. Navy Capt. H.G. Rickover. . . . Rickover talked eloquently about the need for submarine nuclear propulsion. Finally, Hafstad summarized the objectives and justification of the four reactors currently planned.

At 5:05 p.m. the committee met with Adm. Tom B. Hill (Military Liaison Committee), Col. J.H. Hinds (Military Liaison Committee), McCormack, Shields, Warren (director, AEC Division of Biology and Medicine), and Paul C. Fine (nuclear weapons expert) to talk about weapon matters. Hill discussed the present status of the base surge problem and problems such as underground effects, atmospheric effects, thermal radiation effects, ionizing radiation effects, the use of the bomb as a tactical weapon, appropriate height of burst, etc. We agreed that resolution of such problems is important. Warren then discussed the "Gabriel Report" by N.M. Smith of Oak Ridge about the hazards to friendly populations from the use of atomic weapons; Smith concluded that 100 atomic bombs dropped on the USSR would lead to half the U.S. population with plutonium in their lungs. We decided this needed more study, and it was suggested that Joseph Hamilton at Berkeley would be a good consultant. Finally, McCormack described negotiations for a new contractor for Sandia Laboratory with Bell Laboratories and the Western Electric Company. McCormack also told us about the status of the smaller and lighter weapon. We adjourned at 6:10 p.m.

At 8:15 p.m. GAC met in executive session to prepare a draft statement about the present position and performance of the commission and GAC's role in advising the commission; this is in connection with the severe criticism of the commission, especially Chairman Lilienthal, by some members of the Joint Committee on Atomic Energy. We also decided to prepare a draft statement about our views on the investigational procedures in the award of fellowships. Finally, we decided to reaffirm our position on the positive value of the foreign distribution of radioisotopes. This session adjourned at 10:10 p.m.

The next meeting was held in Berkeley:

Thursday, July 14, 1949

At 9:45 a.m. in the Bldg. 8 Conference Room, Oppenheimer, DuBridge,

Rabi, Smith, and I met in executive session, along with our secretary, Manley, and his assistant, Tomei. We approved the agenda, talked a bit about the fellowship program, and asked Manley for copies of the statement we prepared at the last meeting:

> Our intention in recommending that a fellowship program be set up by the AEC was to implement one of the purposes of the Atomic Energy Act, Section 3(a), "to assist in the acquisition of an ever-expanding fund of theoretical and practical knowledge in such fields." The fellowship program, as it has been arranged by the AEC, assists young men and women of unusual abilities in the physical and biological sciences to devote themselves for a period of a year or more exclusively to specialized study and research. . . .
>
> We understand that proposals have been put forward that would require all holders of AEC fellowships to be cleared after an FBI investigation. We should like to register our strong disapproval of any such procedures. Admittedly, the tensions of the times and the secret nature of the atomic energy work require elaborate checks for all who have access to classified material. But to carry over the same security concepts to holders of fellowships who will in no way have access to secret or confidential information seems to us both unwise and unnecessary.

This session was adjourned at 10:05 a.m.

The same GAC members as present at the morning session reconvened at 1:20 p.m., along with Smyth, Pitzer, and George L. Weil (assistant director, Reactor Development). Much of the afternoon was taken up with a discussion of reactors, primarily brought about by Smyth's letter of July 12 to Oppenheimer. Smyth explained that he feels that, at the last GAC meeting, we were uncertain about the support of the reactor program and had doubts about the wisdom of proceeding with the materials testing reactor (MTR). It was explained that GAC consistently supported AEC with regard to each reactor and wants only to defer the question of location of the MTR. Some points brought out included concern about the expensiveness of the program, the rate of progress, and a doubt as to the justification of part of the program on the grounds of military necessity. GAC, we said, is on record as favoring the KAPL reactor, the Argonne reactor, the MTR, the Argonne–Westinghouse program, and procurement of the Arco, ID, site. Smyth said that the staff hopes to learn enough about reactors to enable them to be built in less remote locations in the future and to keep site developmental costs at a minimum. GAC agreed with these points. Oppenheimer said we also agree with Zinn's wish about locating the Argonne reactor at Arco, in order to operate at a higher power than permissible at the DuPage site, with the possible conversion to plutonium as a fuel, and because of the hazards. Smyth asked for a reaffirmation of the GAC's views on these points.

The meeting in September 1949 followed soon after the first Soviet atomic bomb test on Aug. 29, 1949:

Thursday, Sept. 22, 1949

In Washington. I took a taxi from the Mayflower Hotel to the AEC building for the 9:25 a.m. GAC meeting. Present were Oppenheimer, Buckley, Conant, DuBridge, Rowe, Smith, and I, Manley (secretary), and Tomei (administrative assistant). As usual, we first went over the agenda, and then Oppenheimer spoke on several of the items. We went over the evidence that an atomic bomb (their first) had been detonated over Russian Siberia. He briefly described the status of the negotiations on cooperation with the United Kingdom and with Canada. Oppenheimer told us that he has asked Manley to arrange for a presentation for us of the nuclear weapons stockpile data and predictions, along with the plans for expansion. Oppenheimer also pointed out that one matter needed our advice—the policy to be pursued in the award of fellowships and construction contracts since recent legislation limits the freedom of action of the commission in such matters. We looked at background papers until 10:40 a.m., when AEC commissioners Pike, Strauss, Dean, and Smyth came in along with Gen. McCormack (acting general manager). Col. Skaer (Division of Military Application) and Joseph A. Volpe (general counsel) came in later.

McCormack talked about the background of the requirements for fissionable materials that the Joint Chiefs of Staff had presented to the commission,

implying that the balance between plutonium and uranium had not been given detailed attention, but only the total amount of fissionable material, which corresponds exactly to the amount of plutonium that could be made by predicted Hanford operations plus the amount of uranium-235 that could be made by the planned expanded diffusion plant. He said that Dean Acheson (secretary of state), Louis A. Johnson (secretary of defense), and Lilienthal have been named to consider national requirements in more detail. Col. Skaer, at McCormack's request, gave us the figures for the stockpile position as of Aug. 31 and the estimates for 1950. After a brief discussion, Skaer left the meeting at 11 a.m.

Volpe talked about developments in the joint congressional committee hearings and discussions with the Appropriations Committee, which led to restrictive clauses, in the Appropriations Bill, on fellowships and construction. There was a discussion about the restrictive clauses, but we concluded that the commission has an obligation to carry out the fellowship program within the framework as laid down by Congress. We spoke to the commissioners and the General Manager about the question of a Saturday morning meeting; it was concluded that, if there were anything important to discuss, we would hold a session on Friday night in order to avoid a meeting on Saturday morning. The guests of the committee left at 12:05 p.m., and we then talked about the various construction programs.

Our afternoon meeting convened at 1:35 p.m. with Commissioner Dean, McCormack, Fine, Capt. A. McB. Jackson (USN), Col. P.T. Preuss, Capt. J.S. Russell (USN), Col. G.F. Schlatter, J.Z. Bowers, Norris E. Bradbury (director of the Los Alamos Laboratory), and Alvin C. Graves (Los Alamos). AEC commissioners Pike and Smyth came in later. Bradbury reported on the smaller, lighter weapon, the TX-5, discussing the design and the present status of the development work. Then McCormack talked about the use of this weapon to the military, pointing out that the current generation of plane designs is not adaptable to the size and weight of a TX-5 model. The next generation of plane designs will be reached in about three years. Los Alamos will decide on the weapon to be tested in 1951. It appears that if this model is tested in 1951, it will not be available until 1956, when there is a design suitable for use with an airplane carrier. Oppenheimer expressed concern that this will leave a period from 1953 to 1956 in which this nation will not possess what it might have had in the way of coordinated carrier and weapon combinations. There was some discussion about this. McCormack then gave us some information about the development of guided missiles with atomic warheads. Presently, we have four possibilities, all of which have a range of about 100 miles and could carry some form of the TX-5 weapon—the first will be operational in 1954. Bradbury then outlined the present status of the weapons program as applied to those items, in addition to the TX-5 model to be tested in 1951. Graves reported on plans for the 1951 test series and the differences between this and the Sandstone operation. The cost is estimated to be about $50,000,000.

At this point Smyth asked about our thoughts on the Russian atomic bomb, mentioning three things that might become pertinent: (1) possible speed-up of the U.S. weapons program, (2) more intense concern with civil defense, and (3) the overall effect of public reaction on AEC policy. I remarked that I believe the futility of secrecy has now been emphasized and that a reduction of secrecy in certain areas should be considered in order to secure effectiveness of any increase of activity, and that the desirability of cooperation with the British is indicated. Buckley said that this is no occasion to relax secrecy on the art of atomic weapon manufacture and the production of materials. He also thought there may be pressure to increase the production of weapons, although not at the expense of stopping research. Smyth indicated that he feels there is a real danger of such pressures. We discussed whether any real increase in numbers can be achieved by attempting to speed up production and whether it is desirable to try to shift scientific talent from research to applied areas. Smyth left at 3:35 p.m., and we had a short recess.

As indicated earlier a GAC action that gained much publicity

was the recommendation (at a meeting held Oct. 28–30, 1949, which I didn't attend because of a visit to Sweden) that the U.S. not proceed with a high-priority program to develop the super bomb (hydrogen bomb). I had written to Oppenheimer that I had reluctantly come to the conclusion that the United States should proceed with such a program because it was certain that the Soviet Union would do so:

October 14, 1949
Dr. J. Robert Oppenheimer
The Institute for Advanced Study
Princeton, New Jersey
Dear Robert:

I will try to give you my thoughts for what they may be worth regarding the next GAC meeting, but I am afraid that there may be more questions than answers. Mr. Lilienthal's assignment to us is very broad, and it seems to me that conclusions will be reached, if at all, only after a large amount of give and take discussion at the GAC meeting.

A question which cannot be avoided, it seems to me, is that which was raised by Ernest Lawrence during his recent trip to Los Alamos and Washington. Are we in a race along this line and one in which we may already be somewhat behind so far as this particular new aspect is concerned? Apparently this possibility has begun to bother very seriously a number of people out here, several of whom came to this point of view independently. Although I deplore the prospects of our country putting a tremendous effort into this, I must confess that I have been unable to come to the conclusion that we should not. Some people are thinking of a time scale of the order of 3 to 5 years which may, of course, be practically impossible and would surely involve an effort of greater magnitude than that of the Manhattan Project. My present feeling could perhaps be best summarized by saying that I would have to hear some good arguments before I could take on sufficient courage to recommend not going toward such a program.

If such a program were undertaken, a number of questions arise which would need early answers. How would the National Laboratories fit into the

"Although I deplore the prospects of our country putting a tremendous effort into this, I must confess that I have been unable to come to the conclusion that we should not."

program? Wouldn't they have to reorient their present views considerably? The question as to who might build neutron-producing reactors would arise. I am afraid that we could not realistically look to the present operators of Hanford to take this on. It would seem that a strong effort would have to be made to get the DuPont Company back into the game. It would be imperative that the present views of the Reactor Safeguard Committee be substantially changed.

I just do not know how to comment, without further reflection, on the question of how the present "reactor program" should be modified, if it should. Probably, after much discussion, you will come to the same old conclusion that the present four reactors be carried on but that an effort be made to speed up their actual construction. As you probably know, Ernest is willing to take on the responsibility for the construction near Berkeley of a 300-megawatt heavy water–natural uranium reactor primarily for a neutron source, and on a short time scale. I don't know whether it is possible to do what is planned here, but I can say that a lot of effort by the best people here is going into it. If the GAC is asked to comment on this proposal, it seems to me clear that we should heartily endorse it! So far as I can see, this program will not interfere with any of the other reactor-building programs and will be good even if it does not finally serve exactly the purpose for which it was conceived; I have recently been tending toward the conviction that the United States should be doing more with heavy water reactors (we are doing almost nothing). In this connection, it seems to me that there might be a discussion concerning the heavy water production facilities and their possible expansion.

Another question, and one on which perhaps I have formulated more of a definite opinion, is that of secrecy. It seems to me that we can't afford to continue to hamper ourselves by keeping secret as many things as we now do. I think that not only basic science should be subject to less secrecy regulation, but also some places outside of this area. For example, it seems entirely pointless now to hamper the construction of certain types of new piles by keeping secret certain lattice dimensions. In case anything so trivial as the

conclusions reached at the recent International Meeting on declassification with the British and Canadians at Chalk River is referred to the GAC, I might just add that I participated in these discussions and thoroughly agree with the changes suggested, with the reservation that perhaps they should go further toward removing secrecy.

I have great doubts that this letter will be of much help to you, but I am afraid that it is the best that I can do at this time.

Sincerely yours,
Glenn T. Seaborg
GTS/db

The indications are that Oppenheimer did not present my letter for discussion at the GAC meeting. The majority opinion of GAC (Conant, Rowe, Smith, DuBridge, Buckley, Oppenheimer) recommended against the production of such a super bomb:

October 30, 1949

We have been asked by the Commission whether or not they should immediately initiate an "all-out" effort to develop a weapon whose energy release is 100 to 1000 times greater than those of the present atomic bomb. We recommend strongly against such action.

We base our recommendation on our belief that the extreme dangers to mankind inherent in the proposal wholly outweigh any military advantage that could come from this development. Let it be clearly realized that this is a super weapon; it is in a totally different category from an atomic bomb. The reason for developing such super bombs would be to have the capacity to devastate a vast area with a single bomb. Its use would involve a decision to slaughter a vast number of civilians. We are alarmed as to the possible global effects of the radioactivity generated by the explosion of a few super bombs of conceivable magnitude. If super bombs will work at all, there is no inherent limit in the destructive power that may be attained with them. Therefore, a super bomb might become a weapon of genocide.

The existence of such a weapon in our armory would have far-reaching effects on world opinion: reasonable people the world over would realize that the existence of a weapon of this type whose power of destruction is essentially unlimited represents a threat to the future of the human race which is intolerable. Thus we believe that the psychological effect of the weapon in our hands would be adverse to our interest.

We believe a super bomb should never be produced. Mankind would be far better off not to have a demonstration of the feasibility of such a weapon until the present climate of world opinion changes.

It is by no means certain that the weapon can be developed at all and by no means certain that the Russians will produce one within a decade. To the argument that the Russians may succeed in developing this weapon, we would reply that our undertaking it will not prove a deterrent to them. Should they use the weapon against us, reprisals by our large stock of atomic bombs would be comparably effective to the use of a super.

In determining not to proceed to develop the super bomb, we see a unique opportunity of providing by example some limitations on the totality of war and thus of limiting the fear and arousing the hope of mankind.

Two members (Fermi and Rabi), in a kind of minority report, recommended that the United States invite the nations of the world to join us in a solemn pledge not to proceed in the development or construction of weapons in this category:

October 30, 1949

An Opinion on the Development of the "Super"

A decision on the proposal that an all-out effort be undertaken for the development of the "Super" cannot in our opinion be separated from considerations of broad national policy. A weapon like the "Super" is only an advantage when its energy release is from 10-1000 times greater than that of ordinary atomic bombs. The area of destruction therefore would run from 150 to approximately 1000 square miles or more.

Necessarily such a weapon goes far beyond any military objective and enters the range of very great natural catastrophies [sic]. By its very nature it cannot be confined to a military objective but becomes a weapon which in practical effect is almost one of genocide.

It is clear that the use of such a weapon cannot be justified on any ethical ground which gives a human being a certain individuality and dignity even if he happens to be a resident of any enemy country. It is evident to us that this would be the view of peoples in other countries. Its use would put the United States in a bad moral position relative to the peoples of the world. Any postwar situation resulting from such a weapon would leave unresolvable enmities for generations. A desirable peace cannot come from such an inhuman application of force. The postwar problems would dwarf the problems which confront us at present. The application of this weapon with the consequent great release of radioactivity would have results unforeseeable at present, but would certainly render large areas unfit for habitation for long periods of time.

The fact that no limits exist to the destructiveness of this weapon makes its very existence and the knowledge of its construction a danger to humanity as a whole. It is necessarily an evil thing considered in any light.

For these reasons we believe it important for the President of the United States to tell the American public, and the world, that we think it wrong on fundamental ethical principles to initiate a program of development of such a weapon. At the same time it would be appropriate to invite the nations of the world to join us in a solemn pledge not to proceed in the development or construction of weapons of this category. If such a pledge were accepted even without control machinery, it appears highly probable that an advanced stage of development leading to a test by another power could be detected by available physical means. Furthermore, we have in our possession, in our stockpile of atomic bombs, the means for adequate "military" retaliation for the production or use of a "Super."

"We believe a superbomb should never be produced. Mankind would be far better off not to have a demonstration of the feasibility of such a weapon until the present climate of world opinion changes."

GAC members learned from President Truman on Jan. 31, 1950, of his decision that the United States should proceed with the development and production of the hydrogen bomb:

Tuesday, Jan. 31, 1950

In Washington. I went by taxi to the AEC building for the 9:30 a.m. session of the 19th GAC meeting. All members were present except Rowe: Oppenheimer, Buckley, Conant, DuBridge, Fermi, Rabi, Smith, and I, plus Manley (secretary) and Tomei (assistant). The session was preceded by a meeting of the reactors subcommittee with the director of reactor development. We then went over the agenda, and then at 11 a.m. we talked informally with the Security Survey Panel (the session was adjourned).

At 1:30 p.m. we reconvened, and at 2:15 p.m. Oppenheimer read a press release containing the president's statement to proceed with the super bomb development. Pitzer (director of research) came in at 2:20 p.m. to talk about the fellowship program; he and Tomei left when Chairman Lilienthal arrived at 2:55 p.m. to tell us about the president's decision to go ahead with the super bomb and the probable nature of the anticipated directive to AEC. GAC members were stunned and disheartened. Lilienthal left at 3:35 p.m., and Pitzer returned to talk about proposals from the Berkeley Radiation Laboratory for the design of two high-energy, high-current accelerators. Jesse C. Johnson (director of raw materials) spoke with us at 4:10 p.m. on future possibilities of uranium ore supplies. Johnson left at 4:50 p.m., Smith reported briefly on reactor matters, and then I spoke on the status of chemical processing plans and programs for plutonium. We talked about a couple of more items for the next session, and about the significance of the proposed changes in the declassification policy, and then adjourned at 5:35 p.m.

The opinions of GAC members that our country should not give high priority to the development of a hydrogen or super bomb are, in retrospect, provocative and tantaliz-ing. Had I been present at the GAC meeting in October 1949, I suspect that I might have associated myself with the opinion expressed by Fermi and Rabi: "It would be appropriate to invite the nations of the world to join us in a solemn pledge not to proceed in the development or construction of weapons of this category." Such a posture might, as a minimum, have slowed the nuclear arms race and, as a maximum, have mitigated the cold war between the United States and the Soviet Union. Oppenheimer's role in formulating the negative recommendation on the super bomb was key to the suspension of his security clearance a few years later.

During April and May 1954, early in the Eisenhower administration, hearings were held on the matter of security clearance of J. Robert Oppenheimer before the AEC Personnel Security Board (Chairman Gordon Gray, Ward V. Evans, and Thomas A. Morgan). Charges of security indiscretions, dating as far back as the period immediately preceding World War II, were brought by AEC Chairman Strauss and sanctioned by President Eisenhower. A long string of witnesses for the prosecution and defense testified before the board. I was not called as a witness because I would not be helpful to either side—the prosecution, because I maintained that I did not regard Oppenheimer to be (or to ever have been) a security risk, and the defense, because I would need to testify that Oppenheimer had failed to present my letter of Oct. 14, 1949, for discussion at the GAC meeting that month, which I missed. The board, mistakenly I believe, found him to be a security risk and upheld the suspension of his security clearance by a vote of 2 to 1 (Evans voted against the suspension; Gray and Morgan voted to uphold it). This unfortunate action had a devastating effect on Oppenheimer and contributed, I believe, to the deterioration of his health and premature death in 1967.

I attended my last GAC meeting on June 1–3, 1950:

Thursday, June 1, 1950

In Washington. I went by cab to the AEC building for the 21st GAC meeting. My term as a member expires with this meeting. When Chairman Oppenheimer convened the meeting at 9:30 a.m., all members except Rabi— i.e., Buckley, Conant, DuBridge, Fermi, Rowe, and Smith and I—were present, in addition to Secretary Manley and Assistant Secretary Tomei. We went over the agenda and decided to speak informally of possible candidates for the expiring terms of Fermi, Rowe, and me between sessions. We then read background material until 11 a.m.

At 11 a.m. Acting Chairman Pike and AEC commissioners Dean and Smyth (Lilienthal and Strauss had resigned), along with General Manager Wilson, came in (Gen. McCormack joined the group during part of the discussion on custody and all of that on community management). During this morning session we talked about the matter of custody of atomic weapons (and the reservations of Bradbury and Carroll L. Tyler [manager, AEC Operations Office, Santa Fe, NM]), community management (Pike spoke of the concern of the commission on the effect of appropriation bills limiting fees of management contractors at Oak Ridge and Los Alamos), H-bomb statement (Pike mentioned the letter to the president on H-bomb information, which he understood had been submitted to a subcommittee of the National Security Council), international control of atomic energy (Oppenheimer noted that GAC believes this should have some attention), and control of information (Smyth expressed concern about the overstringent restrictions on purely technical information compared with the unsystematic release of news of the atomic energy program). We adjourned at 12:40 p.m.

In the course of the day and informally between sessions, we talked about such men as Norman Ramsey, John Manley, Warren Johnson, Manson Benedict, Franklin Long, Richard Dod-

son, Leonard Schiff, and Fred Seitz as possible candidates for the position of AEC director of research when Ken Pitzer completes his two-year term. Suggestions for replacements for Rowe, Fermi, and me included Robert Bacher, John von Neumann, Isadore Perlman, Warren Johnson, Wendell Latimer, Walter Whitman, and Manson Benedict.

Friday, June 2, 1950

In Washington. I went to the AEC building to attend the 9 a.m. GAC session. Again, all the members except Rowe were present. We first discussed the document describing the international control of atomic energy and decided to mention tritium as a "dangerous" material. Conant, DuBridge, and Oppenheimer were assigned to draft a statement of our views. We agreed to commend the commission on its proposed H-bomb statement, which was forwarded to President Truman, although GAC feels certain matters should be amplified. Additionally, we decided to talk again with the commissioners about the subject of carriers for atomic weapons. Conant left at 10:15 a.m.; after a short recess, we met with Pitzer at 10:35 a.m.

Pitzer went over the progress of the Berkeley neutron-producing accelerator project; he and I estimated that perhaps half of our research personnel might be diverted to this project. Oppenheimer suggested attracting new personnel to replace them. We also talked with Pitzer about the fellowship program, Brookhaven, AEC support of basic science, and the possible role of the National Science Foundation. Pike and Wilson came in at this point, and Pitzer presented data on the atmospheric krypton-85 studies to assess Soviet plutonium production. Wilson described the research committee (Wilson, Pitzer, Hafstad, and Warren), which is looking at the research programs of Brookhaven, Argonne, Ames, and Oak Ridge. Oppenheimer reminded the group about our previous comment on the usefulness of obtaining program statements from these laboratories. Pike, Wilson, and Pitzer left at 12:15 p.m. We went over the wording of the minutes of the 20th meeting, made a few changes, and then approved the draft.

"This unfortunate action had a devastating effect on Oppenheimer and contributed, I believe, to the deterioration of his health and premature death in 1967."

After a discussion of a few more points, we adjourned at 12:40 p.m.

At 6:30 p.m. I joined other GAC members and the commissioners at the Henry Smyth home (1529 29th St., N.W.) for a buffet supper.

Saturday, June 3, 1950

In Washington. I went to the 9:40 a.m. GAC session, during which we first went over the draft paper on International Control and made a few changes in wording.

At 10:30 a.m. Pike, Dean, Smyth, and Wilson joined us for our traditional final session. Oppenheimer gave his usual brilliant review. Oppenheimer, in summarizing the discussion on custody of atomic weapons, stated his view is that the proposals were not as separable from the joint civilian–military control of atomic weapons as might appear. Conant suggested transfer of a fraction of the stockpile to the Defense Department. We understand the need for intimate acquaintance of the Defense Department with atomic weapons and for relief from responsibility for AEC, but we understand the concerns of the manager, Santa Fe Operations Office, and the director of the Los Alamos Laboratory. Oppenheimer commended the commission for the proposed H-bomb statement and suggested that the release of the statement would not jeopardize security. We retiring members were thanked for our work. The session was adjourned at 12:25 p.m.

Oppenheimer prepared the following report of the 21st meeting and submitted it to Pike, acting chairman of AEC.

I. Homogeneous Thermal Reactors

The Director of Reactor Development discussed with us an interesting new proposal for a homogeneous thermal reactor to be built at Oak Ridge. The proposal contemplates a power level for the reactor of about one megawatt and a cost estimate of about two million dollars.

We are interested in this line of reactor development, in part because of the possibility of opening up the thorium breeding cycle, and in part because it may offer substantial power with a smaller investment of fissionable material and greatly simplified chemical processing. We are not, however, recommending to the Commission that the proposal as made be approved. . . .

II. International Control

The Atomic Energy Commission has asked for our views with regard to a letter from the Secretary of State, inquiring about the technical basis of the United Nations plan for control of atomic energy. Specifically, the letter asks "whether any technological changes have occurred or are likely to occur in the United States or abroad which would change the technical assumptions which underlie this plan or which would invalidate it or necessitate changes in its control features."

We have examined the recommendations of the United Nations Atomic Energy Commission, as embodied in Department of State Publication 3646. We have also examined in particular Part IV of the United Nations report entitled "The First Report of the Atomic Energy Commission to the Security Council." This deals with the scientific and technical aspects of the problem of control, and makes explicit the agreed view of the technical problem which underlay the control plan.

We have found (1) that there have been no scientific discoveries known to us which alter the situation; (2) that there may soon be technical developments which have some bearing on control problems; and (3) that, with the passage of time, major changes in the technical situation have occurred which profoundly alter the presuppositions under which the report appears to have been made. We may briefly summarize these points.

1. No scientific discoveries are known to us which open up sources of energy release not publicly known when the reports were written. No discoveries lend support to the view that the large-scale release of atomic energy can be

based on raw materials other than uranium and thorium.

2a. One technical development now underway in this country may, if successful, have an effect on the control plan. This is the electronuclear generation of neutrons. If this turns out to be practical on a large scale, it will mean that atomic energy can be released by converting thorium to U-233 without the use of natural uranium. This would mean that controls of thorium might have to be as strict as those of uranium. This development would also make it possible to produce not only U-233 but tritium and plutonium without the operation of reactors. The success and cost of this development cannot now be foretold; it is unlikely to be realized for a few years.

2b. The development of thermonuclear weapons now underway in this country may also have a bearing on the control plan. If this development is successful, it will mean that tritium must be regarded as a ''dangerous'' material. No development of thermonuclear weapons appears possible which does not start with an atomic explosion using plutonium, U-235, or U-233, and which does not use tritium produced in nuclear reactors, or perhaps by electronuclear neutrons.

3. There are at least three important changes in the technical situation that have occurred since the first use of atomic weapons. One is the production of atomic weapons by the Soviet Union; the second is the great accumulation of stocks of U-235 and plutonium, at least in this country; the third is the fact that the hopes for a rapid development of atomic power have not so far been fulfilled. The first two of these clearly create serious problems with regard to bringing into operation the control provisions of the plan. The third indicates that the development of a large-scale atomic power industry is less certain and may proceed more slowly and on a smaller scale than envisaged in the control arrangements. This point may be relevant in assaying the relative importance of development and control functions of an international authority limited solely to atomic energy.

III. Weapons and Carriers

We have in the past had occasion to discuss with the Commission our concern that the apparently highly successful development of the TX-5 was not being matched by a rapid development of an airplane capable of taking advantage of its reduced size and weight.

A related problem also bearing on the TX-5 and its possible 30-in. successor has come before us during our discussions with the Director of Military Application. From him we learned in some detail of the plans for adopting a howitzer to fire a light-weight gun-type atomic bomb. . . .

We have, therefore, inquired of the Director of Military Application as to whether the Department of Defense had under development a missile for the tactical delivery of atomic weapons suitable for carrying the TX-5 warhead, and with an accuracy and range suitable for this purpose.

As far as could be discovered through these consultations there is no appropriate development underway. It is true that a long-range guided missile with undetermined accuracy is scheduled for development within five to eight years. Neither the characteristics of this missile nor the schedule appears to us to meet the need. We believe that the usefulness of atomic weapons for tactical purposes, and the value of much of the work and much of the investment for which the Commission is responsible, may depend upon the solution of the problem we have outlined.

We recommend to the Commission that it explore the question of the availability of carriers for the best of the weapons developed or planned for development at Los Alamos, and that, in the event that the situation is as unsatisfactory as it now appears to be, the Commis-

"There are at least three important changes in the technical situation that have occurred since the first use of atomic weapons. One is the production of atomic weapons by the Soviet Union; the second is the great accumulation of stocks of U^{235} and plutonium . . the third is the fact that the hopes for rapid development of atomic power have not so far been fulfilled."

sion apprise the Department of Defense of the existence and the critical nature of the situation.

More generally, we have some doubt as to whether the Commission has availed itself adequately of the liaison machinery now in existence for informing the Department of Defense of problems within their jurisdiction arising out of weapon developments in the Commission's laboratories. If the liaison machinery proves inadequate for this purpose, it should be revised.

IV. Custody

We were asked by the Commission to examine AEC 2/21-22-23. This we have done.

We were naturally impressed by the force of the arguments put forward by the Director of Military Application in favor of assigning to the using service full responsibility for the maintenance and handling of weapon components that they must ultimately be prepared to use in combat. It is possible that this requirement could be adequately met by making available a reasonable number of non-nuclear components for practice. We are aware that the Director of Los Alamos Laboratory has strong misgivings with regard to the proposals put forward in AEC 2/23. We have not been completely successful in determining the reasons for these reservations, and hesitate to base our recommendations on conjecture. We do feel that the Commission should not act on the proposals without direct, frank, and complete exploration with the Los Alamos Laboratory of the ground and character of its reservations; and we so recommend to the Commission.

We also recommend to the Commission that it explore the bearing which the acceptance of the proposal of AEC 2/23 would have on the execution of the president's directive with regard to custody of atomic weapons.

V. Community Management

We have learned from the Commission that in their opinion certain legislative restrictions on appropriations now under consideration would have seriously damaging effects on the stability of the communities of Oak Ridge and Los Alamos. We desire to emphasize that both of these installations are now engaged in a critical phase of technical development which it would be contrary to the national interest to jeopardize.

VI. Proposed H-Bomb Statement

We have examined the communication from the Commission to the President (AEC 298/12) with reference to public information on the hydrogen bomb.

We are pleased that the Commission is attempting to have relaxed the severe restriction on official pronouncements on this subject. We regard the proposed public statement as a helpful step in the right direction. We are certain that its release can in no way jeopardize security. We would anticipate that some of the questions touched on in this statement would require further discussion. This seems to us true of the sentence at the end of the statement with regard to the Commission's vigorous prosecution of the peaceful uses of atomic energy. It likewise seems to us true of the deprecation of the possible radiological hazards of the hydrogen bomb. Finally, we believe that further inquiry must naturally be expected with regard to the degree of certainty and probable time scale of the development of thermonuclear weapons.

On Aug. 11, 1950, President Truman wrote to thank me for my service on GAC.

I met Alben W. Barkley, vice president during the second Truman administration (1949–53), on occasions when I visited Washington while I was serving on GAC. One of the occasions when Helen and I met Mrs. Dorothy Barkley was at President Kennedy's Inaugural Anniversary Dinner (Jan. 20, 1962) at the DC Armory. There I had the privilege of sitting between Mrs. Barkley and Maureen (Mrs.

August 11, 1950

My dear Dr. Seaborg:

It is with real regret that I note the expiration of
your term on the General Advisory Committee to the Atomic Energy
Commission.

These few years of our civilian-directed atomic energy
project have brought new and difficult problems and confronted
us with questions of far-reaching import to the country. I am
personally and deeply grateful to you for the time which you have
devoted as a member of the Committee to working out these problems
and to the building of a program which will make us strong in
peace as well as war.

Should there be need in the future, I know you will respond as generously as you have in the past to calls upon your
time and effort.

Very sincerely yours,

Harry Truman

Dr. Glenn Seaborg,
Professor of Chemistry,
University of California,
Berkeley,
California.

President Truman's letter thanking Seaborg for his service as a member of the General Advisory Committee to the U.S. Atomic Energy Commission. Aug. 11, 1950.

Michael) Mansfield, and Helen sat between Senators Carl Hayden and Michael Mansfield (Senate majority leader).

I had contact with or met Harry Truman numerous times following his presidency. During the time of my chancellorship at the University of California, Berkeley, I invited Truman to come to Berkeley to address our students. Unfortunately, he wasn't able to find the time to pay us a visit.

President Harry S Truman and Vice President Alben W. Barkley (left) riding in an open car on the parade route through Washington, D.C., on Inauguration Day, Jan. 20, 1949. Courtesy of the Harry S Truman Library Institute, Independence, MO.

I met him at the time of John F. Kennedy's inauguration as president (Jan. 18, 1961) while he was having lunch at the Mayflower Hotel. Here I renewed my invitation to him to speak to the students at Berkeley. I recall a memorable meeting with him at a stop in Kansas City on June 25, 1965, en route to San Francisco on Air Force One with President Lyndon B. Johnson:

President Johnson had breakfast with former President Harry Truman. I had breakfast with Senator Frank Church in the coffee shop. I participated in a motorcade to the airport. President Johnson presented me to Truman in front of the press and news and television cameras. I recalled to Truman that I had been a member, upon his appointment, of the first GAC of AEC.

My last contact with Harry Truman was a letter received from him, dated Nov. 5, 1969. On Tuesday, Dec. 5, 1972, I made a sad observation in my diary:

President Harry S. Truman was taken to a hospital in Kansas City today because of a serious condition of pulmonary congestion; the prognosis seems to be very bad.

Three weeks later, on Tuesday, Dec. 26, 1972, as I was preparing for a morning departure for Washington, DC, to attend a meeting of the American Association for the Advancement of Science, I learned of his death:

Soon after arising at 5:55 a.m., I heard over the radio that President Truman had died in Kansas City at 7:50 a.m. (5:50 a.m. Pacific time).

Then, two days later, on Thursday, Dec. 28, 1972, I paid tribute to Truman:

I changed into my tuxedo, took a taxi to the Sheraton Park Hotel, and went to the Assembly Room to attend a reception preceding the dinner for past AAAS presidents.

*We then went to Sheraton Hall. . . .
I presided over this program, beginning
with a request that the audience (about
400) stand in a moment of silence in
tribute to President Truman.*

Although my contacts with
Harry Truman were not extensive, I
did receive the impression that he
was a decisive person and deserved
his widespread characterization as a
"no-nonsense" man of action. His
campaign for re-election in 1948,
when he was running against the

President Lyndon B. Johnson (front
center), Seaborg (upper left), and
former President Harry S Truman
(upper right) meet during an *Air Force
One* stopover at the Kansas City, MO,
airport. June 25, 1965. Courtesy of the
Lyndon Baines Johnson Library, Austin, TX.

HARRY S TRUMAN
INDEPENDENCE, MISSOURI

November 5, 1969

Dear Dr. Seaborg:

I was highly pleased to have your letter of October 22, and copy of *Atomic Shield*, 1947-1952, written by Drs. Richard G. Hewlett and Francis Dean. I am vitally interested in the subject as well as the period, and I have scheduled it for reading forthwith.

Thank you for your continuing interest and my appreciation to Drs. Hewett and Duncan.

Sincerely yours,

Harry Truman

Dr. Glenn T. Seaborg
Chairman
U. S. Atomic Energy Commission
Washington, D. C. 20545

Harry S Truman's letter thanking Seaborg for sending the former president a copy of *Atomic Shield, 1947-1952*, the second volume of the official history of the U.S. Atomic Energy Commission. Nov. 5, 1969.

Republican candidate Thomas Dewey and independent candidate Henry Wallace, was a classic in American politics. Early returns seemed to indicate that Dewey, the heavy favorite, had won. However, as the returns continued to come in over the radio, Truman began to gain and pulled almost even. I went to bed with a radio next to me, which I turned on a few minutes every time I woke up. By daylight, it was apparent that Truman had won an enormous upset victory.

I am a long-time member of the Harry S. Truman Institute, in Independence, MO, and both Helen and I are regular readers of their publication, *The Whistle Stop*.

An interesting side note about our thirty-third president is the appearance of his name, Harry S. Truman, without the period following the "S." For most of Truman's life and throughout his presidential tenure, he signed his name as "Harry S. Truman." In 1962, however, Truman joked to some newspaper reporters that the period after the "S" in his name should be omitted since the "S" did not stand for a name but was a family compromise between naming him after his paternal grandfather, Anderson Shippe Truman, and his maternal grandfather, Solomon Young. Many people took Truman seriously, and for a while, Truman himself signed his correspondence as "Harry S Truman." To this day, the Harry S. Truman Library Institute in Independence, MO, a presidential library administered by the National Archives and Records Administration, and the Harry S Truman home (also in Independence), a national historic site administered by the National Park Service, differ in their punctuation of his name. Most published references, however, prefer to use the period after the "S."

Another side note on Truman is that his great grandmother, Nancy Tyler Holmes, was a first cousin of John Tyler, the first vice president to become president by succession; it was Tyler who insisted that the vice president should actually become the president and not just the "acting" president when he assumed office. Truman, of course, succeeded to the office of president after the death of Franklin D. Roosevelt on April 12, 1945.

c h a p t e r 3 DWIGHT DAVID EISENHOWER

1953–1961

THE PRESIDENT'S SCIENCE ADVISORY COMMITTEE

although I had had earlier contacts with Dwight David Eisenhower,

my formal association with his administration began when I was appointed to the newly created President's Science Advisory Committee (PSAC) early in 1959 and attended the monthly meetings in Washington, DC, from April 1959 until January 1961. PSAC was a very influential group. A major component of our advice to President Eisenhower, illustrative of this cold war period, was in the military field, where we advised on antisubmarine warfare, missile and potential anti-intercontinental ballistic missile programs, continental air defense (including early warning against missile attacks), chemical warfare, plans for limited warfare, and especially arms limitation and control. In addition, we gave advice on high-energy and accelerator physics, life sciences, science and foreign relations, space science, and basic research and graduate education. I served as chairman of the Panel on Basic Research and Graduate Education and our report, "Scientific Progress, the Universities, and the Federal Gov-

> "Long before my service on PSAC, however, it was clear that Eisenhower actively sought the advice and participation of scientists."

ernment," which became known as the "Seaborg Report," drew special attention from President Eisenhower and had a substantial impact on federal support of graduate education and science.

Long before my service on PSAC, however, it was clear that Eisenhower actively sought the advice and participation of scientists. An early indication was the Atoms for Peace Program, an idea he introduced at a United Nations meeting in December 1953, which led to creation of the International Atomic Energy Agency (IAEA). His proposal was motivated by a desire to encourage worldwide investigation into the most effective peacetime uses of atomic energy and to begin to diminish the potential destructive power of the world's atomic stockpiles. He suggested that the United States and the Soviet Union and other nuclear-capable states begin and continue to make joint contributions of enriched uranium-235 to such an agency under the aegis of the United Nations for use as fuel in nuclear power reactors throughout the world.

Although initially receptive, the Soviet Union soon opposed such a role for the agency, citing such objections as the potential for increased numbers of nuclear weapons as a result of plutonium production in these reactors. After more than three years of debate, a modified IAEA (without a nuclear fuel repository function), with headquarters in Vienna, Austria, became operative in 1957. As of December 1994 more than 123 countries were participating in IAEA, which administers the safeguards in the Nonproliferation Treaty (see Chapter 5) and is destined to play a key role in the continuing nonproliferation regime; IAEA implements the original objective of providing assistance on the peaceful uses of atomic energy to non-nuclear countries.

Eisenhower's Atoms for Peace Program also resulted in a series of International Conferences on the Peaceful Uses of Atomic Energy, held in Geneva, Switzerland. In August 1955 I attended the first of these conferences and I served as chairman of a session called "Heavy Element Chemistry." Here, I was excited to meet for the first time a number of Soviet nuclear chemists and nuclear physicists, to learn of their experimental results, to meet Otto Hahn (the codiscoverer of nuclear fission), and to talk with many other famous nuclear scientists. Ernest Lawrence and I became acquainted with many of these scientists at evening dinners, and one result of this was a visit by Hahn to Berkeley later that fall.

Niels Bohr lectured on "Physical Science and Man's Position." He spoke in English (the talks were translated into French, English, Russian, and Spanish, and transmitted through earphones to the audience), but his "live" English was so heavily accented that much of the English-speaking audience had difficulty understanding him until some of us, idly turning the dial, successfully tuned in to the English translation. (We all wondered where the organizers found a translator who spoke "Bohrese.")

In September 1958, shortly after I assumed my new position as chancellor of the University of California, Berkeley, I attended the second of these conferences and gave one of the plenary lectures: "Recent Developments in the Field of Transplutonium Chemistry." I proposed that American and Soviet scientists collaborate as modern-day alchemists in the creation of new elements. (See Chapter 4 for a description of my attendance in 1961 at the fifth annual general conference.) I had many contacts with the Atomic Energy Commission (AEC) during Eisenhower's administration. On Oct. 24, 1957, I wrote to Lewis L. Strauss (President

Eisenhower's first AEC chairman) a letter of far-reaching implications for the U.S. program of transuranium element research:

. . . Our phone conversation last week reminded me of our earlier chat about the need for a new very high flux reactor, and I thought that I would drop you a few lines to discuss this and some subsequent thoughts that I have had. . . . The field of new transuranium elements is entering an era where the participating scientists in this country cannot go much further without some unified national effort, which can only be authorized and coordinated by the Atomic Energy Commission itself.

The future progress in this area depends on substantial weighable quantities (say milligrams) of berkelium, californium, and einsteinium. The acquiring of this depends upon our country's entrance into a two-fold program: (1) The irradiation of substantial quantities of Pu^{239} as reactor fuel elements, and the re-irradiation of the products, in the presently available highest flux, high-capacity, reactors to form hundred-gram amounts of Cm^{244} and higher curium isotopes. (The similar neutron irradiation of hundreds of grams of Am^{241} is an alternate method for the faster production of smaller amounts of curium.) (2) The irradiation of the curium in the suggested very high flux reactor. . . . Part (2) of this program (conversion of curium to berkelium, californium, and einsteinium) would require the suggested very high flux reactor. (This, of course, is only one of the needs that such a reactor would fulfill.) You will remember that I suggested to you a reactor in the flux range of 10^{15} to 10^{16}. . . .

I realize that the program suggested here will cost money, but I believe that the suggestion concerning part (1) of the program dovetails very well into the Commission's already planned program and available facilities and would accomplish the desired result at minimum cost, and that the very high flux reactor of part (2) is a much-needed and overdue Commission facility in any case. I would recommend going ahead with part (1) even if an immediate decision is not forthcoming on part (2). Whatever the decision may be, I think

that it is worth recognizing clearly at this time the potentially inferior position which our country may hold in the area of the discovery of new elements a few years hence unless interlaboratory cooperative steps of this type are taken soon. . . .

As a result, the U.S. National Transplutonium Production Program was established. I prevailed upon John McCone (Strauss's successor as AEC chairman during the Eisenhower administration) to sponsor construction of the High-Flux Isotope Reactor (HFIR), which went into operation in 1965 at Oak Ridge National Laboratory (ORNL). Under my administrative leadership during the Kennedy and Johnson administrations, neutron irradiations at the Savannah River Plant and chemical processing of the transplutonium products at a transuranium processing plant at ORNL laid the groundwork for the successful and effective National Transplutonium Production Program.

My first encounter with President Eisenhower came at a White House dinner on Feb. 4, 1958:

. . . Helen and I went to a white tie dinner at the White House, first going to the East Room to assemble with the other guests, prior to passing through the receiving line to be greeted by the President and Mrs. Mamie Eisenhower. (The president was suffering from a slight cold.) We then proceeded to the west end of the White House and the State Dining Room, where Farrington Daniels escorted Helen and I accompanied Mrs. Daniels to our seats. There were about 100 guests at tonight's dinner, three-tenths of whom were military, and the rest scientists and their spouses. Tables were decorated with red carnations in gold tureens. After dinner, the women accompanied Mrs. Eisenhower to the Red Room for coffee, liqueurs, and cigarettes while the men joined President Eisenhower in the Green Room for coffee, liqueurs, cigars, and cigarettes. The president spoke to some of us about the recent launching of our

President Dwight D. and First Lady Mamie Eisenhower on their 40th wedding anniversary. July 1, 1958. Courtesy of Dwight D. Eisenhower Library, Abilene, KS. U.S. Navy Photograph, Public Domain.

space satellite and the importance of science in today's society; he was very personable and approachable. We then converged in the Blue Room, and the president greeted about 150 additional guests, including AEC members, National Advisory Committee in Aeronautics members, the little cabinet of the Defense Department, and others.

Anna Russell (international concert comedian) entertained guests in the East Room with such hilarious interpretations as a lady president of a women's club; singing a passionately sentimental song; a madrigal quartet in which she took all the parts; and a skit entitled "Wind Instruments I Have Known." In the latter she put together a bagpipe with humorous remarks and vocalizing. The Marine Band played throughout the evening, and some of the guests danced to the music.

Later that same month, on Feb. 25, I described a very interesting occasion—with a quite amazing cast of characters—in Washington:

My flight arrived in Chicago at 7:30 a.m., and I transferred to United Flight 710, which left at 8 a.m. for Washington. This arrived in Washington at about 11 a.m. I went directly to the Statler Hotel for the conference on "The Foreign Aspects of United States National Security," arranged at President Eisenhower's request by Eric Johnston (president of the Motion Picture Association of America). The purpose of the meeting was to encourage Congress to continue a strong foreign aid program and to extend the Reciprocal Trade Act, or Tariff-Lowering Act. The meeting was attended by top leaders of both parties in Congress, and speakers included Adlai E. Stevenson (titular leader of the Democratic party), Thomas E. Dewey (former governor of New York), John Foster Dulles (secretary of state), Neil H. McElroy (secretary of defense), Allen Dulles, and Bishop Fulton J. Sheen of New York.

Dean Atcheson, who had served as secretary of state under President Truman, was toastmaster during the luncheon program at which ex-President Harry Truman was the main speaker. President Truman described the meeting as a "good example of a bipartisan approach" and stated that this country's *allies are as vital as our missiles program, adding that our mutual security program actually involves now only one kind of politics: the politics of survival. There were about 1200 people in the main banquet hall. I sat with Ralph Bunche and James P. Baxter, III (president of Williams College).*

In his remarks Adlai Stevenson strongly supported foreign aid but criticized some of its present aspects and suggested that new approaches should be tried.

Vice President Richard M. Nixon didn't make a formal speech but presided over a question-and-answer forum to support the foreign aid program. At the end of this session I went up on stage and spoke with him. I have known Nixon for years, ever since we met at the banquet of the United States Junior Chamber of Commerce in Chattanooga, TN, in February 1948 when we were fellow members of the Ten Outstanding Young Men of the Year for 1947. Nixon suggested that I drop in to see him on one of my trips to Washington, and I agreed to do so.

After the meeting I had cocktails with Ed and Bobbe (Barbara) Pauley. Chief Justice Earl Warren spoke with me and also suggested that I look him up the next time I am in Washington. Other people I met included Mayor Robert F. Wagner (New York) and Robert Murphy (deputy undersecretary of state).

President Eisenhower, the after-dinner speaker, talked of mutual secu-rity as the alternative to chaos. He said that the issue was above partisanship and saluted the Truman Doctrine by which the former president had assisted Greece and Turkey in 1949.

On Jan. 26, 1959, James Killian, chairman of the newly created PSAC, extended to me President Eisenhower's invitation to become a member of the committee. (I was serving as the chancellor at Berkeley.) He told me the committee usually met on the third Monday and Tuesday of each month. I accepted, with the understanding that my first attendance would be at the April meeting (April 20–21, 1959). Missing only a few meetings, I served on PSAC until the end of Eisenhower's term as president; thus, my last meeting was on Jan. 16, 1961. George Kistiakowsky assumed PSAC chairmanship on July 15, 1959, and served until the end of Eisenhower's term.

PSAC members during my time as a member were as follows: George B. Kistiakowsky (chairman and special assistant to the president for science and technology), Robert Bacher, John Bardeen (professor of electrical engineering and physics, University of Illinois), Hans Bethe (professor of physics, Cornell

Photocopies of the invitation cards and diagram of the seating arrangement at an Eisenhower White House dinner. Feb. 4, 1958.

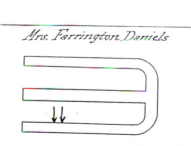

University), William Baker (vice president of research, Bell Telephone Laboratories), George W. Beadle (chairman, Division of Biology, California Institute of Technology), Donald F. Hornig (professor of chemistry, Princeton University), Jerome B. Wiesner (director, Research Laboratory of Electronics, MIT), Walter H. Zinn (vice president, Combustion Engineering, Inc.), Harvey Brooks (dean, Division of Engineering and Applied Physics, Harvard University), Edwin Land (chairman, Polaroid Corporation), Alvin M. Weinberg (director, ORNL), David Z. Beckler, Emmanuel R. Piore (vice president for research and engineering, IBM), John W. Tukey (professor of mathematics, Princeton University), Wolfgang K.H. Panofsky (director, High Energy Physics Laboratory, Stanford University), Detlev W. Bronk (president, The Rockefeller Institute), Robert F. Loeb (Bard professor of medicine, Columbia University), James B. Fisk (president, The Bell Telephone Laboratories), H.P. Robertson (professor of physics, Cal Tech), Paul Weiss (biologist, Rockefeller University), James R. Killian, Jr. (chairman of the corporation, MIT) (chairman), Edward M. Purcell (professor of physics, Harvard University), and Isidor I. Rabi (professor of physics, Columbia University).

PSAC considered a wide range of topics, with some emphasis on military matters. Illustrative of the topics are the names of the many PSAC panels, consisting of PSAC members and additional knowledgeable scientists and engineers: AICBM (Anti Inter-Continental Ballistic Missile), Anti-Submarine Warfare, Arms Limitation and Control, Basic Research and Graduate Education, Chemicals, Continental Air Defense, Early Warning, High-Energy Accelerator Physics, Life Sciences, Limited Warfare, Missiles, Science and Foreign Relations, Space Science, and Ad Hoc Missiles

Study. Members of these panels as of February 1960 are listed as follows.

Official Use Only, The White House, Washington, DC
Feb. 12, 1960

Panels of the President's Science Advisory Committee and to the Special Assistant to the President for Science and Technology

AICBM

Jerome B. Wiesner, chairman; Hans A. Bethe; Lloyd V. Berkner; William E. Bradley; Harold Brown; Lawrence A. Hyland; Wolfgang K.H. Panofsky; Edward M. Purcell; Jerrold R. Zacharias; technical assistant, H.J. Watters

Antisubmarine Warfare

Harvey Brooks, chairman; John E. De Turk; Eugene A. Fubini; William A. Higinbotham; Frederick C. Lindvall; William A. Nierenberg; technical assistant, George D. Lukes

Arms Limitation and Control

J.F. Killian, Jr., chairman; Robert F. Bacher; Harold Brown; Paul M. Doty; Ralph P. Johnson; Oskar Morgenstern; Herbert Scoville; John W. Tukey; Jerome B. Wiesner; technical assistant, Spurgeon M. Keeny, Jr.

Baker Panel

W.O. Baker, chairman; Luis W. Alvarez; Hendrik W. Bode; Richard L. Garwin; Andrew M. Gleason; David A. Huffman; John W. Milnor; John R. Pierce; Nathaniel Rochester; Oliver G. Selfridge; John W. Tukey; technical assistant, Eugene B. Skolnikoff

Basic Research and Graduate Education

Glenn T. Seaborg, chairman; Homer D. Babbidge, Jr.; George W. Beadle; Henry E. Bent; McGeorge Bundy; William B. Fretter; Roy M. Hall; Caryl P. Haskins; E.R. Piore; Roger R.D. Revelle; Frederick E. Terman; Alan T. Waterman; Alvin M. Weinberg; John E. Willard; O. Meredith Wilson; technical assistant, Robert N. Kreidler

> "PSAC considered a wide range of topics, with some emphasis on military matters."

Chemicals Panel

Detlev W. Bronk, chairman; Robert F. Loeb, co-chairman; Edwin B. Astwood; Conard A. Elvehjem; Alfred A. Gellhorn; George J. Harrar; Harold C. Hodge; James G. Horsfall; Cyril N.H. Long; Charles C. Stock; Charles S. Rhyne, consultant; W. Dean Wagner, consultant; technical assistant, Frederic Holtzberg

Continental Air Defense

E.R. Piore, chairman; George C. Comstock; James F. Digby; Daniel E. Dustin; Robert R. Everett; Clifford C. Furnas; Lawrence R. Hafstad; Albert G. Hill; James McCormack; Hector R. Skifter; Jerome B. Wiesner; Brockway McMillan, consultant; Allen E. Puckett, consultant; Julian M. West, consultant; technical assistant, H.J. Watters; staff assistant, Lawrence R. Walters

Early Warning

Jerome B. Wiesner, chairman; Hans A. Bethe; William E. Bradley; Daniel E. Dustin; Albert G. Hill; Edwin H. Land; Brockway McMillan; Edward M. Purcell; Herbert Scoville; W. McC. Siebert; Hector R. Skifter; Herbert Weiss; technical assistant, H.J. Watters

High-Energy Accelerator Physics

E.R. Piore, chairman; Jesse W. Beams; Hans A. Bethe; Leland J. Haworth; Edwin M. McMillan; technical assistant, George D. Lukes

Life Sciences

George W. Beadle, chairman; James Frederick Bonner; Detlev W. Bronk; Graham P. DuShane; George J. Harrar; Robert F. Loeb; Cyril N.H. Long; Colin McLeod; Curt Stern; Paul A. Weiss; W. Barry Wood; technical assistant, Frederic Holtzberg

Limited Warfare

H.P. Robertson, chairman; Robert R. Bowie; John S. Foster, Jr.; Edward H. Heinemann; Albert G. Hill; Charles C. Lauritsen; Don K. Price; Allen E. Puckett; Simon Ramo; Norman F. Ramsey; Glenn T. Seaborg; L.T.E. Thompson; Paul A. Weiss; J.R. Zacharias; technical assistant, George D. Lukes

Missiles Panel

George Kistiakowsky, chairman; Hendrik W. Bode; Harold Brown; James C. Fletcher; Donald P. Ling; Franklin A. Long; Charles H. Townes; Clark B. Millikan, consultant; John H. Rubel, consultant; technical assistant, George W. Rathjens

Science and Foreign Relations

Detlev W. Bronk, chairman; Lloyd V. Berkner; Wallace R. Brode; Joseph B. Koepfli; I. I. Rabi; H. P. Robertson; Alan T. Waterman; technical assistant, Eugene B. Skolnikoff

Space Science

Edward M. Purcell, chairman; Lloyd V. Berkner; Malcolm H. Hebb; Donald F. Hornig; Donald P. Ling; Bruno B. Rossi; Julius M. Schwarzschild; technical assistant, George W. Rathjens

Ad hoc Missiles Study

Donald P. Ling, chairman; Charles S. Ames; Hendrik W. Bode; Harold Brown; Philip J. Farley; James C. Fletcher; W.E. Gathright; Sidney Graybeal; Donald F. Hornig; Franklin A. Long; Harlow Munson; John H. Rubel; Herbert Scoville; John H. Sides; Charles H. Townes; Albert D. Wheelon; technical assistant, George W. Rathjens

PSAC generally met in room 220 of the Executive Office Building, next to the White House, and the meeting was punctuated with lunch in the White House Mess.

Monday, April 20, 1959

I walked to the nearby Executive Office Building, where I went to Room 275 to attend, beginning at 9 a.m., a special security briefing for new members of PSAC, arranged by F.G. Naughten (security officer of the Office of the Special Assistant to the President for Science and Technology).

I then went to Room 220 to attend the opening session, beginning at 9:30 a.m., of the two-day PSAC meeting. We first met in executive session, in which Herb York (director of defense research

> "We discussed the growing complexity in military systems (the president commented that military establishments are, and always have been, obsolete), the importance of arms control (the president agreed emphatically), the cessation of nuclear testing in the atmosphere (the president suggested that the PSAC continue to advocate this), and the strengthening of science in the free world."

and engineering) briefed us on problems of military organization and planning for research and development, a session that lasted about two hours.

This was followed by a report by Kistiakowsky on a recent field trip taken by his missiles panel and the conclusions based on this trip. Following this, H. J. Watters and James M. Mitchell summarized the results of a preliminary look at the expanding activities in the space and missiles fields as they relate to national ranges.

We convened after lunch at about 2:30 p.m. Wiesner summarized the results of a working group meeting on the changing character of warning and the nature of response to warning, and Paul Weiss gave a report on the conclusions of the Ad Hoc Panel on Biological Warfare and Chemical Warfare.

Tuesday, April 21, 1959

I again walked to the nearby Executive Office Building to attend the continuing PSAC meeting. Before the meeting began, a little after 9 a.m., I talked with General Alfred Starbird (director of the Division of Military Application at AEC) about the status of the heavy isotope production program. He said to call on him for help if we ever need it, and I said we will if we develop any crucial bottleneck concerning this program in his area of responsibility.

The morning session began with a briefing by Starbird and F. Shelton (Armed Forces Special Weapons Project, AFSWP) on the status of nuclear weapons testing planning, in the event that nuclear weapons testing resumes. This was followed by a discussion of the pros and cons of an agreement with the Soviet Union banning nuclear weapons testing.

Then we heard a report by Scoville of the Central Intelligence Agency (CIA) on what they have learned about activities in the nuclear weapons field in the Soviet Union. Following this, we heard a report by Governor L. Hough and Gerald R. Gallager (director of research, Office of Civil and Defense Mobilization) on the present status of planning for civil defense in the event of a nuclear attack.

We decided on the following dates for PSAC meetings for the remainder of the year: May 18–19, June 16–17, July 20–21, no August meeting, Sept. 14–15, Oct. 19-20, Nov. 9–10, and Dec.

14–15. These dates are usually the third Monday and Tuesday of the month.

I had to leave immediately after lunch to take a taxi to Washington National Airport, thus missing a short afternoon session at which Killian made a report on the Federal Council for Science and Technology (FCST), a group of science representatives of various departments of government that has just been formed.

Occasionally, we met with President Eisenhower. I recall a meeting in the Oval Office with him on May 19, 1959. We discussed the growing complexity in military systems (the president commented that military establishments are, and always have been, obsolete), the importance of arms control (the president agreed emphatically), the cessation of nuclear testing in the atmosphere (the president suggested that PSAC continue to advocate this), and the strengthening of science in the free world:

Monday, May 18, 1959

I walked to the Executive Office Building after having breakfast in the Statler Hilton coffee shop, to attend the PSAC meeting. The meeting, chaired by Jim Killian, began at 9:30 a.m. with an executive session. He told us that he is resigning as chairman and that George Kistiakowsky will probably take his place.

After some discussion, it was agreed to set up a panel on Limited Warfare, initially to examine the parameter of the problem, to look at the terms of reference for a study, and to report back with suggestions. At 10:30 a.m. we heard a report of the Panel on Science and Foreign Relations, given by Detlev Bronk. During the discussion, Piore observed that science is but one subject area that affects our foreign policies, Rabi emphasized the need to do something about scientific communication with Red China, and Fisk and others observed that there are too many negative statements in the panel report that could be more positively stated.

Then we heard a report of the AICBM panel by Jerry Wiesner. He and York led a discussion on the role of

NIKE/ZEUS missiles in the defense of cities or of hard targets. York observed that it would be cheaper to defend the Minutemen missiles by the deployment of additional Minutemen but cheaper to defend TITAN missiles by hardening, rather than providing more sites. It was agreed that selected members of the AICBM panel would meet with personnel from the office of the Secretary of Defense to discuss the applicability of the U-2 aircraft for airborne infrared early warning. Farley (special assistant to the secretary of state) then summarized a working paper dealing with the problem of establishing a quota for inspections to monitor a nuclear test cessation. Chairman Killian suggested that Bethe, Bacher, and Fisk give careful thought to the quota problem over the next 24 hours to see if they could develop specific ideas.

After lunch in the White House Mess (in the west wing of the White House) we heard reports by Bethe on the evaluation of Soviet tests and on the work of the Panel on Detection of Atmospheric Tests, headed by Panofsky, which studied the detection of nuclear explosions at yields between 50 and 100,000 kilotons. We spent the remainder of the afternoon discussing our program for a meeting with President Eisenhower tomorrow morning.

We discussed the release to yesterday's papers of the report of a special panel appointed by PSAC and the General Advisory Committee of AEC, entitled "An Explanatory Statement on Elementary Particle Physics and a Proposed Federal Program in Support of High-Energy Accelerator Physics." The special panel that prepared the report consists of Piore (chairman), McMillan, Haworth, Bethe, and Beams. The proposed program includes recommendations that a Stanford Electron Linear Accelerator with an initial energy of at least 10 BeV should be built and that, although the technical feasibility and research utility of the specific accelerator recently proposed by the Midwestern Universities Research Association have not been established, the group should be supported on a continuing basis. Furthermore, the research needs for a high-energy accelerator at ORNL should be explored with the laboratory and the southern universities concerned.

At 6 p.m. I attended a farewell cocktail party given by Dr. and Mrs. Killian in the Anderson Room of the Metropolitan Club for PSAC members and their wives. Following this, I had dinner in the coffee shop of the Statler Hilton Hotel, where I spent the night.

Tuesday, May 19, 1959

After breakfast in the Statler Hilton coffee shop, I walked to the Executive Office Building, where I attended the continuing PSAC meeting.

After further discussion in preparation for our meeting with the president, we joined him in the Oval Office from about 11:30 a.m. to 12:15 p.m. Killian opened the session by asking President Eisenhower if there were any matters he wished to discuss with the committee. The president mentioned the Russell Amendment to the Military Construction Bill, which had come up at a meeting with legislative leaders earlier in the day. That amendment would appear to require a special authorization for each missile and each aircraft in the defense program. Eisenhower asked Killian to look into it.

Killian used that request as a way to introduce some comments on the attention PSAC has given to various questions concerned with military technology, among them the trend for growing complexity in military systems and the speed with which they become obsolete. Ike commented that military establishments are, and have always been, obsolete; there is a continual process of weapons replacement.

Killian then initiated a discussion of the importance of arms control, remarking on the growing possibility of accidental war. He suggested that an effort should be made to strengthen our efforts toward arms limitation, especially by developing technical expertise in this area. The president agreed that this is very important and asked how he could help, commenting that the true business of a soldier is to work himself out of a job.

At Killian's invitation, Land continued this discussion by thanking the president for making science as popular as baseball. He feels that, with the president's strong support, PSAC has successfully promoted the importance of science in building our country's strength and providing hope for the *future. He stressed that the president's prestige would be an invaluable aid in supporting efforts toward arms control, especially in discussions with military personnel. He asked that Eisenhower make this scientific effort at arms limitation a part of the American mission, not only by providing funds but also by broadly promoting the idea. The president said he would do what he can.*

Wiesner pointed out that our national goals in the area of arms limitation are not clearly understood. There are many technical details surrounding this question that have not yet been investigated. An example of this, which Wiesner pointed out, was that we [the U.S.] had not seriously considered the question of detection systems before the Geneva talks. Wiesner feels we were ill prepared.

The president said he feels very strongly about some of these issues. He pointed out that a few years ago serious consideration was given to ceasing nuclear testing in the atmosphere, and now that has become only a fall-back position. He thinks perhaps a subcommittee of PSAC and other interested agencies, such as the National Security Council, should be established to deal specifically with disarmament studies.

Killian then introduced the next subject of discussion: the strengthening of science in the free world. He asked Bronk to summarize the principal findings of PSAC in this area, which will appear in a PSAC report in a month or so.

Bronk said that although scientists have always had close associations with colleagues overseas, there are many areas in which these could be strengthened. He suggested that more should be done to publicize opportunities for work in other countries, that universities might place more emphasis on interpretation of foreign cultures and languages, and that agencies like the Foreign Service might create more opportunities for scientists. He emphasized that restrictions on exchange of information limit these opportunities too strictly—our current quid pro quo rule is narrow-minded.

Rabi mentioned his recent visit to the Dubna Laboratory in the USSR and described the international atomic energy laboratory run by the Soviets in cooperation with other communist countries. He raised the question of what we could do to promote such international cooperation, suggesting that Brookhaven or Lawrence Radiation Laboratory might be designated as inter-American laboratories.

The meeting concluded with remarks from Wiesner and the president about the importance of public understanding and support in the areas of both arms control and international cooperation.

After the meeting with the president, we had a postmortem to assess the meeting. We were impressed by the president's views on arms limitation. It was agreed to set up a strong panel of PSAC to look into the broad questions in the field of arms limitation and how they interlock with PSAC activities. It was felt that this could be one of PSAC's principal activities, and I was suggested as a possible chairman for this panel. We also agreed that Rabi will follow up on the idea of Pan American laboratories and report back to the PSAC chairman, who will follow up with Keith Glennan on the possible use of a NASA facility as an inter-American laboratory.

At our June 1959 meeting we considered some follow-up actions to our meeting with President Eisenhower:

Tuesday, June 16, 1959

I attended the PSAC meeting at 9:30 a.m. in the ninth-floor conference room at NASA, at 1512 H St., N.W. We accepted Killian's resignation and elected Kistiakowsky as chairman, effective July 15. In the course of the morning, Kistiakowsky asked that each of us come in with a list of about 12 names as possible replacements for the six members who are leaving: Bacher, Bethe, Baker, Land, Robertson, and Weiss. Names that occur to me now are Kenneth Pitzer, Luis Alvarez, Robert Brode, Simon Ramo, and Warren Johnson. I'll consult with Fretter, Perlman, and McMillan when I get back.

We decided that the PSAC Missiles Panel should help with the problem of Atlas failures (six successive failures), including consideration of whether to embark on further and accelerated tests.

York then joined us to give a technical report on the status of the aircraft nuclear propulsion program. Weiss then presented the final conclusions of his panel on biological warfare and chemical warfare.

PSAC met in the afternoon and again over dinner from 7:15 p.m. to 9:40 p.m. at the Metropolitan Club. We discussed such items as the NIKE/ZEUS system (Bethe is concerned about feasibility), Soviet missile firings (a report by Scoville), and nuclear test ban negotiations (Killian asked Fisk, Bethe, and Bacher to prepare a report on the quota inspection problem). I brought up the difficulties of arranging exchange visits of Russian scientists to the Radiation Laboratory. PSAC will probably emphasize this in a report to President Eisenhower. I also took it up, especially with Wallace Brode (chief science adviser in the State Department), who promised to look into it.

Killian told us about hopeful progress on implementation of a group to study arms limitation (a follow-up of a suggestion I made to PSAC last month, which Land presented to President Eisenhower during our meeting with him). The president is following up on this. I suggested Clark Kerr (president, University of California) as a possible director of this group, which will work full time for six to eight months.

President Eisenhower asked for suggestions regarding Summit Conference topics: He is interested in a massive exchange of students with Russia. I am supposed to prepare a memo to Killian on this.

Wednesday, June 17, 1959

PSAC met at 9 a.m. On the agenda were consideration of the report of the Panel on Science and Foreign Relations and a discussion of it with FCST members. Bronk (chairman of the Panel on Science and Foreign Relations) summarized for the members of the federal council the tentative findings of the panel. Federal council members raised such matters as the training of students from lesser developed areas, the need to generate in the less developed areas the climate for the production of scientists, the question of whether IAEA should have a research laboratory, the activities of the military in extending

overseas support of basic research, the inadequacy of attention given in the document to the life sciences, and the need to develop a greater desire on the part of organizations and persons in the United States to contribute to international science activities.

It was agreed that detailed comments should be sought from federal council members based on the present draft and that the next draft of the report should be circulated for more widespread comment on the part of agency personnel. This was followed by a discussion of possible proposals for a summit meeting. We then held the first meeting of PSAC with FCST members.

Our meeting on Sept. 15, 1959, coincided with a meeting between President Eisenhower and Soviet Premier Nikita Khrushchev, so we all visited the White House lawn that afternoon to watch Khrushchev's arrival:

Tuesday, Sept. 15, 1959

After breakfast in the Statler Hilton Hotel, I walked to the nearby Executive Office Building to attend in Room 220,

President Dwight D. Eisenhower meets with Soviet Premier Nikita Khrushchev. Sept. 15, 1959. Courtesy of the Dwight D. Eisenhower Library, Abilene, KS. Official U.S. Navy photograph.

beginning at 9:30 a.m., the PSAC meeting. We met in executive session until noon to hear a review by Kistiakowsky of the various problems he faces as chairman. He gave us a rather gloomy picture, emphasizing the lack of support from PSAC members, except for a few who do give him some help.

From noon until 1:30 p.m. we had a briefing by Wiesner on air defense, which led to a very vigorous discussion on whether it makes sense to have a defense against airplane attack on the assumption that no missiles will be directed against the target when in fact missiles may be abundant and could actually be the primary threat. It was concluded that a centralization of air defense such as that termed "super-SAGE" makes very little sense.

After lunch we returned to the Executive Office building to continue the PSAC meeting. Jay McRae was scheduled to give a report on the recommendations of his panel on defense matters; however, Kistiakowsky suggested, instead, that we go over to the White House to watch the arrival of Soviet Premier Khrushchev for his meeting with President Eisenhower. (I learned later that AEC Chairman John McCone had objected to presentation of the report on the considerations of the McRae panel in such an open meeting of PSAC and that Kistiakowsky had decided that this was not a suitable subject for briefing in an open session.)

We watched with interest the arrival of Premier Khrushchev at the White House.

During the remaining part of the afternoon session, Kistiakowsky discussed FCST matters, such as materials research and federal policy for provision of facilities at universities, PSAC assistance on analysis of five-year projections, and the need for a committee of the federal council for international science programs.

The first meeting of my panel on Basic Research and Graduate Education was held on Nov. 16, 1959:

Monday, November 16, 1959

The meeting of my PSAC Panel on Basic Research and Graduate Education

began at 9:30 a.m. in Room 220 of the Executive Office Building. Present were: Alan T. Waterman (director, National Science Foundation [NSF]), Caryl P. Haskins (president, Carnegie Institution of Washington), McGeorge Bundy (dean, Faculty of Arts and Sciences, Harvard University), Bill Fretter (faculty assistant to chancellor, University of California, Berkeley), Henry E. Bent (chief, Graduate Fellowship Section, Office of Education, dean, Graduate School, University of Missouri), Frederick E. Terman (vice president and provost, Stanford University), Meredith Wilson (president, University of Minnesota), Homer Babbidge (assistant commissioner for higher education, Department of Health, Education, and Welfare [HEW]), Roy M. Hall (assistant commissioner and director, Division of Statistics and Research Services, Office of Education, HEW), and Paul A. Weiss (The Rockefeller University).

I opened the meeting by describing the two principal items for business: (1) to discuss broadly the panel's assignment and the means for its accomplishment, and (2) to develop advice on a current specific assignment: the allocation of graduate fellowships under Title IV of the National Defense Education Act (NDEA). I noted that this new effort conceivably could be a definitive study leading to the establishment of national policies designed to promote the best in graduate education, recognizing the vital interrelations of the education process to the pursuit of basic research, and suggested that our study include consideration of the following issues:

1. Goals of graduate education
2. Production of college teachers
3. Production of research scientists
4. What is the need? How many graduates for teaching? How many for research?
5. Problems in graduate education
   a. Inadequate undergraduate education
   b. Financial support of graduate education
   c. Financial support of research
   d. Training for teaching; degrees required
   e. Postdoctoral fellowships (much education continues in postdoctoral years)

f. Overcrowding of popular fields
g. Problems forged by large research projects, by large universities
h. Summer support of faculty; of students
i. Interdisciplinary education—broad as well as highly specialized
j. Inadequate graduate student support in social sciences and humanities

I then reported on my conversation with Paul Pearson (Ford Foundation) on Thursday and passed out copies of the two interesting letters I received, one from Conant and another from Beadle. In the course of the discussion, it was mentioned that recent studies and statistics on a number of topics are underway in both government and nongovernment agencies; in the case of the former, notably by HEW and NSF. It would be of great help to the panel if steps could be taken to provide digests of relevant information. We talked at some length about what we hope to accomplish and how we might begin approaching the project.

After lunch, Babbidge presented a rather comprehensive report on the legislative history of NDEA of 1958. He told us that the Office of Education had originally recommended direct matching grants to those institutions desiring to expand graduate education; however, Congress preferred the fellowship procedure now part of the approved act, as a way to avoid the Church–State issue. While the concept of building stronger graduate departments was in the original administration proposal, the language developed in the congressional version calls for "a new program or one being expanded."

The panel's discussion took note of the fact that, although there is virtue in strengthening small schools and marginal producers, it should be equally valid to take advantage of strong existing departmental programs that are underpopulated with students. In this context, the legislative requirement that the program be new or one being expanded discriminated unnecessarily in the availability of fellowships. Succinctly stated, the act has great virtue for those institutions in which the conditions are ready for a graduate school but that, without this fellowship program, cannot develop them. However, the act ought not to create a situation in which good schools lose good students to institutions not yet well equipped to provide an adequate graduate education.

We agreed to submit the following recommendations:

1. The Panel is aware of the defense orientation of the act. Title IV assumes, however, that all graduate education is important to defense and therefore omits categorical references. We, therefore, make our recommendation on the basis of the needs of higher education and the existing distribution of fellowships among disciplines. Our advice is that a larger proportion than present of the fellowships be allocated to the humanities and social sciences.
2. The Panel recommends that more consideration be given to fellowships in strong existing departmental programs with low registration. This would suggest expansion of student enrollment with hope of increase in Ph.D. production and could be associated with a principle of flexibility with respect to the dollar amount of the institutional grant awards.

These resolutions will be transmitted to the secretary of HEW. Before adjourning, we agreed to invite Weinberg (director of ORNL) to join the panel and to hold our next meeting in Berkeley Dec. 18–19, 1959. We also instructed the staff to collect, digest, and plan for presentation to the panel the following information:

1. Relevant information with regard to federal support of graduate programs
2. Relevant information with regard to federal support of basic research
3. Statistics on the effect of basic research on gross national product
4. Statistics on supply and need of scientists and engineers
5. Federal relationships with higher education
6. Need for graduate facilities

We also adopted the following outline:

"While the concept of building stronger graduate departments was in the original administration proposal, the language developed in the congressional version calls for 'a new program or one being expanded.' "

1. *Purpose of study: National objective. Greater qualitative intellectual effort needed. Need to work harder. Reference to effect on nation's international position.*
2. *Need to extend our national effort in basic research: Some statistics related to effect of basic research on our national product. Effect on our defense posture. Argument of justification as an intellectual endeavor per se. Statement on degree of expansion suggested.*
3. *Supply of scientists and engineers: Methods of increasing. Transition of problem of graduate education.*
4. *Problem of graduate education: Unification of different sources of financial aid. Improvement of graduate education. Federal aid: Capital facilities (buildings). Methods of giving financial aid. Financial support of graduate students. Problem of social sciences and humanities.*

The PSAC meeting was held the next day, for which my attendance was interrupted by a meeting with Vice President Richard M. Nixon:

Tuesday, Nov. 17, 1959

The afternoon meeting was devoted to discussing the report of Killian's Panel on Technical Aspects of Arms Limitation. Most of the committee favored the presentation, but Rabi insisted that the plan presented would only increase the armaments race. He indicated that he doesn't believe in the concept of secure deterrence, which is central to the plan.

I left the meeting at about 3:30 p.m. to take a taxi to the Capitol Building for my appointment with Vice President Nixon in the vice president's formal office. I met with him from 4:14 p.m. until 5 p.m. We discussed education. He asked me if there had been noticeable improvements since Sputnik. I said yes, but not enough—it is better than 1900 but not good enough for the mid-20th century in our country's present posture. We need improvement in fundamentals such as English composition and arithmetic, better salaries for teachers, continuing summer teachers institutes, and better textbooks. (I mentioned the UC Chem Project.) He is

interested in the use of TV, so I described NET (National Educational Television) and the Hagerstown experiment (a project to teach at the precollege level via TV in Hagerstown, MD) to him. I said that more federal aid is needed at the university level, that such aid to science in universities has been successful and is now indispensable, and I described the need for federal construction of university buildings (telling him about the recent favorable decision of FCST in this regard). His young brother (29 years old) has recently decided to go into teaching. I got the impression that he will work some of these ideas into his speeches. He said he considers some of Admiral Rickover's ideas to be too extreme.

I urged him to work for the removal of the disclaimer provision of the student loan section of NDEA. He said he would; the administration is united in opposition to this. He urged us to continue to participate in the loan program but not to make public statements (rather, work behind the scenes by talking to the relevant people—public statements in opposition too often antagonize the people we are trying to win over). I also urged that PSAC remain nonpolitical and continue to be so upon the change of administration in 1961, to which he also agreed.

We also talked about the 1960 presidential race. He believes the Republican nomination will go to either himself or Nelson Rockefeller, and both will have to enter the New Hampshire primary before the Feb. 8 deadline. He thinks the Democratic nomination will go to either Stuart Symington or Adlai Stevenson. He said that he has a low opinion of the former but regards the latter as an intelligent and formidable adversary.

I suggested that the government should give out about 10 awards a year, at about $10,000 each, to scientists below the top-level bracket. I said that the returns would more than justify the expense and would have a tremendous effect on morale. He seemed interested and, in thinking about it, said that perhaps $5000 tax-free (and hence worth about $7000 to $8000) would be about right.

He suggested that we get together from time to time on future trips, say in two or three months next time, to dis-

---

"He asked me if there had been noticeable improvements since Sputnik. I said yes, but not enough—it is better than 1900 but not good enough for the mid-20th century in our country's present posture."

cuss things in general, like the U.S. space program, etc.

When I returned to the PSAC meeting in the Executive Office Building, it was just adjourning for the day.

The next meeting of my Panel on Basic Research and Graduate Education was held, as planned, in Berkeley on Dec. 18–19, 1959:

Friday, Dec. 18, 1959

The PSAC Panel on Basic Research and Graduate Education is meeting all day today and tomorrow until noon. We have reserved the Regents Conference Room (Room 221) in Sproul Hall for these meetings. Attending the meeting were members Fretter, Waterman, Terman, Bent, Herbert S. Conrad (coordinator of Research, Division of Higher Education, Office of Education, HEW), Beadle, Hall, Wilson, Bundy, Roger Revelle (director, Scripps Institution of Oceanography, University of California), and staff members Kreidler and Lukes.

I asked Bent about developments in the Office of Education following receipt of our recommendations on allocation of graduate fellowships under Title IV of NDEA. He answered that careful consideration is being given to those recommendations, and it appears that of the 1500 fellowships to be awarded in the current program, two-thirds will go to social science and humanities students and that fellowships in engineering will be doubled. He also noted that heavy emphasis is being given to mathematics, especially in regard to teacher training. We then discussed in some detail the ideal way in which this fellowship program might improve graduate education.

Kreidler then informed the group of the recently revised policy that allows federal agencies to provide financial support to universities for research facilities, including buildings. There followed a long discussion about the historical background of this issue (of which Waterman provided a sweeping view) and of the changing times, which mean that civil service is not as attractive to scientists as it was during the depression, when government jobs provided better salaries, better opportunities, and more security than private

industry. We decided that our report should include particular reference to the role of research facilities in government laboratories and the climate for creative research vis-à-vis the pursuit in universities of basic research in tandem with the education process.

I hosted a luncheon in my conference room in Dwinelle Hall for the panel members, and then we returned to Sproul Hall and tackled the larger question of the intent of our panel report and what areas we would like to cover. It was a fascinating and, I believe, extremely productive conversation in which everyone seemed to be vitally interested.

We adjourned at 5:30 p.m. for cocktails at University House, followed by dinner at 6 p.m. The dialogue about graduate education and basic research continued through dinner, for the most part.

Saturday, Dec. 19, 1959

The PSAC panel meeting continued this morning, beginning at 9 a.m. and concluded after lunch. Most of today's meeting was devoted to reaching agreement on who would be responsible for writing what sections of the panel report. We agreed that the final report should be about 7000 words long and that we will assume responsibility as follows: Terman, introduction; Bundy, the importance of the research process to the educational process; Revelle, the increasing quantity of basic research and the need for a supply of first-rate scientists; Fretter, Waterman, and I; the historical role of the federal government in basic research and graduate education, the need to both recognize and extend this, and the role of large research laboratories; Beadle, questions about the requirements for a Ph.D. degree and the importance of both research and exposure to teaching in this educational process; Bent and Hall, improving stability, continuity, and variety in basic research and graduate education efforts by improving support mechanisms; and Wilson, special problems of middle-sized institutions in basic research and graduate education.

We concluded our meeting, feeling well satisfied with this start on the project, with a luncheon at the Men's

Faculty Club. The campus is very quiet since most students have already left for the Christmas holiday.

The next meetings of my panel and discussions of our report were held in Washington, DC:

Sunday, Jan. 17, 1960

Beginning at 9 a.m. in the Executive Office Building, our panel discussions continued. Chairman Kistiakowsky shared his views of what the panel might accomplish with us. He stated that it is essential for the success of science in the White House and its impact on policy making that the panel not engage solely in lobbying for increased support of science in universities. PSAC deeply believes that science and higher education are essential to national survival and welfare in the short run and to the growth of the economy and leadership in the long run. Paraphrasing Piore's paper of Jan. 12, 1960, written to provoke discussion, Kisty stated that the panel's charter might be as follows: "The report of the panel should not deal with research as a whole but with the problems of academic institutions and the research therein so as to keep them coupled with the nation's needs and the U.S. Government's objectives, even though this coupling may not be wholly consistent with the university's limited point of view."

Kisty suggested that the report should be short, should contain concrete recommendations directed at colleges and universities but, more importantly, should describe ways in which the federal government could contribute to achieving the panel's objectives. He expressed the opinion that if the recommendations simply advocate more funds, then the report will have little effect. There is a need, therefore, to spell out priorities, the forms in which federal support should be given, the alternatives, and any statutory changes that may be required. He then presented us with a fairly detailed outline of the issues, as he sees them. In the discussion that followed, Beadle accepted the assignment of writing a statement on how the research laboratory might provide an environment in which the student researcher and the

teacher come together. Kreidler will try to pull together into one document the papers already prepared by Beadle, Terman, and Bundy. Before adjourning at 1 p.m., we agreed to meet again in Berkeley on March 5 and 6.

Monday, April 18, 1960

The PSAC meeting began at 9:30 a.m. in Room 220 of the Executive Office Building. Before and after lunch in the White House mess, PSAC members discussed in detail the draft of the report of my Panel on Basic Research and Graduate Education, which had been prepared by Kreidler as the result of our meetings and discussions. The discussion focused largely on the text of the report and not very much on the conclusions and recommendations.

Tukey commented on one of the main themes of the draft: Basic research and graduate education should be closely coupled. He pointed out that much basic research is carried on in first-class institutions that are not colleges or universities.

Hornig mentioned the problem of whether universities can give research faculty status to scientists working in large laboratories attached to universities, and he gave the Penn–Princeton accelerator as an example. He indicated that institutional grants are superior to a collection of many small project grants, that the support of graduate students by contracts or project grants is inferior to their support by fellowships or institutional grants, that institutional grants would be a feasible method for the general expansion of graduate faculties, and that grants to support facilities are a critical need in universities.

Weinberg emphasized the increasing importance of basic engineering research, stating that programs aimed at expanding this should be strongly supported. He said it is doubtful that the majority of Ph.D.-trained scientists should go into research because our society is becoming so heavily technological that many administrative jobs now held by lawyers ought to be held by scientists.

Brooks suggested that a substantial proportion of federal support for graduate and postdoctoral education should be in the form of fellowships that permit free choice of institution and carry a much more substantial unrestricted grant to the institution than is now the practice. He stated that our graduate education does not give sufficient attention to the needs of applied science or the more sophisticated kinds of engineering. He expressed the fear that federal support on a massive scale for basic research might drive out private support. He also emphasized the importance of interdisciplinary research.

Bardeen questioned whether the time is right for direct government support for tenured appointments in universities and suggested that there might be political problems involved in direct grants to universities. He suggested, therefore, that it might be better to have a gradual evolution from the project to the grant basis. He suggested that research scientists, that is, those without teaching responsibilities at universities, be confined to temporary postdoctoral appointments. He would like to see the number of universities with quality science departments increased.

Following discussion of the draft report of my Panel on Basic Research and Graduate Education, the remainder of the afternoon was taken up by discussion of international science problems. Fred Seitz (University of Illinois) made a report on the North Atlantic Treaty Organization (NATO) science committee; Wiesner, a report on his visit to the Soviet Union; and Bronk, on the recommendations of the Panel on Science and Foreign Relations, including a program of U.S.–USSR exchange. In addition, the question of attendance of Red Chinese scientists at the Rochester Conference on High-Energy Physics and the proposal for the formation of a Panel on Technical Assistance were discussed.

Tuesday, April 19, 1960

Today is my 48th birthday. I have no complaints; life is treating me very well indeed. I checked out of my hotel room and walked to the Executive Office Building for the continuing PSAC meeting. A number of PSAC members gave me further personal comments on the draft of the report being prepared by my Panel on Basic Research and Graduate Education. There was a discussion of the organization for science in the government, including the role of PSAC, the mission-oriented agencies, NSF, and the arguments for and against

the creation of a Department of Science. The agenda also encompassed reports covering the deliberations of committees and panels, including a report by Bardeen on the National Academy of Sciences Committee on Atmospheric Sciences, by Beadle on the life sciences panel, and by Purcell on the space science panel.

Following a number of additional meetings of my panel, and with critical drafting help from Bundy, we finished our report, "Scientific Progress, the Universities, and the Federal Government," by the November 1960 meeting of PSAC; the report later became known as the "Seaborg Report." When the report was made available to President Eisenhower, he became so interested that he actually edited and made some changes in it. When PSAC members met with the president on Dec. 19, 1960, in the Oval Office, he took special note of my PSAC panel report. Perhaps the report's most famous recommendation was the statement that the basis of general

Members of the President's Science Advisory Committee meet in the Oval Office. Standing, left to right: George W. Beadle, Donald F. Hornig, Jerome B. Wiesner, Walter H. Zinn, Harvey Brooks, Seaborg, Alvin M. Weinberg (in front of Seaborg), David Z. Beckler, Emanuel R. Piore, John W. Tukey, Wolfgang K. H. Panofsky, John Bardeen, Detlev W. Bronk, and Robert F. Loeb. Seated, left to right: James B. Fisk, George B. Kistiakowsky, Eisenhower, James R. Killian, Jr., and Isidor I. Rabi. White House. Dec. 19, 1960. Courtesy of the Dwight D. Eisenhower Library, Abilene, KS. National Park Service Photograph, Public Domain.

> "Furthermore, the report stated that federal support for basic research and graduate education in the sciences should be continued and flexibly increased, so as to support excellence where it already exists and to encourage new centers of outstanding work."

policy should be that basic research and the education of scientists go best together as inseparable functions of universities. Furthermore, the report stated that federal support for basic research and graduate education in the sciences should be continued and flexibly increased, so as to support excellence where it already exists and to encourage new centers of outstanding work.

In February 1960 PSAC met at the U.S. Navy base at Key West, FL:

Saturday, Feb. 13, 1960

I spent the entire day in a meeting of the Board of Directors of the National Educational Television and Radio Center in the Statler Hilton. Present were members Everett N. Case, Norman Cousins, Darwin S. Fenner, Leland Hazard, Richard B. Hull, Lloyd S. Michael, Kenneth E. Oberholtzer, Mark Starr, George D. Stoddard, John F. White, Raymond Wittcoff, and staff members Robert B. Hudson, Warren Kraetzer, and Kenneth Yourd. The business was entirely routine.

I had to leave the meeting before it adjourned, at about 4 p.m., to go by taxi in a very heavy snowstorm to the Andrews Air Force base MATS (Military Air Transportation Service) Terminal, where I boarded a special flight to Key West, FL, for the PSAC meeting. The storm delayed our departure, so we arrived in Key West (Boca Chica) much later than planned, at about 10 p.m. We proceeded to the Seaplane Base BOQ (Bachelor Officers' Quarters), where I shared a room with Pief Panofsky.

Sunday, Feb. 14, 1960

After the miserable weather in Washington, it was a delight to wake up to the Florida sunshine this morning. I am looking forward to swimming and lounging around the pool a bit while I am here in the tropics. Although we Californians do not suffer blizzards like yesterday's in Washington, our winter weather is nonetheless cold and wet, so a bit of summer in the midst of winter is a real treat.

After breakfast at the Seaplane Base Bachelor Officers' Quarters (BOQ), we proceeded to the Fleet Sonar School

Auditorium for a welcome by Rear Admiral Lloyd M. Mustin (commander, U.S. Navy Base Key West). Our meeting began at about 9 a.m. in a conference room with a review of the work of the military panels and then a discussion of government support for science. There was some support, by Rabi and Piore, for a Department of Science, but Chairman Kistiakowsky favored the present system of relying on the National Academy of Sciences, the president's science adviser, and FCST.

We went to the air station for box lunches at noon and then returned to the Fleet Sonar School for the continuing meeting, which included a discussion of the High-Energy Accelerator program. We discussed recommendations of the High-Energy Accelerator Panel and endorsed the panel's recommendations, although Rabi expressed opposition and Weinberg and I had some reservations.

We attended a reception given by Commander and Mrs. Mustin at Quarters A and then had a buffet supper at the Fort Taylor Officers Club.

Monday, Feb. 15, 1960

We again had our breakfast at the Sea Plane Base BOQ and then went to the Fleet Sonar School for a continuation of the PSAC meeting. The main item for discussion was arms limitation and a nuclear test ban. This included a discussion of the verification problem, and there was a good deal of consideration of the concept of a threshold test ban treaty. There was general agreement that the treaty would have to be subject to effective monitoring to win ratification from the Senate. Our meeting ended at about 11:30 a.m.

We all enjoyed the free afternoon and evening in such a lovely climate.

The following PSAC meetings, held in Washington, included some interesting discussions:

Monday, May 16, 1960

After breakfast in the Statler Hotel I walked to the nearby Executive Office Building to attend the PSAC meeting, which began at 9:30 a.m. with an executive session, presided over by Chairman Kistiakowsky.

At the afternoon meeting there were reports by Frank Long on missiles, particularly Minuteman; by Purcell on space science; by Piore on air defense; and by Beadle on biology programs. During the day the bad news came from Paris that, as an aftermath of the U-2 incident, Khrushchev is canceling his summit meeting with President Eisenhower.[1] During his report on plans for an ad hoc meeting to discuss the image abroad of U.S. science, Wiesner, influenced by the news from Paris, made a strong plea that PSAC try to convince the president to set up a strong organization on arms control that would operate outside the Department of State. Kistiakowsky disagreed, saying that if PSAC are members of this mind, they should see the president on their own, individually, and he said that he doesn't think this is the time to do it.

Tuesday, May 17, 1960

Again after breakfast in the Statler Hotel, I walked to the Executive Office Building to attend the continuing PSAC meeting. The entire morning meeting was conducted in executive session.

Today's meeting was conducted under the shadow of cancellation of the summit meeting. The first item was a discussion on problems in the intelligence area, led by Bruce Billings with participation by Baker, Scoville, and others. Discussion of the SAMOS satellite projects didn't lead to any conclusions or any definition of the role PSAC might play in this critical area.

Next there was a discussion of the organization of science activities in the federal government. Kisty led this discussion, which started off with a brief summary by Piore with some suggestions for possible organizational structure. Eugene Wigner (Princeton University) indicated that, in his opinion, the Soviets are doing a better job at this

than we are. He got into an argument with Kistiakowsky, who disagreed with this theme. Wigner identified a number of problems in science organization in the United States but didn't really have any suggestions for remedying them.

The program for the afternoon included a plan to continue the discussion of how PSAC might participate in the intelligence field and the discussion of science in government. (I learned later that Kistiakowsky asked that a paper on this subject be written by Weinberg, Panofsky, and Piore on the basis of a session they will have with Kistiakowsky.)

[Another meeting I attended as a PSAC member took place on Nov. 14–15, 1960; and my penultimate such meeting on Dec. 19–20, 1960, as mentioned earlier, included a wind-up session with President Eisenhower. I had an interesting meeting with Sargent Shriver, President-elect John F. Kennedy's chief recruiter.]

Monday, Nov. 14, 1960

I attended the PSAC meeting of the Executive Office Building today. In the executive session Kistiakowsky told us about the Oct. 25 letter he received from President Eisenhower informing him that the secretary of state has established the U.S. Disarmament Administration to strengthen leadership and coordination of manifold activities of the U.S. government in the field of safeguarded disarmament and arms control. Rabi and Kistiakowsky argued about the correct interpretation of a remark made to Rabi by Vasily Emelyanov of the Soviet Union on the use of nuclear weapons.

We spent most of the afternoon in executive session working on our Fiscal Year 1962 military budget paper, which Herb York and John H. Rubel (Department of Defense) attended and with which they were very helpful.

Late in the afternoon Kisty and I met to make some minor editorial corrections on my Panel Report on Basic Research and Graduate Education.

Tuesday, Nov. 15, 1960

The PSAC meeting continued today with discussions of the NIKE/ZEUS Report; the man-in-space program, led by Hornig; the Armand Report to

---

[1]The "U-2 incident" referred to a U.S. reconnaissance plane, piloted by Francis Gary Powers, that was shot down while flying over Soviet territory on May 1, 1960. Soviet Premier Nikita Khrushchev reacted violently to this event, which led to cancellation of a summit meeting between President Eisenhower and Khrushchev, scheduled for May 14 in Paris. Thus hopes for détente were crushed, and a reversion to a more dangerous level of the cold war ensued.

NATO; and the value of fallout shelters. Keith Glennan and Hugh Dryden of NASA presented their man-in-space report. This raised the question, Is the proposed 15-year program of landing a man on the moon and returning him to Earth worth the $30 billion it will cost? It was pointed out that an unmanned program costing only $5 billion could produce substantially all the information that one is going to get. Townes argued that even with an unmanned program the information so obtained would lead us to want to go on to the man-in-space program. The role of the nuclear rocket (Rover) project was also discussed. Dryden suggested that it isn't necessary for PSAC to formally adopt the man-in-space report. It can be distributed to all PSAC members to ask for their comments, followed by the development of a policy issue paper to be presented to the National Security Council.

Monday, Dec. 19, 1960

The PSAC meeting continued at 9 a.m. in executive session in the Executive Office Building. We discussed the items we would present in our final session with President Eisenhower this morning.

Just before 11 a.m. we walked over to the White House to meet with President Eisenhower in the Cabinet Room (as mentioned above). It was a general briefing on PSAC business, our final meeting at the end of his administration. The president took special note of my PSAC panel report, "Scientific Progress, the Universities, and the Federal Government," which is now being issued for public distribution.

After a lunch break, our meeting continued with a report of the High-Energy Physics Panel, presented by Piore.

In the evening I attended a cocktail party and buffet supper honoring Kistiakowsky, sponsored by his staff. George's term as science adviser to the president has, of course, come to an end as incoming President John F. Kennedy will make his own appointment to this position. We suspect this will be his MIT friend, Jerome Wiesner. Kisty will presumably continue to serve as a PSAC member.

Tuesday, Dec. 20, 1960

I had breakfast in the Mayflower Hotel with Sargent Shriver, President-elect Kennedy's chief recruiter. His main agenda item was to seek my input on who should be appointed AEC chairman. I recommended Jim Fisk (president of Bell Laboratories).

I did my best to promote implementation of our PSAC Panel Report on Basic Research and Graduate Education and to stress the importance of improving science education, citing CHEM Study as an important step in this direction.

Shriver told me something about how JFK's cabinet is developing. It was a rather exciting meeting. One gets the impression of a dynamic, new, young energy in this administration.

I then went to the Executive Office Building for the PSAC meeting. We heard a report by Walsh McDermott (Cornell University) on the technical assistance panel's work and then talked about the scientific and military aspects of the geodetic satellite.

Shortly after the inauguration of President Kennedy, I wrote a letter of appreciation to President Eisenhower:

May I take the liberty of writing to thank you most sincerely for the privilege and pleasure which you have given me of serving as a member of your Science Advisory Committee for the last two years. I have valued very much the Committee's discussions with you and have been most impressed by your views on the importance of basic research and the place and importance of science in the Federal Government.

I am particularly grateful, as chairman of the PSAC's Panel on Basic Research and Graduate Education, for your fine endorsement of our Report, "Scientific Progress, the Universities, and the Federal Government." This launched the Report in such a manner as to give it maximum effectiveness, and I am pleased to say that it has had a uniformly favorable reception.

May I wish for you and Mrs. Eisenhower most joyous and pleasant years in the future.

I met Eisenhower a number of times following the period of his presidency. One notable occasion was on June 7, 1964, at George Washington University, when an honorary doctor's degree was bestowed upon him; at this time we talked about the important role of PSAC and my service during his administration. I met many times with PSAC members during the 1960s when I was serving as AEC chairman.

Shortly before his death, I wrote to congratulate President Eisenhower for his receipt of the Atoms for Peace Award:

March 20, 1969
Dear General Eisenhower:

I just learned that you have been named to receive this year's Atoms for Peace Award. I am very enthusiastic over this choice and feel that it could not have been awarded to anyone more worthy.

It is most appropriate that we honor your role in the formation of our Atoms for Peace program and the establishment of the International Atomic Energy Agency at this time when the Nuclear Nonproliferation Treaty is furthering the efforts you began to assure the peaceful use of nuclear energy around the world.

I was reminded of your leadership in these activities in the course of preparing the Rosenfield Lectures, delivered at Grinnell College in January, which reviewed the important role of the international atom.

Congratulations and best wishes.
Respectfully,
Glenn T. Seaborg

General Dwight D. Eisenhower
Walter Reed General Hospital
Washington, D.C.
GTS:MJ
bcc: Dr. Lee A. DuBridge

This $50,000 award, at his request, was presented posthumously to Eisenhower College in Seneca Falls, NY.

Monday, March 31, 1969—National Day of Mourning

At 4:30 p.m. Helen and I attended the funeral services for former President Eisenhower at the Washington Cathe-

George Washington University President Thomas H. Carroll confers an honorary degree upon former President Dwight D. Eisenhower. Washington, D.C. June 7, 1964. Courtesy of the Gelman Library Photo Archives of George Washington University, Washington, D.C.

```
                    UNITED STATES                      UNCL. BY DOE
              ATOMIC ENERGY COMMISSION                    NOV 86
                 WASHINGTON, D.C. 20545
                      JUL 3 1 1971

OFFICE OF THE CHAIRMAN




     Dear Mrs. Eisenhower:

          The Atomic Energy Commission is observing the
     25th Anniversary of the signing of the Atomic Energy
     Act with a ceremony and reception at the Department of
     State, Sunday afternoon, August 1, 1971.  Recalling
     President Eisenhower's deep interest in atomic energy
     and his historic initiative in establishing the Atoms
     for Peace Program, which has had a pervasively
     beneficial effect on mankind, it is with the deepest
     respect and pleasure that we are sending to you a
     Special Commemorative Medal marking this occasion.

          My colleagues on the Commission join me in
     extending to you our very best wishes.

                              Respectfully,



                              Glenn T. Seaborg

     Mrs. Dwight D. Eisenhower
     Gettysburg, Pennsylvania
```

Atomic Energy Commission Chairman Seaborg's letter to Mamie Eisenhower sending her a special commemorative medal observing the 25th anniversary of the signing of the Atomic Energy Act. July 31, 1971.

dral. Officiating clergy were the Very Reverend Francis B. Sayre, Jr. (dean of Washington Cathedral), and The Right Reverend William F. Creighton (Bishop of Washington). Services were attended by essentially all of present Washington officialdom, some members of the Johnson administration, and state leaders from all over the world. Among those present were several secretaries of the president's cabinet and presidential assistants such as Henry Kissinger, Robert F. Ellsworth, Lee A. DuBridge, Charles B. Wilkinson, Patrick Moynihan, and others. Former President Lyndon B. Johnson and former Vice President Hubert Humphrey also attended the funeral. Heads of state attending included President Charles DeGaulle of France, Prime Minister Mariana Rumor of Italy, Foreign Minister Joseph Luns of the Netherlands, Prime Minister John Gorton of Australia, Chancellor Kurt Kiesinger of West Germany, Prime Minister Chung Il Kwon of South Korea, Prime Minister Marcello Caetano of Portugal, Vice President Nguyen Cao Ky of South Vietnam, the Shah of Iran, Lord Mountbatten of Great Britain, President Habib Bourguiba of Tunisia, President Ferdinand Marcos of the Philippines, Prime Minister Suleiman Demirel

A Chemist in the White House: From the Manhattan Project to the End of the Cold War

of Turkey, and former Prime Minister Nobusuki Kishi of Japan.

On the way out Helen and I spoke to Governor Nelson Rockefeller, Governor Luis Ferre of Puerto Rico, Mrs. Bob Hope, and others.

On July 31, 1971, I wrote to Mrs. Eisenhower to send her a special commemorative medal in observance of the 25th anniversary of the signing of the Atomic Energy Act. Dwight Eisenhower was impressive in many ways, and I particularly admired his dedication to the achievement of an agreement to end nuclear weapons testing. He played a leading role in the initiation of a moratorium on such testing, which began in 1958 and continued until the Soviet resumption of testing in fall 1961 (as described in Chapter 4). I was honored that, in addition to his selection of me to serve as a PSAC member, President Eisenhower approved the choice of me as a recipient of the Fermi Award in 1959 (which, incidentally, gave me the financial freedom to go to Washington when the time came in 1961) and my nomination as a member of the National Science Board. He had a good rapport with his PSAC and encouraged the members to establish a subcommittee to deal specifically with disarmament studies.

Helen and I were pleased to attend, Oct. 9–12, 1990, the observation of the centennial of President Eisenhower's birth at Gettysburg, PA. A description of this impressive event, including the texts of the talks and discussions, has been published as a book, *The Eisenhower Legacy: Discussion of Presidential Leadership*. We have also been pleased to be readers of *Dateline*, published by the Eisenhower World Affairs Institute.

chapter 4   JOHN FITZGERALD KENNEDY

1961–1963

A PASSION FOR ARMS CONTROL

The main theme in the administration of President John

Fitzgerald Kennedy was the pursuit of a nuclear test ban treaty with the Soviet Union. I believe that the achievement of the treaty can be traced in large part to the deep commitment of President Kennedy, to his persistence in pursuing the goal despite numerous setbacks, to his skilled leadership of the forces involved within his administration, and to his sensitive and patient diplomacy in dealing both with the Soviet Union (which meant, basically, with Nikita Khrushchev) and with the United States Senate. In my view, President Kennedy's performance in this matter had qualities of excellence that are worthy of study and emulation.

The failure to achieve a comprehensive treaty ending all nuclear testing by the superpowers is a major disappointment. Despite some near misses, this glittering prize, which carried with it the opportunity to arrest the viciously spiraling arms race, eluded our grasp. It is important, I think, to consider why this happened. Basically, there was deep mistrust between the superpowers, and this mutual suspicion has until recently thwarted the hopes of the largest part of the world community.

A telephone call that changed my life came on Jan. 9, 1961. I was in the Radiation Laboratory of the University of California, Berkeley. Each Monday I took refuge there from my administrative duties as chancellor of the Berkeley campus to follow progress in my own academic field, nuclear chemistry. The call came from President-elect Kennedy. He asked me to accept the job of chairman of the Atomic Energy Commission (AEC). When I asked him how much time I had to make up my mind, he said, "Take your time. You don't have to let me know until tomorrow morning."

That evening I discussed Kennedy's offer with my wife Helen and our six children. It was a big decision, because we would be taking all five of our school-age children out of their schools and away from their friends. The kids did not like the idea of moving from Lafayette, CA, to Washington, DC, and demanded a vote on this issue. The vote was seven to one against moving (with Helen and the six kids, including one-year-old Dianne, voting against). However, I exercised the veto power inherent in the head of a democratic institution and said, "I think we should make the move."

Within a few days I accepted the offer. On Jan. 20, 1961, decked out in morning coat and top hat, I sat with the official party on the stage before the Capitol, waiting for the inauguration of Kennedy as the 35th president of the United States. As Arthur Schlesinger, Jr., describes it, "It all began in the cold." Before the new president came in, I saw several of my friends who were also part of the incoming administration: McGeorge Bundy, national security adviser; Dean Rusk, secretary of state; and Robert McNamara, secretary of defense.

The ceremony began, and after the oath, President Kennedy spoke eloquently and pronounced his now famous "Ask not what your country can do for you" line, which greatly appealed to the crowd.

Later in the day, I met the new attorney general, Robert Kennedy. He greeted me very warmly and brought his brother, the new president, up several rows in the stands to meet me. This was my first face-to-face meeting with Kennedy. He stood beside my seat, and we carried on a cordial chat. After expressing his delight at my willingness to serve in his administration, he gave me my first assignment. "I know there is another vacancy on the Atomic Energy Commission," he said. "Why don't we fill it with another, perhaps young, scientist who could attain experience in the AEC and then go on to other positions in the government?" Having advocated two scientist-commissioners to deal with the commission's increasingly technical problems, I was only too happy to comply with this first order from the new president. In the end, Kennedy's foresight paid off as we recruited physicist Leland Haworth from Brookhaven Laboratory, who, after a few years' service as AEC commissioner, went on to become director of the National Science Foundation (NSF) in 1963. President Kennedy's crisp style and appreciation for science during this first meeting reassured me that I had made the right decision in accepting the AEC chairmanship.

I arrived in Washington on Jan. 31, 1961, and my family joined me at the end of the school year that June. My confirmation hearing was held on Feb. 23, 1961, and I was sworn in on March 1, 1961.

I had had an earlier contact with then-Senator Kennedy when, as Berkeley chancellor, I had invited him to Berkeley to address our students (an invitation that he wasn't able to arrange to accept).

A Chemist in the White House: From the Manhattan Project to the End of the Cold War

I was privileged to be in the audience on July 15, 1960, to hear his acceptance speech for the presidential nomination, as is noted in the following excerpt from my diary.

Friday, July 15, 1960,
Balboa–Los Angeles

This afternoon I drove with Pete, Dave, and Lynne to the Sports Arena in Los Angeles where we met Ed Pauley (University of California regent, prominent businessman, and leader in the Democratic party). The kids were quite thrilled to be invited by Ed to visit with him on the floor of the convention and have soft drinks in his trailer.

We then walked to the nearby Los Angeles Coliseum to hear the acceptance speech of John F. Kennedy, who received the presidential nomination on the first ballot on Wednesday night. He spoke, forcefully and eloquently, in the early evening as the sun was setting, a most exciting occasion. As we were walking outside the Coliseum after his speech, the limousine in which Kennedy was riding inched its way through the crowd and Pete was thrilled to touch his arm, protruding out the window, as the car passed us.

In the exciting early months of Kennedy's "one thousand days," AEC was involved in formulating several major decisions: the nuclear test ban negotiations with the Soviet Union and Great Britain; the man-on-the-moon decision; cancellation of the nuclear-powered airplane; and the funding of two $100 million projects, the Stanford linear accelerator and the New Production Reactor (NPR) at Hanford, Washington State.

I met frequently with President Kennedy and his other advisers and was struck by his informality. He had a tendency, which became famous, to pick up the phone and call officials—even of subcabinet rank—to get first-hand information on issues. To this day, many stories are told about individuals who received such calls and their incredulous reactions, which appar-

JOHN F. KENNEDY
MASSACHUSETTS

COMMITTEES:
FOREIGN RELATIONS
LABOR AND PUBLIC WELFARE
JOINT ECONOMIC COMMITTEE

United States Senate
WASHINGTON, D.C.

February 16, 1960

Glenn T. Seaborg
Chancellor
The University of California
Berkeley 4, California

Dear Chancellor:

Many thanks for your letter of recent date inviting me to address a meeting open to the student body, faculty and staff of the University of California sometime during the coming months.

I certainly appreciate your writing me, but unfortunately I do not have any plans to be in California at the present time which would allow me to visit your university. I will, however, keep your kind invitation in mind and if the opportunity arises for me to come to the university I will certainly get in touch with you.

Thanking you for thinking of me and with every good wish, I am

Sincerely yours,

John F. Kennedy

JFK:el

Massachusetts Senator John F. Kennedy's letter regretting University of California Chancellor Seaborg's invitation to address an open meeting of the University of California–Berkeley. Feb. 16, 1960.

ently delighted the president. Once, he called me at home and one of my young sons answered but delayed informing me that I was wanted on the phone. The president waited patiently for several minutes until I responded.

I served as chairman of the five-member AEC from March 1, 1961, until Aug. 17, 1971. President Kennedy appointed me first to a two-and-a-half year term, the time remaining on the appointment of John McCone, whom I replaced as chairman. Kennedy then reappointed me to a full five-year term when the initial appointment expired in 1963. President Lyndon B. Johnson reappointed me in 1968 but limited the appointment, at my request, to a two-year term. When President Richard M. Nixon reappointed me in summer 1970, it was understood that I would return to my professorial post at Berkeley a year later. The termination date of this appointment occurred while I was in the Soviet Union leading a group of U.S. nuclear scientists, engineers, and administrators on visits to nuclear establishments and laboratories. The president asked me to continue with the visit and to serve in September as head of the U.S. delegations to the Fourth UN Conference on the Peaceful Uses of Atomic Energy in Geneva and the Fifteenth General Conference of the International Atomic Energy Agency (IAEA) in Vienna.

Many people recall AEC only, and perhaps ingloriously, for its development and testing of nuclear weapons and its sponsorship of nuclear energy as a source of electricity. Although these were two of its principal functions, the agency also had major programs for the production of nuclear materials; reactor research and development for the armed services, including the nuclear navy; research in high- and low-energy nuclear physics and in chemistry and biology; production and sale of radioisotopes for

use in medicine, agriculture, industry, and research; licensing of the use of nuclear materials for power plants and other peaceful purposes; and international cooperation in developing the peaceful atom.

I played a central role in managing these activities. The composition of the commission, its officers, and my staff changed throughout my tenure. A total of 13 commissioners served as my colleagues and, although they were accorded equal authority under the Atomic Energy Act, in practice I—as chairman—was more "equal" than the others.

The commissioners operated pretty much as a collegial body, but we did use a system of "lead" commissioner, in which individual commissioners paid special attention to certain areas of AEC's program. For example, John S. Graham and Robert E. Wilson specialized in civilian nuclear power; Loren K. Olson on regulation; Leland J. Haworth, Gerald I. Tape, and Clarence E. Larson on weapons and research (and attended meetings of the Federal Council on Science and Technology, FCST); James T. Ramey on regulation and civilian nuclear power; John G. Palfrey on international activities; Samuel M. Nabrit and Polly I. Bunting on life sciences and education; Theos J. Thompson on weapons and civilian nuclear power; and Wilfrid E. Johnson on civilian nuclear power.

Congressional oversight was a very serious fact of life for AEC. In our case it was exercised primarily by the Joint Committee on Atomic Energy (JCAE), a unique body established by the Atomic Energy Act. Under the statute, we were required to keep JCAE "fully and currently informed" on all our activities. In addition, AEC's budget had to be authorized in detail by JCAE before it could be acted upon in the normal appropriations process. Much of my time and that of the other commissioners and prin-

A Chemist in the White House: From the Manhattan Project to the End of the Cold War

cipal staff was spent testifying at hearings held by the joint committee on various aspects of the agency's program. The record of these hearings provides a valuable source of information on the agency's programs throughout its history. According to a custom established by the committee itself, its chairmanship alternated each congressional session between a House member and a Senate member. During my tenure the post was filled alternately by California Congressman Chet Holifield and Rhode Island Senator John Pastore. By and large, we had smooth relations with JCAE and the White House, although it was sometimes a difficult balancing act.

Soon after I came, I initiated informal information meetings between commissioners and staff to deal expeditiously with day-to-day operational and administrative matters. These meetings, during which we sometimes dealt with as many as 30 or 40 agenda items, were held in addition to the long-established more formal commission meetings in which the commissioners and staff dealt with policy matters and more long-range business, usually with the help of staff papers submitted by the general manager stating a problem, possible solutions, and recommending an action. During my tenure I presided over some 1700 information meetings and some 850 commission meetings. About 500 of the information meetings and 100 of the commission meetings dealt exclusively with regulatory matters.

Since 1957 AEC's official headquarters had been in Germantown, MD, some 30 miles from downtown Washington. This distance was inconvenient for those of us who needed to transact business at the White House, the Executive Office Building, and with Congress and government departments and agencies. Therefore, an alternative headquarters had been established at 1717 H St., NW, two blocks from the White House, where my fellow commissioners and I, secretarial and key staff, spent most of our time. Still, we regularly held forth in Germantown as well. This gave rise to serious logistical problems because all of our files had to accompany us as we moved from one office to the other. Adding to the cumbersome arrangement was the fact that the regulatory people were quartered in still a different location: Bethesda, MD.

At one of my first meetings with Budget Director David Bell, he suggested that we should try to move toward replacing the five-member commission with a single administrator, a position that I would presumably fill. The other commissioners were amenable, and on May 16, 1962, we sent him a letter. We argued that, due to changed circumstances, the initial concern over concentration of too much power in a single individual had become relatively less important than the need for a more efficient decision-making process. This was a remarkable step—a government administrative body was recommending its own demise. An additional reason that the White House wanted this change was to reduce the leakage of confidential administrative information to JCAE. There had been many such leaks.

Attempts to get support from JCAE, which would have had to provide the necessary legislation to effect the change, were without success. Congressman Holifield, a powerful force in JCAE, was adamantly opposed. Several later attempts, including some during the Johnson administration, were similarly unsuccessful. I was not too disappointed with this result. I found the commission form of administration, although somewhat cumbersome, to have many advantages for attacking the numerous knotty problems we faced. Five minds were potentially better than one.

I shall describe my involvement with a number of councils and committees, my progress reports sent to President Kennedy, his visits to AEC installations, my visits to the Soviet Union and attendance at the annual IAEA's Geneva conferences, the Cuban Missile Crisis, Kennedy's man-on-the-moon program, and the problems of atmospheric testing of nuclear weapons.

As AEC chairman, I belonged to a number of interagency committees that existed for all or part of my tenure. Foremost of these was the Committee of Principals, which advised the president on arms control policy. Established by President Eisenhower, this group expanded and achieved new prominence under President Kennedy, continued to be important in the Johnson administration, and then was abandoned by President Nixon

in favor of more closely held White House control. Other committees whose meetings I, or my designated representative, attended included FCST (1961–71, composed of scientific representatives of federal agencies that had a science component in their operations); the U.S. Intelligence Board; the Federal Radiation Council (1961–69); the President's Committee on Equal Employment Opportunity (1961–65); the President's Science Advisory Committee (PSAC), as an observer and as an alumnus of this committee; the National Aeronautics and Space Council (1961–71); and the National Council on Marine Resources and Engineering Development (1966–71). Vice Presidents Lyndon Johnson, Hubert Humphrey, and Spiro Agnew served as chairmen of the space council, and Humphrey and Agnew of the marine council—I first became well acquainted with Lyndon Johnson because of his service as chairman of the space council. I also attended meetings of AEC's General Advisory Committee, as an observer and as an alumnus of this committee. I had many meetings with the congressional JCAE.

I include here extracts from my diary covering the first of a number of these meetings as illustrative of their nature and content. My first meeting with PSAC was especially interesting because it was attended by President Kennedy:

Friday, February 10, 1961

At 4 p.m. I attended a conference with President Kennedy and PSAC. This was a very informal meeting, and we discussed a wide range of topics. Rabi opened the discussion with a statement on the importance of international scope in science and, in particular, he mentioned the desirability of an international meeting in Geneva in summer 1962 to discuss the applications of science and technology to strengthen underdeveloped areas; he also mentioned the necessity for strengthening NATO science; the president said he would look into both of these matters.

Meeting of the National Aeronautics and Space Council in the Executive Office Building. Left to right: Ragnar Rollefson (Department of State), Seaborg, U. Alexis Johnson (Deputy Undersecretary of State), Edward C. Welch (Executive Secretary), Vice President Lyndon B. Johnson (Council Chairman), Robert S. McNamara (Secretary of Defense), and James E. Webb (Administrator, National Aeronautics and Space Administration). Washington, D.C. July 17, 1963.

We discussed a number of ways in which science could be helpful, and the president was especially interested in sea water conversion. It was pointed out to him that at present this is economically unfeasible; the cost is about ten times greater than the normal costs for fresh water, and there is no presently available scientific principle that holds out much hope for economical water from this source.

In the course of the discussion, the status of the economics of atomic power was raised. Zinn pointed out that only the very largest reactors were economically competitive, and these only because of the government subsidies of the rules of the game, as he put it. I pointed out the value of nuclear power at remote stations and for nuclear propulsion of naval vessels such as submarines. I also pointed out that the aim of the commission was to have competitive nuclear power by 1968.

The president expressed the wish to work closely with, and to rely on, PSAC for advice in a number of matters. He referred specifically to his difficulty in getting unbiased advice from the various departments in the Department of Defense because of the competition between them.

A number of other areas in the general fields of science and its approach to human welfare were discussed. I mentioned the proposal for a U.S.– USSR joint accelerator as a means of scientific cooperation; Wiesner also mentioned my idea of a joint reactor to produce certain isotopes.

Illustrative of Kennedy's interest in science was his unique appearance on April 25, 1961, to address members of the National Academy of Sciences at their annual meeting. Here he emphasized the importance of the academy, and of scientists in general, to the future welfare of the United States in an outstanding extemporaneous speech.

The first commission meeting at which I presided was held on March 1, 1961. The main item of discussion was AEC's position on test ban negotiations preparatory to my meeting with the principals on the following day. Details of this meeting are given in the following memorandum.

United States Government

Memorandum Date: March 1, 1961
To: A.R. Luedecke, General Manager
From: W.B. McCool, Secretary
Subject: Action Summary of Meeting 1707, Wednesday, March 1, 1961, 3:00 p.m., Room 1113-B, D.C. Office
Symbol: SECY:DCR
Commission Decisions

1. Minutes of Meetings 1701, 1702, 1703, and 1704
   Approved as revised.
2. AEC 374/69—Joint AEC-DOD Nuclear Weapon Vulnerability Program
   Approved (Betts)
3. AEC 154/11—Australian Request to Transport Fuel Elements Across the United States
   Approved as revised.
   Mr. Graham requested the Australian Government to be informed by letter this decision does not commit the Commission to approval of the return shipment of irradiated fuel elements. (Wells)
4. Ernest Orlando Lawrence Award-Approved. (Secretariat)
5. Test Cessation Negotiations and AEC 226/276—Draft Instructions for Guidance of U.S. Delegation at Geneva Conference
   Approved as revised. (Betts-English)

OTHER BUSINESS

1. Hanford Labor Contract Negotiations
   Mr. Seaborg requested you provide him with a progress report on the negotiations by Friday.
   Mr. Seaborg said he would report to Senator Jackson on the status of negotiations on Friday. (Secretariat)
2. Rover-SNAP Hearings
   Mr. Olson requested a report on the Commission consideration of the Rover Project. (Secretariat)
   Mr. Seaborg requested that Mr. Olson be provided a copy of Mr. Dryden's testimony on Project Rover. (Donovan)
   You said you would schedule a staff paper and a briefing on Project Rover for next week.

ITEMS OF INFORMATION

1. CIA Briefing of JCAE
2. Senator Anderson's Letter to the Secretary of the Navy
3. Hearings on the Proposed Military Agreement with Italy

I presided over my first Information meeting on Friday, March 3, 1961, for which I include the following notes.

United States
Atomic Energy Commission
Washington 25, D.C.
March 3, 1961

Memorandum for the Commissioners

Subject: Morning meeting notes, March 3, 1961, Chairman's Office, D.C.

Commissioners' Morning Meetings—Dr. Seaborg discussed briefly the continuing arrangements for the meetings.

Report on Regulatory Study—The General Manager said this paper would be issued today.

Chairman's Report on Meeting of the Principals, March 2, 1961

The Chairman's Meeting with Dr. Weisner—Dr. Seaborg said he would meet with Dr. Weisner on Monday afternoon in preparation for the meeting with the president and Congressman Holifield.

Letter to Professor Emelyanov—The letter was approved by the Commissioners. The General Manager is to remind the staff that this involves no Commission commitment to a joint accelerator project. The chairman said he would mention the matter to Congressman Holifield.

Letter to the Interstate Commerce Commission on Radiation Levels re: Shipment of Radioactive Material—Commissioners approved Dr. Wilson's proposed letter.

National Science Foundation on Byrd Reactors—In a telephone conversation with Mr. Olson, Dr. Waterman said the Foundation had no objection to the installation of a reactor at Byrd Station.

Materials Research Laboratory at University of Illinois—Mr. Graham reported on calls from the University of Illinois representatives and Congressman Cannon's letter of February 28, 1961. The General Manager will give the chairman a report on the meeting including the names of the University of Illinois persons concerned with the matter.

Commission Letter to the U.S. Chamber of Commerce (Committee on Atomic Energy)—The general Manager said this matter was in staff and Mr. Graham requested Dr. Wilson to assure timely Commission action.

Draft Bill on Absolute Liability re: Gnome Shots—Mr. Graham Claytor will discuss a draft bill with Mr. Parks. The General Counsel will keep the Commissioners informed.

Joint Committee Letter on Section 91.b (See AEC 1041/9)—General Loper's testimony on February 27 will be circulated.

Southwest Graduate Research Center—The Commissioners will meet with Mr. Stohl and Associates in mid-March contingent on development of a definitive agenda.

Senator Jackson's Letter to the President re Antarctic Reactors—The General Manager is preparing a reply which will be coordinated with the DOD.

Fourth French Nuclear Shot—The General Manager reported on the matter.

Chairman's Report to Senator Jackson on Hanford Strike—The General Manager is provided a talking paper.

Press Release on Hanford Strike—Commissioners approved the proposed press release for Mr. Travis' use. The Joint Committee is to be informed. The chairman is to meet with Secretary Goldenberg and Congressman Holifield, 5:00 p.m., March 6 on this matter.

Hearing Examiners' Move—Mr. McCool reported that the General Manager's staff had agreed to proposed space allocation for Hearing Examiners in A Wing, Third floor.

Attendance
Dr. Seaborg
Mr. Graham
Dr. Wilson
Mr. Olson
Gen. Luedecke
Mr. Naiden
Mr. Oscar Smith
Mr. McCool
W.B. McCool
Secretary
cc: General Manager
General Counsel
Mr. Hollingsworth

AEC commissioners John S. Graham (Washington lawyer), Loren K. Olson (Washington lawyer), and Robert E. Wilson (former chairman, Standard Oil Company of Indiana) were present. I described my philosophy for this daily meeting as follows:

1. *I will attend from 9:30 to 10:15 a.m., when Mr. Graham will take over if it is not finished.*
2. *Inasmuch as the meeting is primarily for the exchange of information, no regular commission business will be transacted; McCool will take brief action minutes.*
3. *Gen. Luedecke and McCool will attend the whole meeting except for the first few minutes, which will be confined to private business if there is any.*

Luedecke and McCool then joined the meeting and I explained my plan to them. Then I described yesterday's principals' meeting. We okayed a letter from me to Vasily S. Emelyanov (chairman of the State Committee on the Utilization of Atomic Energy of the Soviet Union) regarding the U.S. visit to the USSR on waste disposal problems, on the exchange of reports, and on a joint U.S.–USSR accelerator for 300–1000-BeV energy; I sent the letter later in the day:

March 3, 1961
Dear Professor Emelyanov:

Since my nomination by the president to be chairman of the Atomic Energy Commission, I have reviewed the progress that has been made under the Memorandum on Cooperation in the Utilization of Atomic Energy for Peaceful Purposes Program of our countries, and have discussed the subject within the new administration. I have since given careful consideration to the discussions which took place with respect to our exchanges of visits.

It has long been my view that scientific and technical exchanges increase understanding and friendship between our two nations and I am looking forward to further contacts which I am certain will be mutually beneficial. In this regard, I should like to submit for your consideration my thoughts on several matters which have been under discussion, looking toward early consummation of them.

The Atomic Energy Commission is prepared to implement an exchange of visits by U.S. and USSR scientists to radioactive waste disposal installations. I hope that these visits can take place at an early date. For our part, we are prepared to proceed before the end of March, if that were convenient to you and the details could be worked out in time. These visits, we feel, should come about prior to any conference on the subject because they would provide the basis for discussion of information on our respective programs which would be both current and meaningful. In this connection, I am referring to discussions which we understand you had with Mr. Sterling Cole regarding an Agency conference on the subject to which we would be agreeable. Also, I would like to indicate our continued interest in including visits to operational waste disposal facilities if this can conveniently be arranged by you.

It also seems appropriate that we renew consideration of the scheduling of exchanges of visits by our scientists in the fields of fast breeder and nuclear superheat reactors later in the spring of 1961.

Details of the exchanges in the area of waste disposal and in the two reactor fields can be arranged through the usual diplomatic channels. However, I should appreciate any suggestions you might have now for scheduling these visits.

I understand that Dr. Victor Spitsyn may be among the Soviet scientists who would comprise a waste disposal delegation to the United States. In this connection, please convey my continued good wishes to Dr. Spitsyn and my hope for the success of his published, translated papers on radiation chemis-

"It has long been my view that scientific and technical exchanges increase understanding and friendship between our two nations and I am looking forward to further contacts which I am certain will be mutually beneficial."

try. I hope that I shall have the opportunity to see him again in the near future.

The Commission is prepared to initiate the exchange of information provided for in the Memorandum on Cooperation. I have directed that the first group of the Commission's abstracts of unclassified current work in atomic energy be dispatched to you within the next two weeks. We shall welcome receipt of abstracts from the Soviet Union. In addition, we are prepared to provide copies of reports in the reactor field which you may request from our abstracts.

I have reviewed the report of the U.S. and USSR scientists on their New York meeting in September 1960 pertaining to their discussion of the scientific feasibility of constructing a large and novel accelerator. I am hoping that scientists in our country will be able to proceed with the studies recommended by the group on a time schedule that will afford the opportunity of discussing their results at the time of the 1961 International Accelerator Conference. I would be interested in learning of the action you are taking in connection with the group's recommendations and of your views on the best means of coordinating our respective studies and arranging for informal exchange of results.

I am looking forward to meeting you again in the near future and hearing from you on the matters discussed above. I should like once more to express the hope that both of our countries will continue to benefit from the contact arising from the Memorandum on Cooperation. I am sure we will have the opportunity to consider other matters of cooperation in the field of the peaceful uses of atomic energy in the course of the further implementation of the already agreed upon program.

My best wishes.

Sincerely,
Glenn T. Seaborg
Chairman

I later pursued the suggestion of building a large joint U.S.–USSR accelerator. At the time of the U.S.–USSR confrontation over access to Berlin (the "Berlin Crisis"), I wrote the following letter to President Kennedy.

August 11, 1961
Personal and Confidential
Dear Mr. President:

My purpose in writing is to advance an idea that could possibly be useful in bringing about some relief to the Berlin situation. Conceivably the idea could be expanded to a point where it might represent a possible solution to the dilemma.

The idea is this: International science may well have reached the point where a portion of the globe should be set aside for large-scale international scientific projects. We are already exploring with the Soviet Union the possibility of entering into a bilateral arrangement for the joint construction and operation of a large-scale accelerator and possibly a high flux nuclear research reactor. These discussions have not reached the point of considering sites for these projects—but we anticipate that there may be some difficulty in finding neutral sites which also meet technical requirements for adequate space.

If Berlin and enough territory in that portion of East Germany which separates West Germany from Berlin could be dedicated to international scientific activities under the United Nations, it might afford a dignified solution to the Berlin crisis and, at the same time, represent a dramatic and constructive step toward world peace and toward the advancement of science. One could conceive of this neutral zone as being another "Switzerland" for world science.

There are numerous other scientific projects of international interest which could be undertaken in such an area. They could include projects in radio astronomy, experiments in the use of radioisotopes, and even joint activities in the exploration of space. While they might be initiated on a bilateral basis with the Soviet Union, I would envisage that they would soon be expanded to participation by many other nations throughout the world. Possibly the International Atomic Energy Agency would become the appropriate organ to administer this multilateral program for the United Nations.

I commend this idea to your attention. I believe that it would have the support of scientists throughout the world, and it may well be the kind of

A Chemist in the White House: From the Manhattan Project to the End of the Cold War

idea that statesmen and diplomats—now desperately seeking just one fresh idea that might contribute to the relief of the Berlin situation—would welcome and support.

Respectfully submitted,
Glenn T. Seaborg
The President
The White House

Unfortunately, any movement toward such a joint U.S.–USSR endeavor very soon became impossible when the Soviets further shattered the fragile U.S.–USSR relationship with the resumption of atmospheric nuclear weapons testing, as described below.

Leland J. Haworth (director, Brookhaven National Laboratory) was sworn in as the fifth AEC commissioner on April 17, 1961, bringing the commission to full strength. James T. Ramey (executive director, JCAE) and John G. Palfrey (dean, Columbia College) were sworn in on Aug. 31, 1962 to replace John S. Graham and Loren K. Olson.

My first meeting with GAC took place on Feb. 1, 1961 (my first day on the job):

At 2 p.m., we met with GAC. Chairman Kenneth S. Pitzer presided. Members John H. Williams (physicist, University of Minnesota), Willard F. Libby (chemist, University of Chicago), John C. Warner (president, Carnegie Institute of Technology [now Carnegie Mellon University]), Norman F. Ramsey (physicist, Harvard University), Philip H. Abelson (Department of Terrestrial Magnetism, Carnegie Institution), and Manson Benedict (nuclear engineer, MIT) were present. In addition to the commissioners, General Manager Luedecke, Robert A. Charpie (technical secretary), and Anthony Tomei (secretary) were present. Dr. Pitzer gave us GAC recommendations from a three-day meeting, including recommendations for the Lawrence Awards—Dr. Leo Brewer, Dr. Henry Hurwitz, Jr., Dr. Conrad L. Longmire, Dr. W.K.H. Panofsky, and Dr. Kenneth E. Wilzbach. They reaffirmed their choice of Dr. Hans Bethe as the recipient of the Enrico Fermi Award. They recommended more work on materials and low-energy physics so that the funds for these could compete with those for high-energy physics.

My first meeting with FCST took place on Wednesday, March 29, 1961:

Swearing-in ceremony for Leland J. Haworth, newly appointed AEC commissioner. Left to right: Woodford B. McCool (AEC Secretary), William Vitale (AEC Administrative Secretariat), and Haworth. Washington, D.C. April 17, 1961. Courtesy of the U.S. Department of Energy, Germantown, MD. Photo by Elton P. Lord.

The five commissioners of the Atomic Energy Commission. Left to right: Loren K. Olson, John S. Graham, Seaborg, Robert E. Wilson, and Leland J. Haworth. Washington, D.C. July 1961. Courtesy of the U.S. Department of Energy, Germantown, MD. Photo by Elton P. Lord.

At 1:30 p.m. I attended my first FCST meeting with Presidential Science Adviser Jerome Wiesner (in chair), Herb York (director of defense research and engineering, Defense Department), Walter Whitman (science adviser to the secretary of state), Frank Welch (assistant secretary, Department of Agriculture), Edward Gudeman (undersecretary of commerce), Boisfeuillet Jones (special assistant to the secretary, Department of Health, Education, and Welfare [HEW]), James Carr (undersecretary of the interior), Alan Waterman, Jim Webb (administrator, NASA), Elmer Staats (deputy director, Bureau of Budget), and Bob Kreidler (secretary).

Wiesner gave the history, background, and philosophy of the federal council, mentioning competition for funds, needs for priorities, the use of science for civilian (nonmilitary) purposes, and so on. It was decided to meet the fourth Tuesday of each month and to work on broad interagency problems. We then discussed George Kistiakowsky's paper on support of science by the federal government. Gudeman reported on a study of natural resources. Wiesner reported on the work of the PSAC life sciences panel and also emphasized the need to investigate the problem of large scientific salaries paid (with U.S. government funds) by industrial contractors in competition with the smaller salaries paid to government and university scientists. Waterman reported on the Meteorological Center, supported by NSF, to be built at Boulder, CO, and Whitman

reported on the work of the Panel on International Cooperation.

Following amendment of the Aeronautics and Space Act to place the vice president at the head of the space council (April 25, 1961), and the directive of the president to have the space council undertake the necessary studies and government-wide policy recommendations for bringing into optimum use at the earliest practicable time operational communications satellites (June 15, 1961), I attended my first meeting with the space council:

Friday, July 14, 1961,
Washington, D.C.

At 4 p.m. I attended a meeting of the National Aeronautics and Space Council, which was presided over by Vice President Johnson. Also present were Secretary of State Dean Rusk, Ros Gilpatric (for Secretary of Defense Robert McNamara), Jim Webb (administrator of NASA), Newton Minow (FCC), Fred Alexander (for Frank B. Ellis, Office of Emergency Planning), Attorney General Robert Kennedy, Ed Welch (executive secretary), Arnold Fritsch (my special assistant), Captain H.E. Ruble, Philip Farley (special assistant for atomic energy, State Department), and others. The purpose of the meeting was to prepare a recommendation of policy and public statement by President

Kennedy on regulations governing Communication Satellites (including international aspects, role of private industry, urgency, etc.).

As a result, the president issued a statement on Communication Satellite Policy.

299 Statement by the President on Communication Satellite Policy. July 24, 1961

Science and technology have progressed to such a degree that communication through the use of space satellites has become possible. Through this country's leadership, this competence should be developed for global benefit at the earliest practicable time.

To accomplish this practical objective, increased resources must be devoted to the task and a coordinated national policy should guide the use of those resources in the public interest. Consequently, on May 25, 1961, I asked the Congress for additional funds to accelerate the use of space satellites for worldwide communications. Also, on June 15, I asked the Vice President to have the Space Council make the necessary studies and policy recommendations for the optimum development and opera-

tion of such system. This has been done. The primary guideline for the preparation of such recommendations was that public interest objectives be given the highest priority.

I again invite all nations to participate in a communication satellite system, in the interest of world peace and closer brotherhood among peoples throughout the world.

The present status of the communication satellite programs, both civil and military, is that of research and development. To date, no arrangements between the government and private industry contain any commitments as to an operational system. . . .

I have asked the full cooperation of all agencies of the government in the vigorous implementation of the policies stated herein. The National Aeronautics and Space Council will provide continuing policy coordination and will also have responsibility for recommending to me any actions needed to achieve full and prompt compliance with the policy. With the guidelines provided here, I am anxious that development of this new technology to bring the farthest corner of the globe within

Swearing-in ceremony at the White House for newly appointed AEC commissioners John G. Palfrey and James T. Ramey. From left to right: Palfrey, Ramey, Seaborg, President John F. Kennedy, and Leland J. Haworth. Washington, D.C. August 31, 1962. Courtesy of the U.S. Department of Energy, Germantown, MD.

reach by voice and visual communication, fairly and equitable available for use, proceed with all possible promptness.

Two noteworthy events are recorded in my diary at the end of June 1961:

Wednesday, June 28, 1961— Washington, DC

I left the office around 3 p.m. to go to Friendship Airport, where my family was arriving at 4:30 p.m. on United Flight No. 808 from San Francisco. After they arrived I brought them to our home at 3825 Harrison St., and then we all went to the Howard Johnson restaurant at Chevy Chase Center on Wisconsin Ave. for dinner.

Thursday, June 29, 1961— Washington, DC

Transit satellite 4-A was successfully launched from Cape Canaveral at 12:25 a.m. this morning. Two transmitters, powered by a 2.7-W SNAP (Systems for Nuclear Auxiliary Power) device using two different wavelengths, are transmitting navigational data back to Earth. The orbit is such as to suggest a long life.

I held a press conference at noon, attended by about 40 people from the press and TV and about 50 others, at which I answered numerous questions, described the SNAP device, emphasized its safety, and described the SNAP device at the Martin Company in Maryland that is transmitting weather data to our H St. headquarters.

I also described our entire SNAP program, including the development of compact reactors for use in such things as communication satellites for worldwide TV.

The press conference seemed to go well, and the Transit 4-A received national publicity as an example of U.S. science applied for the benefit of mankind. This was the first use of a nuclear power source (using Pu-238) in a satellite, a very important first in the U.S. space program.

After the press conference I went home with a severe migraine headache.

President Kennedy had an especially close relationship with his science and technology adviser, Jerome Wiesner, and his special assistant on national security affairs, McGeorge Bundy. I met with Kennedy many times, during some periods on an almost daily basis, when such items as nuclear weapons testing and the question of a test ban were paramount issues. My contacts were on an individual basis, but more often, whenever some item of interest to AEC was discussed, during meetings of the cabinet or the National Security Council (or one of its subcommittees), sometimes including JCAE members. The cabinet meeting on Oct. 18, 1962, is illustrative of such meetings.

From 10 a.m. to 10:30 a.m. I attended a cabinet meeting at the White House. The president opened the meeting by saying that, because of the rising costs of government, especially those of the space program, defense, the pay bill, and so on, and because of the tax cut, severe economies must be exercised to keep the budget in hand; and he has asked Budget Director David Bell to make cuts in the budget. Bell said that the bureau is aiming for a total that is $2 billion below planning figures and will require cutbacks in programs already started and committed. He said that, although most of us have already submitted our budgets, our staffs should work on possible further reductions preparatory to the review procedure with the bureau. Although every agency must go below planning figures, the Departments of Agriculture and HEW must each cut another few hundred million dollars below the target figures. (HEW Secretary Anthony J. Celebreeze interposed some objection to this, pointing out that this will mean that they will have to cut their budget a total of $600 million below their submitted figures.) Bell indicated that this program shouldn't be interpreted to mean that there will be no new starts of projects. The president observed that the 160,000 increase in government employees since the beginning of his administration is probably too much. Ted Sorensen then discussed next year's legislative program, urging us to con-

sider this now, even though decisions will not be made until after the election. He also suggested that we reconsider the items that didn't pass in this session. He said we should discuss our programs with his office and with the Bureau of the Budget (BOB). Item 3 of the agenda, Review of Foreign Situation, by the secretary of state, was postponed because Secretary Rusk was not present.

I had a good deal of success in my contacts with President Kennedy whenever I requested the inclusion of items of special interest to me in AEC's budget. I recall an early meeting with him at which Wiesner and I convinced him to retain funds in the AEC budget for the construction of the Stanford Linear Accelerator Center. In addition, I sent weekly (later biweekly) informal reports to the president describing developments of import and interest in the atomic energy program.

February 16, 1961
Personal and Confidential
Dear Mr. President:

This memorandum constitutes the first of a series of informal weekly reports to you on developments of importance and interest in the atomic energy program. I will, of course, keep you or your staff informed of significant developments which occur between these weekly submissions.

1. Joint Committee on Atomic Energy Study of Civilian Control of Nuclear Weapons.

In the Fall of 1960, the JCAE made a trip to military bases in seven European countries. A report is now being prepared by the JCAE on the manner in which nuclear weapons are being integrated by the NATO Defense. It is believed that this report will raise basic questions on the organization and purpose of NATO. The JCAE also has under preparation a more encompassing study of the present viability of the concept of civilian control of the manufacture and storage of nuclear weapons, as well as other atomic energy applications.

2. International Activities.

a. The U.S. Representative to the International Atomic Energy Agency—

John F. Kennedy Cabinet meeting, White House Cabinet Room. Clockwise: Bernard L. Boutin (Administrator, GSA) (back), James Webb (Administrator, National Aeronautics and Space Administration), Edward Day (Postmaster General), Ted Sorensen (hidden), Robert S. McNamara (Secretary of Defense), Donald F. Hornig (Director, Office of Science and Technology), William Wirtz (Secretary of Labor), Anthony J. Celebrezze (Secretary of Health, Education, and Welfare), Luther H. Hodges (Secretary of Commerce), Seaborg, (unidentified individual), Kennedy, Douglas Dillon (Secretary of the Treasury), John Carver (Undersecretary of the Interior), Robert F. Kennedy (Attorney General). Behind Dillon is Najeeb Halaby (Administrator, Federal Aviation Administration). Washington, D.C. Oct. 18, 1962. Courtesy of the John Fitzgerald Kennedy Library, Columbia Point, MA.

Mr. Paul Foster—left Vienna last week to return to the United States. Mr. Foster carries the personal rank of Ambassador. It is expected that he will resign in the coming months and I am currently working with the Department of State and members of your staff to find an appropriate candidate for his replacement.

b. The term of Sterling Cole as Director General of the IAEA expires in 1961. The Agency Board of Governors must decide upon his successor at its meeting in June.

It does not appear that the Commission and the Department of State will recommend that Mr. Cole seek a second term. He has not been informed of this. I am currently working with Dr. Wiesner and Dr. Whitman to find a candidate whom the U.S. could support and who would be acceptable to the Soviet Union.

3. Congressional.

The JCAE has not yet organized. It has not held its internal organization meeting and, until this occurs, the Committee cannot hold formal hearings, including my confirmation hearings.

4. Congressional.

The JCAE has indicated its desire to hold a weapons briefing in the near future. There is a strong possibility that at such a briefing the JCAE will want views on our continued test moratorium, as well as on nuclear weapons dispersal.

5. Study of AEC Regulatory Organization and Operations.

A study of the AEC Regulatory function has been completed and forwarded to the JCAE. The study recommends a separation of the promotional activities of the Commission from the Regulatory activities by having both the General Manager and a Director of Regulation report directly to the Commission.

As a result of discussions with Budget Director Bell, we are not submitting requests for legislation to provide for these proposed changes. Rather, we propose to handle them internally and within our present authority.

6. Possibility of Work Stoppage at the Hanford Plutonium Operation.

On October 25, 1960, the Atomic Energy Labor Management Relations Panel intervened in negotiations between the General Electric Company and the Hanford Atomic Metal Trades Council (AFL-CIO). In December and in February the Panel convened and issued recommendations. The Council is prepared to accept all of the recommendations of the Panel and to conclude a new agreement on this basis. General Electric, however, has indicated that the recommendations are not acceptable to it.

If the Panel recommendations are not accepted by both parties by March 12, a work stoppage could occur. The Commission is actively exploring possible means of avoiding or meeting this eventuality.

7. Congressional Hearings Under Section 202 of Atomic Energy Act.

Under Section 202 of the Atomic Energy Act, hearings are held annually to receive information concerning the development, growth and state of the Atomic Energy Industry. Testimony for this hearing has been prepared and forwarded to the JCAE. The testimony indicates that considerable technical progress has been realized during the past year, but that there have been delays in construction of most of the civilian power reactor projects.

8. Prime Minister Menzies' Interest in Project Plowshare.

Prime Minister Menzies of Australia has expressed an interest in AEC's Project Plowshare—use of nuclear explosives for peaceful purposes. Information received from Dr. Mark Oliphant, Member of the Australian Parliament, indicates that Mr. Menzies intends to discuss the program with the president during his forthcoming visit and to raise the possibility of cooperation.

Glenn T. Seaborg

Of special interest is my reference to IAEA in section 2 (International Activities) of the report to President Kennedy. IAEA, a United Nations sponsored organization with headquarters in Vienna, Austria, was established in 1957 on the basis of the suggestion by President Dwight Eisenhower that there was need for an international authority to promote the peaceful uses of atomic energy (Atoms for Peace) while at the same time exercising some control to prevent the diversion of fissionable material from peaceful to nuclear weapons purposes. I attended the annual General Conference of IAEA, held in the fall of each year, 11 times (1961–71) during my tenure as AEC chairman. Representative of these general conferences is my first one in 1961:

Tuesday, Sept. 26, 1961—Vienna

I presided over a meeting of the American delegation to the conference to discuss plans for meetings. Then we went to the Neue Hofburg to attend the opening session, held in the Festsaal. As I entered the hall, I met the president of the Austrian parliament, Leopold Figl, Chancellor of Austria Alfons Gorbach, and Austrian Minister Rudolph Renner. I spoke at the opening session in support of Adm. Oscar A. Quihillalt of Argentina, for president of the confer-

ence, and also in support of admitting the Republic of the Congo to IAEA. I had lunch at the Sacher Hotel with a group hosted by Adm. Quihillalt. I then visited the Swedish ambassador to Austria, Sven Allard, and IAEA headquarters, where some 500 people work.

I attended a reception given by Director General Sterling Cole at the Palais Schwarzenberg, followed by dinner at a Hungarian restaurant (gypsy music) with Haworth and Daniel Wilkes (my assistant).

Wednesday, Sept. 27, 1961—Vienna

I heard the statements of various delegates in the morning. I then had lunch at the Russian embassy, where I sat between Emelyanov and A.I. Alexandrov (member of the USSR Foreign Office). I had an interesting and friendly conversation with Emelyanov; we agreed on continued investigation of a joint accelerator project, exchange of visits, and so on. He renewed the invitation to me to visit Russia.

I gave my speech to the general conference at 2:25 p.m., which opened with the following words, including a message from President Kennedy:

"I am honored to appear before the General Conference of the International Atomic Energy Agency. First of all, I wish to express on behalf of my country the deep sense of loss which it feels as a result of Dag Hammarskjold's death. Words cannot come near to expressing appreciation for his service to all mankind.

"Mr. President, your election as president of the conference is a tribute to the impressive role you have played in advancing the agency's objectives, and I congratulate you.

"I should like to pay tribute to the present director general, Sterling Cole, for his notable work in leading this organization so successfully through its formative years.

"I am privileged to read the following message to the delegates from the president of the United States:

Presidential Message

The General Conference of the International Atomic Energy Agency is a welcome event to all people who value peace. Your meeting accentuates the enormous potential of the atom for improving man's well-being.

We already know the atom can help place more food on our tables, provide more light in our

At the opening session of the International Atomic Energy Agency Fifth General Conference, AEC Chairman Seaborg chats with high-ranking officials of the Austrian government at the Festsaal, Vienna. Left to right: Alfons Gorbach (Austrian Chancellor), Rudolph Renner (Austrian Minister), Seaborg, and Leopold Figl (President, Austrian National Assembly). Vienna, Austria. Sept. 26, 1961. Courtesy of the U.S. Information Service, Vienna, Austria. Photo by Rainer.

homes, fight disease and better our health, and give us new technical and scientific tools. The exploitation of this force for human welfare is just beginning. The International Atomic Energy Agency can assume a position of leadership in bringing the peaceful uses of atomic energy to the people of the world.

Moreover, the intangible benefits of your work are no less than the material rewards. When people from different countries work together in a common cause, they help to maintain a bridge of understanding between nations during times of tension and build firmer foundations for a more stable and peaceful world of the future. I applaud your efforts and assure you that they have the full support of the United States.

John F. Kennedy

President, United States of America

"I am here in the spirit of President Kennedy's message to advance, on behalf of my government, the high principles of the agency, which in the words of the statute are to accelerate

and enlarge the contribution of atomic energy to peace, health, and prosperity throughout the world. My remarks will be confined to that purpose. I shall not pretend that we can ignore world tensions. It is, of course, difficult to be optimistic that these tensions will vanish at an early date. We can hope, as I earnestly do, that this condition will be limited to the tension phase, however difficult this may be. We can do much more: We can resolve, here, to dedicate ourselves to the pursuit of the objectives of the agency and, through mutual goodwill and the positive accomplishment that is within our power, to strive to build solid international foundations that will diminish tensions in the longer future.

"There are good reasons why this organization is especially suited for such accomplishment. Man has many arts which can be applied to the building of a better world order, including diplomacy, law, economics and others. I believe science stands high among these activities."

My speech seemed to be well received. Afterward, in the evening at the reception, I learned from Emelyanov's speech writer, and also from Emelyanov himself, that as a result of my speech, Emelyanov's speech tomorrow morning is being revised tonight, mainly to shorten it. (My speech was about 25 minutes; his apparently was scheduled to be more than an hour long.)

I co-hosted the joint Western Hemisphere reception with representatives of Argentina (Adm. and Mrs. Oscar Quihillalt), Brazil (Mr. and Mrs. Helio F.S. Bittencourt), Canada (Ambassador Max Wershof), El Salvador (Juan Contreras-Chavez), and Mexico (Mr. Jose Maria Ortiz Tirado) at the Palais Pallavicini (North and South American members of the IAEA Board of Governors).

There was a great deal of controversy at this conference over the election of the director general of IAEA to succeed the incumbent American, Sterling Cole. I championed the election of my Swedish friend, Sigvard Eklund, over the violent objections of Vasiily Emelyanov, head of the USSR delegation, and Homi Bhabba of the Indian

North and South American members of the International Atomic Energy Agency Board of Governors co-host with Seaborg a joint Western Hemisphere reception at the IAEA General Conference in the Palais Pallavincini, Vienna Left to right: Admiral Oscar Quihillalt (Argentinian Delegate), Mrs. Quihillalt, Seaborg, Mrs. Bittencourt, Helio F. S., Bittencourt (Brazilian Alternate Delegate), Ambassador Max Wershof (Canadian Delegate), Juan Contreras-Chavez (Salvadoran Delegate), and Joe Maria Ortiz Tirado (Mexican Delegate). Sept. 27, 1961. Courtesy of the U.S. Department of Energy, Germantown, MD.

A Chemist in the White House: From the Manhattan Project to the End of the Cold War

delegation. I described this altercation in my biweekly report to President Kennedy:

October 10, 1961
Dear Mr. President:

I have the honor to submit our regular biweekly report to you. I would like to take this occasion to report to you briefly on my recent visit to Europe.

I. Fifth General Conference of the International Atomic Energy Agency.

A. Confirmation of Director General:

1. By all odds, the confirmation of Dr. Sigvard Eklund (formerly Deputy and Managing Director of the Division of Reactor Development of the Swedish Atomic Energy Commission) as Director General of the IAEA was the highlight of the Fifth General Conference. Dr. Eklund was finally confirmed on October 3 by a vote of 46 countries for, 16 against, 5 abstentions, and 10 "not voting." He was inducted into office on October 6.

2. Dr. Eklund's confirmation was preceded by, and accompanied with, attacks on the Western Powers by the Soviets and Soviet Bloc. Much of this was precipitated by the Indians, who alleged that Dr. Eklund's candidacy was a failure to recognize the African-Asian geographic regions, and that there had not been adequate consultation prior to his nomination by the Board of Governors.

3. The Indian objections were all too transparent. It was obvious that they hoped to create enough doubt that Dr. Eklund himself would withdraw and that in the confusion a "neutral" candidate, such as Mr. Arthur Lall, Indian Ambassador to Austria and member of the IAEA Board of Governors, who was much in evidence, might be persuaded to occupy the position.

It should be emphasized that in all of the numerous speeches by the Soviets and Soviet Bloc countries, and the Indians, there was only one brief and mildly critical reference to Dr. Eklund's ability; namely, alleged inadequate administrative experience. No one challenged his integrity as an individual nor his competence as a scientist, and his administrative ability was generally acknowledged, except in the one instance mentioned above.

4. Professor Vasily B. Emelyanov, head of the Soviet delegation to the Fifth General Conference and member of the IAEA Board of Governors, received a good deal of attention in the U.S. newspapers following the confirmation of Dr. Eklund. In statements to the press and in official statements at the conference, he said that he personally would not cooperate with Dr. Eklund, that he *personally* would recommend to his government that it withdraw from the IAEA. And then, he *personally* stalked out of the conference.

I emphasize "personally" because the Soviet Delegation itself did not retire from the conference with the Professor. In addition, the Professor told us shortly after our arrival in Vienna that he personally would not be returning to Vienna. There was no other explanation, but we assume that he has another assignment. Accordingly, his tactics in Vienna may well have been intended to mask the fact that he would not be returning anyway. I assume they were also part of the global effort to intimidate members of the United Nations into believing that unless the Soviet gets its way it will also withdraw from the United Nations.

I don't think the Soviets will withdraw from the IAEA, although I believe they may for a while be even less cooperative than in the past. However, the Soviets pledged 40,000 rubles to the Agency's voluntary fund.

In contrast to Professor Emelyanov's "official" behavior, my contacts with him were cordial, if somewhat reserved. He made several references to his readiness to cooperate with the U.S. in scientific exchanges of information and personnel, and in large-scale, joint scientific ventures.

Professor Emelyanov's official position in the Soviet Government is chairman of the State Committee on the Utilization of Atomic Energy, Council of Ministers. . .

Respectfully submitted,
Glenn T. Seaborg

Actually, the Soviets and Indians didn't withdraw from IAEA; they were completely won over by Eklund, who remained as director general for five four-year terms, until 1981.

Seaborg gives the opening address to the International Atomic Energy Agency Seventh General Conference in the Hofburg. Vienna. Sept. 25, 1963. Courtesy of the U.S. Information Service, Vienna, Austria.

The flavor of the IAEA general conferences, held in the Hofburg, Vienna, can be gleaned from, where I am shown addressing the Seventh General Conference on Sept. 25, 1963.

Shortly after the inauguration, on Feb. 16, 1961, President Kennedy made an historic visit to AEC's Germantown headquarters, traveling from the White House by helicopter. On the way, I told him a little about the AEC program. He had coffee with me in my office before meeting with the AEC staff. Following this, I gave him a short introductory briefing on the fundamentals of atomic and nuclear structure.

Feb. 16, 1961

After this, we proceeded to the commission meeting room, where all present rose to their feet. I introduced Dwight Ink (assistant general manager), the person nearest at hand, to the president, and Bob Hollingsworth (deputy

general manager), then introduced in turn the staff members occupying front-row seats as the president proceeded down the row. These included Col. Allan Anderson (Division of Military Application), Brigadier Gen. Austin Betts (director, Division of Military Application), Dr. Frank Pittman (director, Division of Reactor Development and Technology), Dr. Paul McDaniel (director, Division of Research), John Hall (International Activities), Dr. Charles Dunham (director, Division of Biology and Medicine), and Hugo Eskildson (manager, Idaho Operations Office). As I escorted the president over to his seat at the far side of the oval table, the members of the president's staff—Gen. Chester V. Clifton (military aide to the president), Dr. Wiesner, and Mr. Bundy—passed down the front row, introducing themselves and shaking hands with each person in that row.

All persons promptly took their places. The president's chair was in the middle of the far side of the oval table, directly opposite the lectern to be used by the speakers. Eight other officials occupied places at the table. Viewed from the lectern we were seated, left to right, as follows: Gen. Alvin R. Luedecke (general manager), General Clifton, Commissioner Olson, Acting Chairman Graham, the president, myself, Commissioner Wilson, Dr. Wiesner, and Mr. Bundy.

On the left side of the lectern and facing the president was the easel for charts; on the right side was a translucent screen for the projection of slides and transparencies. A cutaway model of a Polaris submarine, about two feet long, was on the table to the right of the lectern. Speakers and watchers were seated in three rows on the opposite side of the room from the president and behind the lectern, easel, and screen, which served to obstruct their view and render them less conspicuous.

I opened the discussion at 10 a.m. with a few brief remarks on fundamental physics. I contrasted chemical energy, released through rearrangement of planetary electrons, and nuclear energy, derived from rearrangement of particles within the nucleus. To explain both, I then contrasted fission reactions and fusion reactions and described in the simplest terms the uses, potential and actual, that man makes of these

A Chemist in the White House: From the Manhattan Project to the End of the Cold War

reactions with weapons, reactors, and thermonuclear devices. I pointed out that the limit to the usefulness of nuclear energy is man himself, because of the effects of radiation, and in this connection emphasized the commission's concern for greater understanding of radiation effects and for protecting public health and safety. I brought out that the fission process is self-initiating and self-perpetuating at normal temperatures, whereas the fusion process requires the creation of an extremely high-temperature environment that can be maintained only with great difficulty. I also pointed out that a very great potential advantage of the fusion reaction is the absence of residual radioactivity such as is created by fission products.

*I then gave a broad-brush picture of the principal AEC activities, including military weapons, propulsion for ships and aircraft, peaceful uses, civilian power reactors, nuclear propulsion and auxiliary power for rockets, production and use of radioactive isotopes, research program in physics, chemistry, materials and biology and medicine, the regulatory role of the commission, the operating budget, the value of capi-*tal plant, the number of employees, and so on.

Next, AEC staff members provided briefings that took until about 11:15 a.m. I then presented to the president a number of policy questions. This took until about 12:10 p.m., much longer than planned. The president seemed to be very interested throughout and asked many questions in the course of the briefings and policy discussions. We returned by helicopter to the White House, where we arrived at about 12:40 p.m.

To the American public, perhaps the most important science decision made by Kennedy was the Apollo Project. AEC got involved in the U.S. space program mainly through two nuclear space projects, called SNAP and Rover. SNAP used radioisotopes or minireactors to produce long-lasting power for spacecrafts and space instruments. It proved, in the case of radioisotopes, a very successful program. In Apollo's second trip to the moon,

Seaborg and President John F. Kennedy stride into the Atomic Energy Commission headquarters after helicopter flight from the White House. Germantown, MD. Feb. 16, 1961. Courtesy of the U.S. Department of Energy, Germantown, MD. Photo by Elton P. Lord.

Seaborg and Kennedy enjoy cookies and coffee in the AEC chairman's office. Seaborg slyly places the chocolate cookies, which he does not much like, in front of the president. Feb. 16, 1961. Courtesy of the U.S. Department of Energy, Germantown, MD. Photo by Elton P. Lord.

astronauts left there a data-gathering device whose sole source of energy derived from a SNAP battery fueled by plutonium-238. I took special pride in its success because I had been a codiscoverer of this element (and this isotope) back in 1941.

In contrast to SNAP, Rover (which was aimed at producing desk-size nuclear reactors to propel rockets in space) proved far more ambitious but less successful. The nuclear-powered rocket would have been much more powerful and efficient than any chemical rocket. Because of this advantage, it held much hope in the 1960s for lunar and interplanetary space flight. But the obstacles were formidable, and the project was terminated during the Nixon administration.

SNAP and Rover defined AEC's roles in the national space program and brought me into the highest level space policy deliberations. As a member of the space council chaired by the vice president, I developed a close relationship with Lyndon Johnson and NASA Administrator James Webb.

Conflict arose early between AEC and BOB over the costly Rover project. In 1961 AEC wanted to put $7 million in its FY 1962 supplemental budget for an early flight test, but BOB wanted to cut it. On March 22, 1961, BOB Director David E. Bell, his deputy, Elmer Staats, Jerome Wiesner, Fred Dutton, and I met with the president to iron out the AEC–BOB disputes. Kennedy settled some differences but decided to defer decision on the $7 million Rover increase. He wanted to wait for the NASA budget meeting that afternoon.

The meeting with NASA representatives proved a key step toward Kennedy's moon shot decision.

From 5:55 p.m. to 7 p.m. I attended another meeting with President Kennedy and Bell, Staats, Wiesner, Bundy, Vice President Johnson, James Webb (administrator of NASA), Hugh Dryden (deputy administrator of NASA), Willis Shapley (BOB), Bob Seamans (NASA), and Edward Welsh (executive secretary, space council). Webb and Dryden described the forthcoming Mercury flight experiment in which a man would be projected into space by ballis-

A Chemist in the White House: From the Manhattan Project to the End of the Cold War

tic means for a relatively short time—not into orbit. The point was made that the responsibility for the decision to do this would lie with NASA, and not with the president, so that the president could investigate the situation if something went wrong.

Dryden gave a résumé of the arguments in favor of having our expensive space program, such as increase in human knowledge, practical military value for satellites and reconnaissance, buildup of national technological capability, and prestige value. There was considerable discussion about the relative value of Rover, and it seemed to be agreed that this development would be important only for missions in the distant future, such as taking a man to the moon or even further into space. At one point the president asked Webb and Dryden what their choice would be as between Saturn C-2 and Rover, and they indicated that it would have to be Saturn C-2. It was pointed out to the president that a decision to go ahead with Rover and some of the other NASA projects meant that he was taking steps to go into the very expensive advanced man-in-space projects. The president asked for further comparisons of costs with the Eisenhower budget, and then said he will meet again tomorrow with BOB before he makes a decision.

The next day, March 23, Kennedy, Johnson, Wiesner, and I met with JCAE leaders on the AEC budget. Senator Clinton Anderson (D-New Mexico) spoke strongly in favor of the $7 million increase for Rover, which was conducted mainly in the Los Alamos Scientific Laboratory in his state. After the meeting, Kennedy told Bell to add the money for Anderson for Rover. But later Bell was able to change Kennedy's mind by arguing that the $7 million Rover increase for AEC would require adding $23 million for NASA. The fate of the Rover increase was again left uncertain.

In the meantime, a crash space program was suggested. On April 19, Webb, in a phone conversation, alerted me to a secret Trevor Gardner report on space that advocated everything going to the military.

Webb, obviously opposed to this change, commented that President Eisenhower was probably right in warning against the military–industrial complex. I asked him to talk this over with Mac Bundy. In the meantime, Holifield vowed JCAE would try to add the $7 million for Rover to the AEC authorization bill.

Space soon received added momentum from a series of international events. On April 12 Maj. Yuri A. Gagarin of the Soviet Union stunned the world by becoming the first man in space. Five days later came the Bay of Pigs fiasco, with furor abroad and political fallout at home. When the National Security Council met on April 22, 1961, its prescheduled main agenda was on the nuclear test ban. But in the last part of the meeting, it turned to space:

Saturday, April 22, 1961

At 11 a.m. I attended a meeting of the National Security Council at the

Seaborg briefs Kennedy on some fundamentals of atomic and nuclear structure before meeting with staff members for a discussion of individual AEC programs. Feb. 16, 1961. Courtesy of the U.S. Department of Energy, Germantown, MD. Photo by Elton P. Lord.

White House. The president called on John J. McCloy (President Kennedy's special adviser on disarmament), who gave out a memorandum dated April 22, 1961, on possible courses of action relating to negotiations with the Soviets regarding a test ban. The president asked about the significance of the Soviet insistence on a three-man council in place of a single administrator, and it was agreed that this is a key point.

The president asked Defense Secretary McNamara how important it was to resume testing, and McNamara said it was important, especially in three areas: (1) to increase the yield-to-weight ratio of a number of nuclear warheads, (2) to study the effect of antiweapons on nuclear warheads, and (3) to develop the radiation or neutron bomb.

The president suggested that McCloy prepare a position, emphasizing the two or three points that can be dramatized, which could be used by Arthur H. Dean (McCloy's assistant) and David Ormsby-Gore (British minister of state for foreign affairs) when they return to Geneva to make a last attempt to achieve a test ban. This might be followed by a communication from the president to Soviet Premier Nikita Khrushchev.

During the last five minutes, the president got into the space problem. He said he had written to the vice president, asking him and the space council to study a number of problems in the area of space missions, such as what is needed in order to put a man on the moon, a comparison of chemical versus nuclear rockets, and so on. He asked that a memorandum be prepared responding to these questions.

After the meeting I talked to Webb, and he said they are probably going into the big space program, already planning for NOVA (the proposed successor to SATURN, never developed), the project beyond SATURN (the launch vehicle for the man-on-the-moon project), so as to put a man on the moon by 1967.

Tuesday, April 25, 1961

Webb called me and told me that as a result of last Saturday's National Security Council meeting, the president sent a memorandum to the vice president, asking a lot of questions on space, such as: Do we have a chance of beating the Russians by putting a laboratory into space? Using rockets, can we get a man to the moon and back? What about the interspace program? How much would this cost? Is the program on a 40-hour work week? What could be done to move us ahead?

Webb told me he talked with McNamara last Friday about the man on the moon program, and it seems that he is willing to undertake development of a large solid first-stage NOVA and a second-stage NOVA by a redistribution of his funds. There is now no doubt but that we have a real military requirement for a big booster. The solid propellant route is good, in addition to Rover and SATURN.

Saturday morning (April 22) Webb met with the vice president on the man on the moon program, and he asked for answers by 2 p.m. to the president's memorandum, which NASA furnished. In the reply, NASA said that the Eisenhower 10-year program was funded on the basis that it would cost $17.9 billion, and that it would raise about $2 billion by 1968. The program was underfunded by about $5 billion, in terms of the work incorporated in the plan; instead of $17.9 it should have been about $22 billion. The actions taken by President Kennedy to speed up the booster program and go beyond Mercury would add about another $5 billion, bringing the figure to about $26 billion to $27 billion over a 10-year program.

In answer to the president's question about speeding things up, Webb told me he said that the $26 billion to $27 billion rate over 10 years would be about the most economic rate at which to spend the money. The $17 billion to $22 billion is really an uneconomic one, because you spend the same amount of money but over a longer period of time. However, if they really wanted a speed up, a program of about $33.5 billion could be envisaged; that would mean, for instance, driving ahead on making a lunar landing by 1967, which is now under discussion. In a meeting the vice president included Frank Stanton of CBS, George Brown of Texas, Don Cook of New York, Wehrner von Braun, et al., to confer. Adm. John T. Hayward and someone from the army also attended. Webb told the president that the most sensible thing is not a space czar, but for him

A Chemist in the White House: From the Manhattan Project to the End of the Cold War

(Webb) and me and McNamara to work these things out together.

This led to Kennedy's famous message to Congress announcing his program to place a man on the moon before the end of the decade:

Thursday, May 25, 1961

At 12:30 p.m. I was present as President Kennedy gave his message to a packed joint session of Congress (there wasn't a seat available) on "Urgent National Needs." It included a program for placing a man on the moon in this decade and also a request for AEC supplementary funds for Rover.

The president's special message to Congress included the following passages:

. . . Finally, if we are to win the battle that is now going on around the world between freedom and tyranny, the dramatic achievements in space which occurred in recent weeks should have made clear to us all, as did the Sputnik in 1957, the impact of this adventure on the minds of men everywhere, who are attempting to make a determination of which road they should take. Since early in my term, our efforts in space have been under review. With the advice of the vice president, who is chairman of the National Space Council, we have examined where we are strong and where we are not, where we may succeed and where we may not. Now it is time to take longer strides—time for a great new American enterprise—time for this nation to take a clearly leading role in space achievement, which in many ways may hold the key to our future on earth. . . . I therefore ask the Congress, above and beyond the increases I have earlier requested for space activities, to provide the funds which are needed to meet the following national goals:

First, I believe that this nation should commit itself to achieving the goal, before this decade is out, of landing a man on the moon and returning him safely to the Earth.

No single space project in this period will be more impressive to mankind, or more important for the long-range exploration of space; and none will be so difficult or expensive to accomplish. We propose to accelerate the development of the appropriate lunar space craft. We propose to develop alternate liquid and solid fuel boosters, much larger than any now being developed, until certain which is superior. We propose additional funds for other engine development and for unmanned explorations—explorations which are particularly important for one purpose, which this nation will never overlook: the survival of the man who first makes this daring flight. But in a very real sense, it will not be one man going to the moon—if we make this judgment affirmatively, it will be an entire nation. For all of us must work to put him there.

Secondly, an additional 23 million dollars, together with 7 million dollars already available, will accelerate development of the Rover nuclear rocket. This gives promise of someday providing a means for even more exciting and ambitious exploration of space, perhaps beyond the moon, perhaps to the very end of the solar system itself.

Third, an additional 50 million dollars will make the most of our present leadership, by accelerating the use of space satellites for world-wide communications. . . .

President Kennedy continued the cooperation with the United Kingdom in the nuclear weapons program. As a result of an initiative that I instigated, the president asked AEC to "take a new and hard look at the role of nuclear power in our economy:"

March 17, 1962
Dear Mr. Chairman:

The development of civilian nuclear power involves both national and international interests of the United States. At this time it is particularly important that our domestic needs and prospects for atomic power be thor-

> "Now it is time to take longer strides—time for a great new American enterprise—time for this nation to take a clearly leading role in space achievement, which in many ways may hold the key to our future on earth."

June 21, 1961

UNCL. BY DOE
NOV 86

Dear Mr. Chairman:

Reference is made to your letter to me of June 20th concurred in by the Secretary of Defense, concerning proposed cooperation with and transfer of certain non-nuclear parts of atomic weapons to the United Kingdom pursuant to the "Agreement Between the Government of the United States of America and the Government of the United Kingdom of Great Britain and Northern Ireland for Cooperation on the Uses of Atomic Energy for Mutual Defense Purposes."

I note that, pursuant to Executive Order 10841 dated September 30, 1959, the Atomic Energy Commission and the Secretary of Defense, acting jointly, have determined that the proposed cooperation and the proposed transfer arrangement for the non-nuclear parts of atomic weapons as set forth in your letter will promote and will not constitute an unreasonable risk to the common defense and security of the United States. I hereby approve the program for the transfer of the types and quantities of non-nuclear parts of atomic weapons as set forth in your letter.

By copy of this letter I am informing the Secretary of Defense of this action.

Sincerely,

The Honorable Glenn T. Seaborg
The Chairman
U. S. Atomic Energy Commission
Washington 25, D. C.

President John F. Kennedy's letter to Seaborg approving transfer of non-nuclear parts of atomic weapons from the United States to the United Kingdom. June 21, 1961.

nuclear power development program in the light of the nation's prospective energy needs and resources and advances in alternate means for power generation. It should recommend appropriate steps to assure the proper timing of development and construction of nuclear power projects, including the construction of necessary prototypes. There should, of course, be a continuation of the present fruitful cooperation between Government and industry— public utilities, private utilities and equipment manufacturers. Upon completion of this study of domestic needs and resources, there should also be an evaluation of the extent to which our nuclear power program will further our international objectives in the peaceful uses of atomic energy. The nuclear power plants scheduled to come into operation this year, together with those already in operation, should provide a wealth of engineering experience permitting realistic forecasts of the future of economically competitive nuclear power in this country.

As you are aware, two major related studies are now or will soon be under way. The study being conducted at my request by the National Academy of Sciences on the development and preservation of all our national resources will focus on the nation's longer-term energy needs and utilization of fuel resources. The other study to be launched soon by the Federal Power Commission will determine the long range power requirements of the nation and will suggest the broad outline of possible programs of growth for all electric power companies—both private and public—to meet the great increase in power needs. Your study should be appropriately related to these investigations.

The extensive and vigorous atomic power development programs currently being undertaken by the Commission should, of course, be continued and, where appropriate, strengthened during the period of your study. I urge that your review be undertaken without delay and would hope that you could submit a report by September 1, 1962.

Sincerely,
Dr. Glenn T. Seaborg
Chairman
Atomic Energy Commission
Washington 25, D.C.

oughly understood by both the Government and the growing atomic industry of this country which is participating significantly in the development of nuclear technology. Specifically we must extend our national energy resources base in order to promote our nation's economic growth.

Accordingly, the Atomic Energy Commission should take a new and hard look at the role of nuclear power in our economy in cooperation with the Department of the Interior, the Federal Power Commission, other appropriate agencies, and private industry.

Your study should identify the objectives, scope, and content of a

Following eight months of evaluation of the nuclear power situation by AEC and its advisers, I submitted a report to President Kennedy:

November 20, 1962
Dear Mr. President:

I am pleased to submit herewith the report resulting from our "new and hard look at the role of nuclear power in our economy," as requested by you on March 17, 1962. In preparing this report, we have had the benefit of comments and advice from interested offices and individuals within and without the Government. However, the Commission takes full responsibility for the conclusions and recommendations of the report.

The Commission, of course, has concentrated on issues related to the development and use of nuclear power; it has not attempted to appraise the possible effect of major research efforts on the economics of non-nuclear energy sources or on improved transmission methods for either source of energy. However, the study has been greatly aided by the information furnished by the Department of Interior, the Federal Power Commission, and the National Academy of Sciences' Committee on Natural Resources.

Those who have participated in the study you requested are agreed that it has proved to be very timely. While the Commission has been proceeding on a considered course in general accord with its 10-year civilian power program adopted in 1958, that program is now on the threshold of attaining its primary objective of competitive nuclear power in high-fuel-cost areas by 1968. However, it became evident with the passage of time that our attention had probably for too long remained focused narrowly on short-term objectives. This restudy made it apparent that, for the long-term benefit of the country, and indeed of the whole world, it was time we placed relatively more emphasis on the longer-range and more difficult problem of breeder reactors, which can make use of nearly all of our uranium and thorium reserves, instead of the less than one percent of the uranium and very little of the thorium utilized in the present type of reactors. Only by the use of breeders would we really solve the problem of adequate energy supplies for future generations.

We believe that it still is necessary for the Government as an interim measure to maintain a substantial program of research and development on advanced types of reactors other than breeder reactors, which are some years away. It appears from the projections made that efficient converter reactors will be required in conjunction with breeder reactors to meet the rapidly growing national demands for electrical power. This Government program over the next several years is also important since it provides the national means for "bridging the gap" between the infancy and maturity of nuclear power. This interim aid will allow the consolidation of the gains made to date and will permit the national nuclear program to proceed in an efficient and sensible manner toward the development of more efficient and economical converter reactors and eventually breeder reactors.

Furthermore, a vigorous national nuclear power program can be pursued without interfering with a growing coal industry; in fact, all our projections indicate that, even assuming an optimistic forecast of nuclear power development, the use of coal by the rapidly expanding electric generating industry will increase several fold over the next 40 years.

It should be recognized that, largely as a result of early optimism, we have, in a short space of time, developed a competitive nuclear equipment industry which is over-capitalized and under-used at the present time. This optimism has had some good results in terms of bringing many able technical men, manufacturers, and utility executives into the field, and assuring Congressional and industrial support during the development years.

The optimism has also brought about some difficulties in that unless there are new starts on atomic power plants, the atomic equipment industry will probably dwindle down to fewer manufacturers than would be desirable for a healthy and competitive nuclear industry. Fortunately, it now appears that only relatively moderate additional governmental help will be necessary to insure the building of a substantial number of large, water-type power reactors that will be economically competitive in the high-fuel-cost areas of this country and the world. This would increase public acceptance, keep the nuclear industry healthy, and help to furnish the plutonium necessary for a breeder reactor economy as soon as it can be adequately developed.

In summary, nuclear power promises to supply the vast amounts of energy that this Nation will require for many generations to come, and it probably will provide a significant reduction in the national costs for electrical power.

The Commission unanimously concurs in this report.

Respectfully yours,
Glenn T. Seaborg
Chairman
The President
The White House
Enclosure

Although the report erred in predicting a substantial future role for breeder and other advanced types of reactors, the following years in the 1960s saw a surge in the building of water-cooled reactors; however, beginning in the next decade nuclear power development in the United States came to a stop because of a combination of low electric power load growth, environmentalist pressure, regulatory uncertainties, and increased costs. Many of these same pressures are now affecting nuclear power worldwide.

Kennedy was the first president to make a personal presentation of AEC's highest honor, the Enrico Fermi Award; he made these presentations in 1961 and 1962 and had planned to do so again on Dec. 2, 1963. On the first of these occa-

John F. Kennedy presents the 1961 Enrico Fermi award to Hans Bethe (center) in the presence of Seaborg. Kennedy is the first president to make a personal presentation of the Atomic Energy Commission's highest honor. White House Cabinet Room. Washington, D.C. Dec. 1, 1961. Courtesy of the U.S. Department of Energy, Germantown, MD.

sions—presentation of the Fermi Award to Hans Bethe in the White House in 1961—the camera caught all of us smiling when I somewhat tardily handed over the $50,000 check for the president to present to Dr. Bethe, the president having just read the citation in tones of convincing finality. He presented the Fermi Award to Edward Teller in a White House ceremony on Dec. 3, 1962.

Helen and I flew with President Kennedy and his entourage on Air Force One to California on March 23, 1962, when he gave the University of California Charter Day address in Memorial Stadium. We flew from Andrews Air Force Base to Alameda Naval Air Station. Others in the party included California Senator Clair Engle and Congressmen George Miller and Jeffrey Cohelan. During the flight Kennedy came back from the presidential quarters in the plane to where Helen and I were seated to talk to me about the University of California; I briefed him on the university, the Berkeley people serving in his administration, and so on. I participated in a parade (with the president, Ed Pauley, and Governor Pat Brown in the lead car) through Alameda, Oakland, and Berkeley. The presidential party then proceeded to the Radiation Laboratory where, in Building 70A, we briefed the president using a display of model weapons and command-and-control devices. Robert McNamara, Harold Brown, Norris Bradbury, John Foster, Edwin McMillan, Edward Teller, Roger Batzel, Carl Haussman, Marvin Martin, Ted Merkle, Duane Sewell, Duncan McDougall, Robert Frohn, and Michael May were present.

We then went to University House on campus, where Kennedy met UC President Clark Kerr, the regents, chancellors from other campuses, deans, and their wives. After lunch, we went to Memorial Stadium, where after the inauguration of Ed Strong as chancellor of

Kennedy visits the AEC-supported Lawrence Radiation Laboratory. Left to right: Norris Bradbury, (Director, Los Alamos Scientific Laboratory), John S. Foster (Director, LRL–Livermore), Edwin M. McMillan (Director, LRL–Berkeley), Seaborg, Kennedy, Edward Teller, Robert S. McNamara (Secretary of Defense), and Harold Brown (Director of Defense Research and Engineering, Department of Defense). In front of building 70A. Berkeley, CA. March 23, 1962. Courtesy of the Los Alamos National Laboratory, Los Alamos, NM.

UC Berkeley, the president spoke to a crowd of 85,000 to 90,000 people, occupying all the seats and covering the floor of the stadium. He gave an inspiring address, speaking largely extemporaneously, though he had a prepared text to which he could refer. He opened with personal references to a number of the Berkeley people in his administration about whom I had briefed him during the airplane ride on the way out in the morning.

Helen and I were privileged to attend the famous black-tie dinner for Nobel Prize winners held in the White House the evening of April 29, 1962. There were 49 Nobel Prize winners, most with their wives and including one female Nobelist, Pearl Buck, among the 175 guests at this black-tie dinner. The winners of the Nobel Prize the previous year were the featured guests: Melvin Calvin (chemistry), Robert Hofstadter, and Rudolf Mössbauer (physics), and Georg von Bekesy

Kennedy gives the University of California's Charter Day address before a packed Memorial Stadium. Berkeley, CA. March 23, 1962. Courtesy of the University Archives, The Bancroft Library, University of California–Berkeley.

(physiology or medicine). The assembled group of 49 Nobel Prize winners had our pictures taken with President and Mrs. Kennedy, and including such people as Pearl Buck, Mrs. Ernest Hemingway, and Mrs. George C. Marshall.

The president spoke in the State Dining Room where he was seated at one of 14 festively decorated tables, where I was also seated. A loudspeaker beamed his words to the neighboring Blue Room where First Lady Jacqueline Kennedy presided over five additional tables, and where Helen was seated.

It was on this occasion that Kennedy made his famous extemporaneous remark, "I think this is the most extraordinary talent, of human knowledge, that has ever been gathered together at the White House, with the possible exception of when Thomas Jefferson dined alone."

Among the 175 people present, including many spouses, were

Prime Minister Lester B. Pearson (of Canada), Robert F. and Ethel Kennedy, astronaut John Glenn, John Dos Passos (writer), Robert Frost (poet), Lee DuBridge, James Killian, George Kistiakowsky, Robert Oppenheimer, Alan Waterman, Jerome Wiesner, Clark Kerr (president of the University of California), Nathan Pusey (president of Harvard University), and Norman Cousins (writer). Gold cloths covered the tables in the State Dining Room and the Truman China with green and gold borders was used. In the Blue Room, the cloths continued the blue theme while the Harrison China with blue and gold bands was used there.

In a conversation with Robert Oppenheimer, I asked if he would accept the Fermi Award if offered, and he replied that he would. In reply to another question, he indicated that he wouldn't want to go through another hearing process to get his security clearance renewed. (Oppenheimer's security clearance had been revoked in 1954 as the result of an infamous hearing process instigated by AEC, as I have described in Chapter 2.)

After dinner, when we had assembled in the East Room, we were privileged to hear Frederic March perform. He read three works by past Nobel Prize winners. One was a short foreword by the late novelist Sinclair Lewis. March followed this by reading the stirring words spoken by Gen. George C. Marshall when, as secretary of state, he spoke at Harvard University upon the inception of the Marshall Plan. Lastly, March recited a moving chapter from an unpublished novel of the sea by the late Ernest Hemingway.

On occasion the White House dinners were at the white-tie level. Two of these, which Helen and I attended, when Vice President Lyn-don B. Johnson, House Speaker John McCormack and Chief Justice Earl Warren were honored, were held on Feb. 20, 1962, and Jan. 21, 1963.

On Oct. 17, 1963, I was privileged to attend a luncheon at the White House in honor of Yugoslav President Tito, in connection with his meeting with Kennedy (one of Kennedy's last such affairs of state).

In December 1962 I visited Los Alamos Laboratory, Sandia Laboratory, and the Nevada Test Site with President Kennedy. I recall flying over the Sedan Crater with him and remember his fascination with the size of the hole (1500 feet in diameter and 400 feet deep). He wanted his helicopter pilot to land on the crater's edge for a closer look. When the pilot, after some difficulty, persuaded the president that the dust might be deep and cause a

Dinner honoring the 1961 U.S. Nobel prize winners at the Kennedy White House. A total of 49 American Nobelists are in attendance. Front row, seated left to right: Pearl Buck (Literature, 1938), Rudolf L. Mössbauer (Physics, 1961), Mrs. Ernest Hemingway, Kennedy, Mrs. George C. Marshall, Melvin Calvin (Chemistry, 1961), First Lady Jacqueline Kennedy, and Robert F. Hofstadter (Physics, 1961). Seaborg (Chemistry, 1951), is standing in the back row (left center of picture). George C. Marshall won the 1953 Peace prize; Ernest Hemingway won the 1954 Literature prize). Washington, D.C. April 29, 1962. Courtesy of the John Fitzgerald Kennedy Library, Columbia Point, MA.

Reception line at a white-tie dinner at the Kennedy White House, honoring Vice President Lyndon B. Johnson, Chief Justice Earl Warren, and Speaker of the House of Representatives John McCormack. Seaborg is fifth in line (from left) and Helen Seaborg (behind Seaborg) is greeting Johnson (left of Jackie Kennedy). Washington, D.C. Jan. 21, 1963. Courtesy of the John Fitzgerald Kennedy Library, Columbia Point, MA.

problem in taking off, he asked that a low-level flight be made around the crater, and this was done.

Friday, Dec. 7, 1962

I accompanied President Kennedy on his tour of AEC and Air Force installations. We left Andrews Air Force Base

at 8:30 a.m. in the presidential plane. Others on board were: Deputy Secretary of Defense Ros Gilpatric, Secretary of the Air Force Eugene Zuckert, Gen. Curtis E. LeMay (chief of staff, U.S. Air Force), Adm. George W. Anderson, (chief of Naval Operations), Gen. Earle G. Wheeler (chief of staff, U.S. Army), Gen. David M. Shoup (special assistant to President Kennedy), Jerry Wiesner,

President John F. Kennedy meets with Marshall Josef Tito (bottom left corner), president of Yugoslavia and his aides. Kennedy is flanked by U.S. Under Secretary of State W. Averell Harriman (left) and U.S. Assistant Secretary of State for European Affairs (William Tyler (right). White House. Washington, D.C. Oct. 17, 1963. Courtesy of the John Fitzgerald Kennedy Library, Columbia Point, MA.

Harold Brown, Spurgeon Keeny (Wiesner's White House office), Don Hornig, Mac Bundy, Pierre Salinger (press secretary to President Kennedy), Mrs. Evelyn Lincoln (personal secretary of President Kennedy), Maj. Gen. Chester V. Clifton, Brigadier Gen. Godfrey T. McHugh (President Kennedy's air force aide), Capt. Tazewell Shepard (naval aide to President Kennedy), Kenneth O'Donnell (special assistant to President Kennedy), and Dr. George Burkley (the president's physician).

We arrived at Offutt Air Force Base in Omaha about 10 a.m., where we visited Strategic Air Command headquarters and heard briefings on operations, the latest intelligence on Russian missile and Anti Inter-Continental Ballistic Missile (AICBM) sites, and so on. Vice President Johnson and party (including my technical assistant Arnold Fritsch) followed us in another big 707 jet.

At 1 p.m. our group, minus most of the Defense Department people but including Fritsch, Vice President Johnson, and General Wheeler, flew to Santa Fe (arriving at 2:15 p.m.), where a huge crowd greeted the president. After being joined by Senator Clinton Anderson we flew to Los Alamos, with the president, Senator Anderson, O'Donnell, Shepard, and me in the front compartment of a helicopter and Wiesner, Bundy, Salinger, and Secret Service men in the rear compartment. We were met by a huge crowd at Los Alamos.

We then continued by helicopter to Los Alamos Scientific Laboratory and landed at the Chemical and Metallurgical Research Building. Raemer Schreiber, Norris Bradbury, Harry Finger, and I briefed the president on the Rover program with the aid of models.

The president and entourage, including New Mexico Congressmen Joseph Montoya and Thomas Morris (who had joined us when we landed at Los Alamos), then drove in a motorcade to the football stadium, where the president gave a marvelous impromptu speech. The motorcade then proceeded to the airstrip. I flew with President Kennedy and Senator Anderson in the front compartment of the helicopter to Albuquerque (Kirkland Airport). There the president received another big reception. We proceeded by motorcade

to Sandia Corporation. Kenner Hertford (area manager, Albuquerque Operations Office) and Monk Schwartz (president, Sandia Corporation) briefed the president on nuclear weapons and command and control with sample of weapons, old and new, and much of the component parts.

Saturday, Dec. 8, 1962

We flew in the presidential plane to Indian Springs Air Force Base. We left Albuquerque at 9 a.m. and arrived at 9:15 a.m. I flew in the front compartment of a helicopter with President Kennedy, Al Graves (Los Alamos physicist), and Nevada Senators Howard Cannon and Alan Bible (who flew with us in the plane from Albuquerque) over the Nevada Test Site.

We flew over the spot that is planned for a community, the site of SMALL BOY (a July effects test), the Los Alamos Scientific Laboratory and the Lawrence Radiation Laboratory underground test sites, the SEDAN crater (which the president requested that we circle), and LRL testing tunnels. We landed at the Nuclear Rocket Development Site (NRDS) where Finger and I escorted the president through the Maintenance and Disassembly Building (MAD). I then rode in an open car with the president for a tour of NRDS. We ended up at Engine Test Site #1 where, with the help of Erickson (of Aerojet General), we showed the president the huge stand for testing the NERVA (Nuclear Engine for Rocket Vehicle Application) engine (22 ft. by 40 ft.). Reporters and photographers (some 40 or 50) met us at the MAD building and Engine Test Site #1 and at points in between to take numerous pictures and movies of the president with Finger and me.

I had a chance to discuss many things with the president during the helicopter trip and the ride in the open car: the Rover schedule, the civilian nuclear power program, Plowshare excavation potential, basic research leading to the discovery of fission and plutonium, the need for a community near NRDS, the PLUTO (Nuclear Ramjet Missile Propulsion System) project, SNAP-50 and the $25 million start he gave it, the 10,000-MW nuclear desalination project, and many other items.

Kennedy and Seaborg tour the Nuclear Rocket Development Site at the Nevada Test Site. Dec. 8, 1962. Courtesy of the Office the Naval Aide to the President, Washington, D.C. Photo by R. L. Knudsen.

In May 1963 I led a 10-man delegation to the Soviet Union for a tour of nuclear facilities. The president made Air Force One available to our delegation, and we arrived in Moscow on May 19, after a record-breaking 8 hours and 39 minutes of nonstop flight from Washington, DC. Our tour of facilities, some not shown before to Western scientists, and the cordial treatment we received helped to affirm my belief that science can successfully serve as a common meeting ground and a common language between the East and West:

Saturday, Sunday, May 18–9, 1963

We departed Dulles International Airport at 9:35 p.m. on the Air Force One plane. I slept in President Kennedy's bed aboard the plane from midnight to 4 a.m. (Washington time), or 7 a.m. to 11 a.m. (Moscow time). We arrived in Moscow at 1:10 p.m., making the flight from Washington, DC, to Moscow in 8 hours, 38 minutes, and 42 seconds, which established a record. I saw Riga, Latvia, and many small Russian towns on the way. My delegation consists of the following: Gerald F. Tape (recently appointed AEC commissioner, replacing Leland J. Haworth, and outgoing president, Associated Uni-

versities, Inc.), Manson Benedict (chairman, GAC, AEC; professor of nuclear engineering, Massachusetts Institute of Technology), Alvin R. Luedecke (general manager, AEC), Albert V. Crewe (director, Argonne National Laboratory), Albert Ghiorso (scientist, Lawrence Radiation Laboratory), Alexander Zucker (scientist, Oak Ridge National Laboratory), Algie A. Wells (director, Division of International Affairs, AEC), Arnold R. Fritsch (my technical assistant, AEC), Cecil St. C. King (my staff assistant, AEC).

In addition, on the flight to Moscow, we had an unexpected last-minute companion: Vitalii I. Goldanskii (chief of the Radiation and Nuclear Chemistry group at the Chemical Physics Institute of Moscow). Goldanskii, an outstanding scientist with whom I was well acquainted, had been visiting U.S. nuclear research facilities, including Brookhaven National Laboratory. On learning of our projected trip, he had asked me only a few days earlier whether he could hitch a ride home with me. I had agreed without hesitation. While officials of both the U.S. State Department and the USSR embassy in Washington may have been somewhat taken aback by this impromptu arrangement, neither side raised any objection.

We were met at the Sheremetyevo airport (north of Moscow) by U.S. Ambassador Foy Kohler, State Committee on Atomic Energy (SCAE) Chairman Andronik M. Petrosyants, V.S. Kandaritskiy (director of the International Affairs Office of SCAE), and many newsmen and photographers. A Russian newsman asked if I brought greetings from President Kennedy to Khrushchev, and I said the president sent his best wishes and hopes that a successful Agreement on Peaceful Uses of Atomic Energy would be concluded. I was driven to Spaso House in a SCAE car. Fritsch also is staying at Spaso House, and the others are staying at the Sovietskaya Hotel. We had lunch at Spaso House with Kohler, Embassy Second Secretary Herbert Okun (who is to be our escort officer), John McSweeney (minister), Godfrey T. McHugh (President Kennedy's air force aide, who came on the plane with us), Fritsch, Wells, and King.

After lunch, Fritsch, King, Okun, and I visited the Kremlin and took movies and pictures. After sandwiches at Spaso House, the entire group (including those staying at the Sovietskaya

The American scientific delegation boards *Air Force One* at Washington's Dulles International Airport to Moscow, USSR, for a tour of Soviet nuclear facilities. Left to right: Seaborg, Alvin Luedecke (AEC General Manager), Manson Benedict (General Advisory Committee Chairman), Alexander Zucker (Scientist, Oak Ridge National Laboratory), Gerald F. Tape (recently appointed AEC Commissioner), Albert Ghiorso (Scientist, Lawrence Radiation Laboratory), Albert V. Crewe (Director, Argonne National Laboratory), Cecil King (AEC Staff Assistant), and Arnold R. Fritsch (AEC Technical Assistant). Algie A. Wells, AEC Assistant General manager for International Activities, is not present.) Washington, D.C. May 17, 1963. Courtesy of the U.S. Department of Energy, Germantown, MD.

Hotel) attended a show that began at 7 p.m. at the magnificent Palace of Congresses in the Kremlin. The capacity is 6000, and it was full. We watched The Red Army Chorus (including dancers). The singers were good, but many songs were somewhat militant.

Monday, May 20, 1963—Moscow

The whole group went to SCAE headquarters, where we held discussions with Andronik M. Petrosyants, Igor D. Morokhov (first deputy chairman), Valentin A. Levsha (deputy chairman), Nikolay M. Sinev (deputy chairman), Vasily Emelyanov (deputy chairman), and V.S. Kandaritskiy (director of the International Affairs Office, SCAE). We agreed on a schedule of our stay. They then showed us a model of their 70-BeV accelerator and injection linear accelerator being constructed at Serpukhov.

Tuesday, May 21, 1963—Moscow

Petrosyants and I signed four copies (two in each language) of the Agreement for Cooperation in Peaceful Uses of Atomic Energy. Tass, AP, and UPI photographers as well as CBS and ABC TV representatives were present. I gave special interviews to CBS and ABC TV.

After signing, we spent about 40 minutes in conviviality around the table. Our whole delegation and Ambassador Kohler were present, plus all the Soviet officials who were present yesterday at the discussion of the program. The whole group was driven to Arkhangelskoe, former estate of Duke Golitsyn and later Yusupov (killer of Rasputin) of early and late 19th century, respectively—a beautiful park and museum (closed today). We were then driven back to Moscow (the park was approximately 30 km from city center) and had lunch at the National Hotel.

Then all of our delegation with Morokhov went to the USSR Academy of Sciences, where I introduced our group, and President Mstislav V. Keldysh (identified some years later as a chief architect of the Soviet Union's space program) introduced his group. Benedict and I presented to N.N. Semenov a certificate of membership in the U.S. National Academy of Sciences on behalf of NAS President Frederick Seitz. Ghiorso and I presented a Mendelevium Folio to Keldysh. Crewe and I presented CP-1 (Chicago Pile-1) graphite pieces to Keldysh, Mikhail D. Millionshchikov (vice president of the USSR Academy of Sciences) and Morokhov (of SCAE). Academicians I.Y. Tamm, Aleksandr P. Vinogradov, V.I. Spitsyn, Dimitrii V. Skobeltsyn, and Lev A. Artsimovich were also present.

After this, we attended a reception at the Academy. Pictures were taken here as well as previously at Arkhangelskoe. We attended the U.S. Army–Navy diplomatic reception at Spaso House (hundreds of guests, including Russian military and also the head of Aeroflot). Our whole group attended the Bolshoi ballet (Tchaikovsky's Swan Lake) at the Palace of Congresses, with Petrosyants, Kandaritskiy, Okun, D.P. Filippov, and others.

During the following week our delegation, as a whole or in subgroups, visited the Lebedev Physics Institute, the Institute of Chemical Physics, Moscow State University, the Kurchatov Institute of Atomic Energy in Moscow, the Obninsk Physical–Technical Institute (for nuclear reactors, 65 miles south of Moscow), the Scientific Research Institute of Atomic Energy Reactors

Seaborg and Andronik M. Petrosyants (Chairman, USSR State Committee on Atomic Energy) sign the Memorandum on Cooperation in the Field of Utilization of Atomic Energy for Peaceful Purposes. State Committee on Atomic Energy headquarters. Moscow, USSR. May 21, 1963. Photo by Arnold R. Fritsch, Pittsburgh, PA.

Seaborg and Petrosyants shake hands after the signing of the memorandum on cooperation in the peaceful uses of atomic energy. Left to right: Seaborg, Tilchonow (Interpreter), Igor D. Morokhov (State Committee on Atomic Energy First Deputy Chairman), Valentin A. Levsha (SCAE Deputy Chairman), Petrosyants, and Nikolay M. Sinev (SCAE Deputy Chairman). May 21, 1963. Photo by Arnold R. Fritsch, Pittsburgh, PA.

at New Melekess (85 km from Ulyanovsk, where we visited Lenin's birthplace, childhood home, and school), the historic Khlopin Radium Institute, the Ioffe Physical–Technical Institute, the Yefremov Scientific–Technical Research Institute (and the famous Hermitage) at Leningrad, the Novovoronezh Nuclear Power Station, the Kharkov Physical–Technical Institute, the Dubna Joint Institute for Nuclear Research, and the site of the 70-BeV Proton Synchrotron at Serpukhov. Many of these sites had never before been visited by Western scientists.

I met and had a long talk on Wednesday, May 29, 1963, in Moscow, with Soviet President Leonid I. Brezhnev:

I had an appointment for an hour and a quarter (10 a.m. to 11:15 a.m.) with Leonid Ilyich Brezhnev, chairman of the Presidium of the USSR Supreme Soviet. (Two days before, my Soviet hosts proudly told me they had made an appointment for me with an important official of the Soviet government, namely, the president. When I indicated some hesitation or lack of knowledge of this individual, they hastened to assure me that this presented a very unusual opportunity for me, because the president, a man named Leonid Brezhnev, was destined in their opinion to play a very important role in the future of the Soviet government.)

The meeting took place in the Kremlin office building (their White House), where Khrushchev also has his office. Petrosyants was present, also Anatoliy Belov (SCAE), who acted as interpreter; I.V. Tikhonov (same committee), who took notes for SCAE; and Herbert Okun, who took notes for me and the U.S. embassy.

Brezhnev seemed a personable man of pleasant appearance, about 55 or so, with more of a Western manner than most Soviet officials. He first asked me if this was my first visit to the Soviet Union. When I replied that it was, he said he thought it was a good start and added, "Good relations require frequent visits." He agreed with me that this was particularly true in the field of science because of the international aspect of scientific research, which makes science an excellent vehicle for continuing contact and the development of good relations.

I noted that my visit to the Soviet Union was a short one—too short, according to Petrosyants—and that I was sorry not to have the opportunity to visit the new scientific center at Novosibirsk. Brezhnev replied that he had visited the establishment on two occasions, the last time, three years ago. He said that a great scientific center is being built there, on a large scale and in a very picturesque location. He added that the state has spared no expense in this endeavor and that Academician Mikhail Alekseyevich Lavrentyev had reported to him that work was almost finished on the site. I asked when the work had begun on the center, and Brezhnev said it had begun six years ago. Petrosyants noted that such prominent scientists as Lavrentyev, Guri I. Marchuk, and Vladimer S. Sobolev now work in Novosibirsk. Brezhnev pointed out that the center carries on a very broad program of scientific research, and said that I should include Novosibirsk in the itinerary of a future visit. I said that I hoped Petrosyants could visit the United States this year, perhaps in October or November.

Answering Brezhnev's question whether I found the atmosphere open during my visit, I said that I thought it was. He said that the chairman of the state committee (Petrosyants) was the type of man who wanted to open things for me and that the government had given him instructions to show me everything. When I commented that the delegation had been well fed on its trip, Brezhnev noted that Russian cuisine was varied and heavy. He described in detail the manifold aspects of the cuisine of the various regions and Republics of the USSR. We discussed sports in the USSR, and Brezhnev remarked that the Soviet people are passionate soccer fans.

Noting that I had visited only the Russian republic during my travels in the USSR, Brezhnev spoke about the other interesting national republics, with special stress on the Central Asian republics. He declared that before the revolution they were semicolonies of the Czarist regime and very backward, whereas now they have everything: agriculture, industry, and an active cultural life. He said that agriculture, in particular, was on a very high level in the Central Asian republics and that Uzbekistan alone produces more than 3 million tons of cotton a year.

Brezhnev asked for my impressions of Moscow, Leningrad, and other cities I had visited. I said it was hard to judge since I had never been in the USSR before, but that there was a lot of construction going on. Brezhnev agreed, noting that in Moscow 3.7 million square meters of new housing are being built a year, in addition to the reconstruction of the city. He said that new types of buildings—skyscrapers and open-air housing projects—were being built and that most Soviet housing construction used prefabricated panels of prestressed concrete. I observed that the quality of construction in other Soviet cities seemed superior to that in Moscow. Brezhnev replied that he would not conceal the fact that after the war the Soviet government had to work on its industry, and this had held back housing construction and caused difficulties. Several years ago, he continued, a housing program had been adopted with the principle of one family per apartment. Brezhnev said that now the USSR was sending its architects abroad to study the construction of housing and office buildings. He said Soviet architects had already visited France, Italy, and Austria but that he was not sure whether any had gone to America. Brezhnev said that only yesterday a government decision had been made to send more Soviet architects abroad to study advanced techniques. He noted that a site was going to be set aside in Moscow where foreign architects would be invited to design buildings on an experimental basis and that the best designs would be adopted for use in the USSR.

I commented on Brezhnev's apparent scientific–technical background, and he discussed his career. He agreed that he had had technical training as a metallurgist but pointed out that for more than 25 years he has been in party work. "In my party work, I ve always been concerned with industry," he remarked, citing his work in Zoporozhye, Dnepropetrovsk, Moldavia, and Kazakhstan. He said that for several years he had been secretary of the Central Committee, where he was responsible for industry and construction work. He concluded this outline of his career by saying that he was now in the government. Brezhnev further noted that metallurgy seemed to run in his family, for his grandfather, father, brother, sister, and son were all metallurgists, and all worked at the same plant. Brezhnev also noted that his father-in-law had been a railway engineer for 45 years and now was pensioned.

We agreed that society today lives in a scientific world. He said that he has been working in Moscow for a relatively short time, less than 10 years, and in that span he has seen sputniks and spaceships. He stated that he had visited plants here such equipment is manufactured (it is possible for members of the government to visit such plants) and that he had been astonished by what he saw.

I remarked that several of my books had been translated into Russian and were used in Soviet schools. He said that the USSR translates many books, both fiction and science, and has a special publishing house for such work. I asked whether the USSR pays royalties to foreign authors, and Brezhnev said, "Yes, I think so."

He asked me for my opinion of what I had seen and asked me to criticize boldly. I said I had seen a lot, both Soviet work in my own scientific specialty of the transuranium elements and work in the reactor field. After reviewing briefly the work being carried on at various Soviet installations, I remarked that a solid base was being built. I commented that the future would depend on the people concerned, and the Soviet government was giving them a lot of support. Brezhnev said jocularly that they ask for a lot of money, and Petrosyants interjected quickly, "And thank God we get it." On the subject of funds, I said that I had asked Petrosyants what his total budget was but that I had not received any answer. Petrosyants said, rather apologetically, that because of the vertical structure of his state committee, "We couldn't let you know without lying, and we don't lie."

Brezhnev said that the USSR expected a lot from radiochemistry. I remarked that at the Khoplin Institute in Leningrad a solid base was being built, but the results would come in the future. Brezhnev said, "It's very important that you don't stay too far ahead of us," and Petrosyants added that this work was not all peaceful. Brezhnev noted that as long as the United States was peaceful, things would be all right.

Brezhnev agreed with my statement that the Soviet government appears to understand the importance of basic science and cited the Soviet chemical industry as an example. Momentarily confusing the functions of Baibakov and Tikhomirov, he said that the Soviet Union was investing heavily in this field.

He asked me about Soviet atomic power plants and whether I thought that Soviet scientists were working in the proper direction. I replied that the Soviet approach was very similar to ours and that similar problems were being encountered, such as corrosion. I said that cooperation in solving such problems would help both countries, and Brezhnev replied that he would welcome such cooperation. He agreed with my statement that there was a place for atomic power now in the economies of both countries, particularly in areas where other sources of power are scarce and expensive.

As a memento of my visit, I presented him a square, transparent paperweight, containing (as I explained) a small piece of the original graphite taken from the reactor (CP-1) in which the world's first self-sustaining chain reaction had been achieved on Dec. 2, 1942, in Chicago, and Brezhnev responded to this gift with great warmth. After thanking me, Brezhnev said that he wanted to leave the subject of science. He said that I would doubtless meet with President Kennedy upon my return and that he wanted me to tell the president that Khrushchev means what he says in his speeches, addresses, and documents sent to the president about peaceful coexistence and peaceful cooperation. "This is not propaganda," he said, "it is the sincere desire of our government, our people, and of our party, which leads the country. I can't say any more than that. I hope that this area will be as successful as your scientific contacts. Please tell President Kennedy this, even though I don't know him. And give my best wishes to him and his family." I replied that I would give this message to the president, who is a fine man. I said that Brezhnev would feel this way if they ever had the opportunity to meet someday. Brezhnev ended this conversation by declaring that if they met he thought the president would like him because of his candor and openheartedness.

Brezhnev's manner was warm and friendly; he seemed to display a

"Brezhnev noted that as long as the United States was peaceful, things would be all right."

good understanding of science and technology when he spoke of the details of the work in various institutes. All in all he made a favorable impression of a man who wanted to get along with the United States. My talk with him was perhaps even more interesting in retrospect, since his replacement of Khrushchev as first secretary of the communist party of the Soviet Union occurred less than a year and a half later. I think it is worth mentioning that although at the time of our meeting a number of people regarded him as a mere figurehead, the opinion was growing among certain experts (including Ambassador Kohler) that he was assuming a position of increasing importance and that (I was told prophetically) he might actually be Khrushchev's successor. I had the impression, which I could not document, that he spoke as though he anticipated his future role in government.

It is symptomatic of the extreme insularity of Russian leaders that, as I was told later, I was the first non-Communist American to meet Brezhnev. The only American to meet him before me had been Gus Hall, head of the Communist party in the United States.

Our trip was profoundly interesting and satisfying to all of us as scientists and simply as human beings, and we were glad of this chance to become somewhat acquainted with the land and the people of the Soviet Union. It was also deeply encouraging with respect to our hopes for peaceful nuclear cooperation between our countries and constructive exchanges in general. Excerpts of our published trip report are included here.

> . . . My visit to the Soviet Union with the other members of the U.S. Atomic Energy delegation affirms my belief that science can successfully serve as a common meeting ground for East and West. The basic principles and methods of science are invariant; they are the same for the East and the West. Isaac Newton's third law of motion works as well for Soviet space flights as it does for those of the United States.
>
> Both sides need a beginning—a beginning which can hopefully lead to mutual trust and which will facilitate the freer exchange of information and ideas. Science is as good a place as any from which to start since it is measured against the criteria of nature, rather than the judgment of persons, and therefore is more removed, perhaps, than other fields from the arena of political and social emotions.
>
> I am hopeful that the fine spirit that existed between our delegation and its Soviet hosts is indicative of a growing desire for constructive cooperation, and that this attitude will continue so that exchanges between our countries may broaden and contribute to a peaceful settlement of our differences. . . .

All members of the U.S. delegation to the Soviet Union felt that the trip was quite rewarding and worthwhile. A considerable amount of new information and insight into the Soviet Union's programs in nuclear energy were gained. Not only were a number of sites visited which never before were seen by Western groups, but tours of institutes and installations previously visited were in general more extensive and complete.

If, as expressed by the Soviet hosts, this visit was indicative of the cooperative attitude that the USSR proposes to take toward the Memorandum on Cooperation in the Peaceful Utilization of Atomic Energy, signed by Chairman Seaborg and Chairman Petrosyants in Moscow at the start of this trip, then in fact it would appear reasonable to anticipate that both parties to the agreement will profit greatly from it. It was clearly stated by several of the hosts during the trip that the USSR had broken precedent and had shown, in many

"It is symptomatic of the extreme insularity of Russian leaders that, as I was told later, I was the first non-Communist American to meet Brezhnev. The only American to meet him before me had been Guy Hall, head of the Communist party in the United States."

cases, reactors and apparatus still under construction. Generally, the Soviets prefer to withhold visits to sites until work is completed so as to give a more positive impression.

As was evident, the exchange of scientific ideas and information is welcomed by the Soviet scientific community. The enthusiasm with which the U.S. scientific delegation was greeted by Soviet scientists and Soviet people met during the visit was notable. As has been observed by other delegations and visitors to the Soviet Union, science appears to represent a positive link which can be advantageously used to bridge the gap that now exists between the two societies.

It is deemed beyond the province of this report to consider in detail the social life which the delegation encountered during the trip. Briefly, besides the innumerable excellent banquets, the group also attended a concert of the Red Army Band and chorus and saw the Swan Lake ballet in the new Palace of Congresses within the Kremlin. In Leningrad, the delegation had the opportunity to hear the opera, "Boris Godunov," in the former Royal Opera House. Members of the group also were guests at an international soccer match in Lenin Stadium, Moscow. Finally, there were several social receptions which were notable for the warm spirit of friendship evidenced by all the guests.

One sidelight on the delegation's trip, which was a measure of the friendliness of the visit, was the innumerable exchanges of gifts. At each site appropriate mementos of the U.S. delegation's visit were given by Chairman Seaborg and other members of the party. In exchange, in many instances, the group received books or scientific journals concerning either the institute being visited or a Russian translation of one of several of the books written by various members of the U.S. delegation. Highlights of these exchanges of gifts were the presentation of the folio commemorating the discovery of ele-

ment 101, mendelevium, to President Keldysh of the USSR Academy of Sciences and the exchange of gifts with Chairman Petrosyants at a reception given by the State Committee during the final day of the visit.

Through the kind auspices of Dr. Edwin Land of the Polaroid-Land Corporation, the delegation had obtained a Polaroid camera with an ample supply of color film. Numerous one-minute color photographs were taken throughout the tour which were then presented to the Soviet scientists at the various sites. These always made a fine impression. Upon departure, Dr. Seaborg presented Chairman Petrosyants with the camera for his personal use, together with a supply of color film. In return, Chairman Seaborg and the other members of the delegation received a large quantity of photographs taken of the party throughout its visit to the Soviet Union. Generally, in the many institutes and sites visited, the delegation was not permitted to take pictures. However, the State Committee provided an adequate number of photographers to cover the visit.

A few specific conclusions regarding the delegation's visit to the Soviet Union's nuclear energy sites can be made. Contrary to the U.S. program, there is little biological work being done in the Soviet Union nuclear program (or at least little was shown). Although this was not one of the main interests of the members of the delegation, the only indication that biological work in the atomic energy program was being conducted was at the Kurchatov Institute of Atomic Energy where they had recently begun a biology program. The only other biological work that was seen was at the Chemical Physics Institute of the Academy of Sciences.

Noticeably different in approach from this country was the tendency to consolidate in several large institutes all the research in a scientific field. For example, the Kurchatov

Institute of Atomic Energy is quite large and equal to the largest national laboratories in the U.S. In the U.S., although there are several national laboratories, there is also a strong effort to diversify nuclear energy research into the many universities and colleges of the country in order to conduct research in conjunction with the training and education of new scientists. This did not appear as evident in the USSR.

With few exceptions, the lack of experimental equipment in the large experimental facilities visited by the delegation was noticeable. It was the group's general impression that the experimental gear had been removed for the visit in an attempt to give an impression of orderliness. This was unfortunate, since it would have been of more scientific benefit to see the equipment and discuss the experiments in more detail.

The U.S. delegation had known prior to the trip that the Soviet attitude regarding nuclear safety was somewhat different from that in the U.S. Whereas the Soviet Union is as sensitive to the inherent dangers connected with the handling of radioactive isotopes either in fission product studies, radiochemical work, or hot cell studies, they take a decidedly different attitude toward nuclear reactor safety. During the visit to the nuclear reactors in the Soviet Union, no air-tight containment structures were seen. The prevalent attitude apparently is that once a reactor has been safely designed it is safe and reactor accidents cannot happen.

[They were wrong, as I pointed out to them at the time of my visit; the Chernobyl accident followed in 1986.]

Also, reactors are built extremely close to one another, such as at Novovoronezh and New Melekess, in order to take advantage of common facilities such as water lines or ventilation systems. On the other hand, the Soviets are concerned with exposure of workers to radiation. The normal work week for such employees in the Soviet Union is 35 hours, compared with a 40-hour average elsewhere in the USSR. Paradoxically, of course, at no point in the delegation's visit were the members of the delegation given film badges or other dosimeters, hard hats, or safety shoes, although numerous construction sites and radiation areas were visited.

The final conclusion of the delegation was that the Soviet Union's nuclear energy program is competent and that in many areas a very ambitious and aggressive attitude is in evidence, such as in high energy accelerators, controlled thermonuclear reactions, and transuranium research. It was clear that the Soviet scientists are quick to assess the value of any newly-discovered device, theory, or principle, and then to attempt to improve upon the original discovery. In this way they are among the leaders in such fields as alpha spectroscopy, neutron capture gamma ray work, and the discovery of new transuranium elements and isotopes. The Soviet Union will for a time have the world's highest energy electron linear accelerator at Kharkov and the world's highest energy proton accelerator at Serpukhov. There was no evidence to show that the Soviets will not continue to hard press the U.S. in years to come. . . .

These were the themes I emphasized when I reported to President Kennedy on our trip. I met him at the White House on June 14, 1963. After first giving him the message Brezhnev had asked me to relay from Khrushchev, I described my meeting with Brezhnev and discussed the speculation that he might be Khrushchev's successor. I went on to give an account of our trip and my impressions:

I said that their work in nuclear power is similar to ours, but they are not developing as many varieties and do not have nearly as much power on the line. They believe, however, just as we do, that nuclear power has a role to play and that it will be competitive in high cost areas where the price of coal is high. One reason they are not so far along as we is probably that they haven't pressed so hard, because of their ample sources of conventional power. I said, however, that they are building on a broad base and will have a good future program in this and all other areas of peaceful nuclear applications and that their budget for the peaceful uses of atomic energy is probably about equivalent to ours.

I said that they have a program in controlled thermonuclear reactions that is at least equivalent to ours, and their budget in this field is probably larger than ours, but I reminded the president that this is a field in which the solutions probably will not come for another 20 or more years. I said that they are making an extremely large effort in the field of transuranium elements, probably at least partly with a view to the prestige that would be gained through discovery of a new element.

The president asked if the use of his airplane had been helpful. I said that it had been, and that it made quite an impression when we arrived in Moscow. He brought the meeting to a close by saying that he was very happy that our trip to the Soviet Union had been so successful. I believe that my visit played a role in setting the stage for the successful negotiations that led to an agreement on the Limited Test Ban Treaty during Averell Harriman's ''Twelve Days in Moscow'' some two months later.

Let me go back now to the beginning of the Kennedy administration to trace his vigorous, persistent, and ultimately successful pursuit of a nuclear test ban treaty with the Soviet Union.

Very early in the Kennedy administration, on March 7, 1961, I attended a lunch in the Gold Room at the White House with President Kennedy. Other guests at the luncheon were Vice President Johnson, Secretary of State Rusk, Secretary of Defense McNamara, Mr. Bundy, Mr. McCloy, Mr. Dean, Dr. Wiesner, Gen. Lemnitzer, Sena-

tor Jackson, Senator William Fulbright, Senator Hubert Humphrey, Senator Clinton Anderson, Senator Bourke Hickenlooper, Senator John Pastore, Congressman Chet Holifield, Congressman James Van Zandt, Congressman Thomas Morgan (chairman of the Foreign Relations Committee), Congressman Albert Gore, Congressman Melvin Price, Mr. Ramey, and Mr. Adrian Fisher. At the meeting McCloy and Dean briefed those present on the current status of agreement on the aims of the test ban negotiations. A very spirited discussion followed, which was concluded when the president rose and said, with a bit of a twinkle in his eye, that he was glad to see that there was agreement on the U.S. side, and now we would see if we couldn't get the Russians to agree, too. My journal entry describing the conclusion of this meeting reads as follows:

As we were leaving the White House, a group consisting of the president, Senator Jackson, Mr. Rusk, Mr. Bundy, Mr. McCloy, Mr. Dean, Dr. Wiesner, and I gathered in the main White House entrance and continued the discussion. Senator Jackson emphasized the need for a clear, intellectually honest position. The president agreed and emphasized again the fact that more was at stake here than just the test ban agreement. We have to think of the other path, the alternatives: the increasing (without limit) stockpiles of weapons, not only in the U.S. and the USSR but in other countries, and the possible consequences of this. He also mentioned, as an example, Israel. He emphasized again that we should make a really serious effort so that, even if we fail, in the eyes of the world we will be in the position of having done the best we could.

My own predilections were strongly on the side of arms control in general and a test ban in particular. I felt that the future of mankind required such steps to arrest the arms race. Yet I was not inclined to put down those in the

"I believe that my visit played a role in setting the stage for successful negotiations that led to an agreement on the Limited Test Ban Treaty during Averell Harriman's "Twelve Days in Moscow" some two months later."

AEC community who held opposite views, usually because of strong feelings about the requirements of national safety. There seemed little doubt that there were elements of risk in seeking a test ban agreement with the Soviet Union. The question that weighed heavily, however, was whether the risks of not reaching an agreement might not be greater. The opposing views of the administration leadership and much of the AEC community continued to wage war in my head and conscience for the entire period of the test ban negotiations under Kennedy, more than two and a half years. I tried to play the honest broker between them, calling to the attention of each what seemed to be valid points raised by the other. At all times, for example, I tried to counsel the administration against positions that might alienate JCAE, whose strong influence in Congress could block the approval of a test ban treaty. I was in effect playing a double game in a way that I thought served the national interest. Overall, my hope is that I struck approximately the right balance in attempting to nudge the government apparatus toward policies and practices that favored a test ban while still according with the technical and national security realities.

After many meetings of the Committee of Principals (including Secretary of State Rusk, Secretary of Defense McNamara, the chairman of the Joint Chiefs of Staff Lemnitzer, the special assistants to the president for Disarmament, McCloy, National Security Affairs, Bundy, Science and Technology, Wiesner, and I), an American position on a nuclear weapons test ban was hammered out. The president participated in a number of these meetings and invited the participation of key members of Congress.

On April 18, 1961, the United States and the United Kingdom introduced at the Geneva conference a complete draft treaty embodying agreements previously reached by the conference plus the Western position on all the contested issues. The treaty would have banned all nuclear tests in the atmosphere, in outer space, and in the oceans, and all tests underground except those producing signals of less than 4.75 seismic magnitude. (Tests below that threshold were thought to be too small to be detected and identified consistently by the proposed verification system.) It was proposed by the West, though not set forth in the draft treaty, that there would be a three-year moratorium on underground tests below the threshold. The moratorium would be renewable annually while further research was carried out to improve detection techniques. A voluntary moratorium on testing of all kinds had been in effect since fall 1958.

The treaty was to be policed by a worldwide detection system operated by a single administrator and an international staff, headquartered in Vienna. Policy direction was to be provided by a Control Commission of eleven members, four from the Soviet side, an equal number from the U.S.–U.K. side, and three neutrals. Commission decisions were to be by a simple majority. The commission was to appoint the administrator. The detection system was to utilize 180 manned detection stations on land—19 on Soviet soil—and on ships at sea. There were to be a maximum of 20 inspections per year on the territory of each of the three original parties. Inspection teams were not to include nationals of the country being inspected, but that country could escort the inspectors and determine the route they traveled.

One month later the Soviets formally rejected the draft treaty. S.K. Tsarapkin (USSR Ministry of Foreign Affairs) asserted that the number of inspections sought by the West was artificially high. The real reason Tsarapkin was not interested in any new U.S. effort to accommodate the Soviet point of view by reducing the number of on-site inspections became evident on Aug. 30, 1961. I spent that day at home, supposedly on my vacation. About 6 p.m. I received a phone call from Mac Bundy saying that a Tass transmission of items for provincial papers, monitored earlier in the day, had contained a statement, scheduled for release at 7 p.m., that the Soviet Union had decided to resume nuclear testing! The announcement did come at 7 p.m., and the Soviets conducted their first test, a 150 kiloton atmospheric test ending the three-year moratorium, on Sept. 1, 1961.

When the Soviets broke the moratorium on nuclear weapons testing, I had many meetings with Kennedy to consider our course of action. Kennedy decided that the United States would resume underground testing, and the timetable for these tests had to be worked out. Our first nuclear test took place in Nevada on Sept. 15, 1961. We also held discussions on whether the U.S. should resume atmospheric testing. I appeared on the NBC television program, "Meet The Press," on Oct. 29, 1961. In preparation for this, the president briefed me on an answer that I might give to the inevitable question concerning whether the United States had decided to resume atmospheric testing. The president told me to be very forthcoming but not to reveal anything of substance.

The opening of the program (in which I used the guidance from President Kennedy) was as follows:

> Spivak: Dr. Seaborg, I know it hasn't been announced, but can you tell us—has a final decision been made yet as to whether or not the U.S. will test in the atmosphere?
>
> Seaborg: The final decision has not been made yet, Mr. Spivak.

Seaborg appears on the NBC television program "Meet the Press." Top, left to right: Ned Brooks (Moderator), Seaborg. Bottom, left to right: Lawrence Spivak (Panelist), Marquis Childs (Syndicated Columnist), John Finney (New York Times), and Peter Hackes (NBC News). WRC television studios. Washington, D.C.

> Spivak: Can you tell us when it will be made?
>
> Seaborg: No, I can't tell you when it will be made—if it will be made.
>
> Spivak: Can you tell us what the final decision will be based on?
>
> Seaborg: Well, the final decision will be made by the president, of course, and I suppose that he would want to base it in large part on the result of the analysis of the Soviet tests. I would also suppose that he would never take this decision to test in the atmosphere on the basis of political or terroristic considerations, such has been at least part of the reason for the Russian testing, but would base his decision entirely on the technical need for the information in the interests of our national security.

I spent a very interesting half-hour responding to the persistent questions of Lawrence Spivak (the permanent member of the "Meet

OCTOBER 16, 1961 25c

Newsweek

For War – THE ATOM – For Peace

Seaborg of the AEC: A Scientist in Command

spheric tests, the dangers of fallout, and so on. The Soviets exploded their 57-megaton bomb the next day, Oct. 30.

The national concern over nuclear weapons testing led to my appearance in a cover story in *Newsweek* magazine and in *Time* magazine. The *Time* photographer also took a picture, which did not get used with the article, of my family.

President Kennedy was driven toward a decision to resume atmospheric testing. For this we needed a place to conduct the tests, and Christmas Island, a British possession about 1000 miles south of Hawaii, seemed to serve the purpose. To obtain permission to use this island, a summit meeting, which took place at Bermuda, was necessary.

The historic meeting of Kennedy with U.K. Prime Minister Harold Macmillan at Bermuda was preceded by a meeting at the presidential summer home in Palm Beach, FL, on Dec. 20, 1961. On this occasion, we ate dinner in the family dining room where I had the pleasure of sitting next to Jacqueline Kennedy, whom I found to be a very interesting conversationalist. The next day I flew with the president and his party to Bermuda for a two-day meeting in Bermuda with Macmillan and his advisers, including Sir William Penny (chairman of the United Kingdom's Atomic Energy Authority). My memory of Kennedy in these early meetings, an image strengthened and reinforced as time went on, is of a man remarkable for his immediate grasp of ideas, his ability to arrive at the gist of a discussion, and his eloquence in summarizing the main points at issue:

Thursday, Dec. 21, 1961
    . . . arriving in Bermuda at 1 p.m. (their time).
    We had lunch at the Bermudiana with Bermudans Sir Norman Brook and Sir Evelyn Shuckburgh, Secretary Dean

the Press'' panel), Marquis Childs (a widely syndicated newspaper columnist), John Finney (the aggressive *New York Times* reporter), and Peter Hackes (NBC News).

The rest of the interview involved questions about the 50-megaton bomb that Khrushchev had announced the Soviets planned to test, the possibility of Soviet superiority in nuclear weapons if the U.S. didn't have atmo-

A Chemist in the White House: From the Manhattan Project to the End of the Cold War

Rusk, David Ormsby-Gore (British test ban negotiator), U.K. Foreign Secretary Earl Home, Gen. Chester V. Clifton (aide to President Kennedy), Sir William Penney, and William Tyler (deputy secretary of state). We discussed the Berlin and Congo problems, the Indian invasion of Goa, the Dutch–Indonesia problem, and the status of the U.K. possessions throughout the world.

Between 5:00 p.m. and 6:30 p.m., I attended the meeting of President Kennedy and Prime Minister Macmillan at Government House and participated with Rusk, Home, Penney, and Harold Brown (director of research and engineering, Defense Department) in a discussion of the atmospheric testing question and our possible use of Christmas Island.

Penney opened with a technical summary. He believed the U.S. was still slightly ahead. He felt both sides might now be forced to attempt to set up an antimissile defense. This would be a fantastically difficult and costly effort. The prime minister was concerned about piling up bigger and bigger bombs if the arms race proceeded unchecked. He felt that the U.K. would probably drop out, but the two superpowers would continue piling up sophisticated weapons. Meanwhile, "everyone else" would have simple Hiroshima-type bombs within 25 years. This was an intolerable prospect. Mankind could not go on this way. We had to make another effort to reach an agreement.

Macmillan noted that both the United States and Great Britain had made great efforts in the test ban talks. David Ormsby-Gore had spent three years in Geneva, the dullest city in the world. The talks had almost succeeded and then somebody (he asked me to forgive the allusion) unveiled the big hole. (He was referring to the decoupling method of evading detection of underground tests under a treaty, as a result of which the need for controls on underground tests got out of hand. People were saying that underground tests were not very useful.) We might have had an agreement; the failure to get one was a great pity.

Now a decision had to be made. Should there be a new test series for which the British would allow the use of Christmas Island? Or should we make

a serious new disarmament effort? Could not he and the president and Khrushchev get together and make a great new effort to break the cycle of the arms race? It might fail, but if so we would have lost only a few months. He had been reading Russian novels in order to learn what he could about them and felt that they might come around.

The president commented that Soviet behavior over the last nine months suggested that they did not

Nov. 10, 1961, cover of *Time* magazine. Copyright *Time*, New York.

The Seaborg family at home. Clockwise: David, Lynne, Glenn, Dianne (in Glenn's arms), Helen, Pete, Steve, and Eric. 3825 Harrison Street, Washington, D.C. Nov. 2, 1961. Copyright *Time*, New York.

want an agreement. Was it not true, for example, that they had been preparing since February the test they initiated in September?

The prime minister asked what a 100-megaton bomb would do to people. Penney replied that it would burn up everyone in even the largest city. The prime minister next asked how many large bombs it would take to destroy England. Penney estimated that eight of the existing multimegaton weapons would be enough to make a terrible mess of England. The prime minister continued that he had asked Khrushchev what would happen if all the bombs in the world went off, and Khrushchev had said there would be nobody left but the Chinese and the Africans.

The prime minister observed that a large proportion of the U.S. strategic weapons were based in England, and then added: "Every time you lift the phone, Mr. President, I think you may say that you intend to go, and I wonder what answer I would give."

President Kennedy felt that before long, the nuclear arms race would come to a standoff where neither side could use these weapons, because it would be destroyed if it did. Rusk said this would work only if it were a standoff in which both sides believed. Brown agreed, saying the standoff would have

to be psychologically and technologically stable.

Macmillan thought the position of the Russians might be changing. Their economic structure was really not very different from that of Western Europe. In both cases railroads and mines were nationalized. They had a ruling class that enjoyed special privileges, including sending their children to schools like the British public schools. In other words, the forces of humanity were operating the same in all countries. Couldn't we allow these forces to bring us closer together instead of persisting in the arms race?

Foreign Minister Home observed that the new U.S. plan for general and complete disarmament was basically very similar to the Soviet plan except in the important area of inspection. This being the case, could we not make the impending meeting of the Eighteen-Nation Disarmament Committee a moment of major effort? Perhaps it could be kicked off by the 18 heads of state.

I remarked that the test ban negotiations might have to be linked with general disarmament, because the problem of secret Soviet preparation made it harder and harder for the West to accept a test ban treaty as a separate measure. The president adverted again to the discouraging experiences of recent months. We could not get taken twice. Even though he was a great antitester, he felt we ought to make preparations for a test series and then carry out the tests unless we got something substantial in some other field that helped our security in the world. The president added that there was only one serious issue: the balance of missile/antimissile capabilities.

Lord Home repeated that there must be a major effort to get on with disarmament and stop the arms race. The president replied that the timing was difficult. We could not start an atmospheric testing just when we were kindling new hopes for disarmament.

The president summarized the discussion by saying that there appeared to be three questions: (1) Should the U.S. prepare to test and then test? (2) Should there be a parallel effort on disarmament? (3) Would the U.K. help the U.S. testing program by making Christmas Island available?

Macmillan then said that he thought the discussion had been most helpful and that he felt better.

I had dinner at Government House with the governor-general of Bermuda Sir Julian and Mrs. Gascoigne, President Kennedy, Gascoigne's son and his wife, Bundy, Shuckburgh, Brook, Rusk, Charles (Chip) Bohlen (former U.S. ambassador to the Soviet Union), Ormsby-Gore, Home, Clifton, Penney, Tyler, and others. I sat one place removed from President Kennedy at the table, which gave me an opportunity to talk to him. He invited me to go along (and I accepted) on his trip to California to give the Charter Day address at Berkeley on March 23, 1962; he also accepted my invitation to visit Livermore Laboratory at that time. I also had the opportunity to talk to President Kennedy and Prime Minister Macmillan about nuclear testing; they both feel that the U.S. overemphasized the value to the Soviets of their cheating on underground testing and were unhappy that the U.S. made so much out of the "Big Hole" method of evasion by decoupling, because it is now apparent that it would have taken the USSR many, many, years to catch up through underground testing. They feel that if the test ban treaty had been consummated, without worrying so much about clandestine underground testing, we might be in a much better position

today so far as the arms race is concerned, and we would also be farther ahead of the USSR.

The trip to Bermuda took me away from the first annual reception that Helen and I gave at our home for members of the chairman's and commissioners' staff in recognition of the Christmas season. In my absence, Helen went ahead with the reception. Howard Brown (my executive assistant) was very helpful in that he introduced to Helen many AEC staff members she hadn't yet met. In the course of the evening I telephoned home from Bermuda. Howard Brown and my son, Pete, tried to set up an amplifier system so that I could talk to the assembled guests, but they were not successful.

Friday, Dec. 22, 1961

I attended a continuation of the meeting between President Kennedy and Prime Minister Macmillan at Government House. The same themes were evident as the day before: the British wanted assurance that there would be one more try for an agreement; the

Meeting between President John F. Kennedy and British Prime Minister Harold Macmillan held in Government House, Bermuda. Dec. 21–22, 1961. Left to right: Dean Rusk (U.S. Secretary of State), Kennedy, Macmillan, and Lord Sir Alec Douglas-Home; known as Lord Home (Foreign Minister). Courtesy of the John Fitzgerald Kennedy Library, Columbia Point, MA.

Americans wanted to get down to cases about Christmas Island.

The prime minister asked how long a time would be needed for preparations on Christmas Island. Brown answered that we would have to start early in January to be ready for testing by April 1. The prime minister remarked that he must have cabinet consent to any decision on these matters. (This came as a bit of surprise and disappointment; we were hoping to wrap things up in Bermuda.)

Macmillan repeated his question: Couldn't we make a new effort to reach an agreement? On one side was the Berlin question; it could be settled if people wanted to settle it. On the other side was this testing competition, which seemed to be a travesty of the purposes of human life.

Kennedy insisted that the United States had to decide to test, but he was willing to couple the decision with another statement of our disarmament proposals. Also, he was willing to hold back any announcement of our decision to test. We would state only that we were making preparations. Kennedy then asked Macmillan if he could agree on Christmas Island now. The prime minister answered that the two countries were partners and the United Kingdom would back up the United States. But could we not announce our plans so that they would seem less a threat than a hope? Kennedy said we could do this if we were careful not to use words that might trap us in the future.

Home asked if the president intended to make the testing decision dependent on Berlin. Kennedy answered that if a really satisfactory settlement could be worked out on Berlin, he felt that it would be easier for the United States to forego testing at this time. He cautioned that this position should not be published.

The president and the prime minister took note of the new situation created by the massive series of atmospheric tests conducted in recent months by the Soviet government after long, secret preparations. They agreed that it is now necessary (as a matter of prudent planning for the future) that, pending the final decision, preparations should be made for atmospheric testing to maintain the effectiveness of the deterrent.

Meanwhile, they continue to believe that no task is more urgent than the search for paths toward effective disarmament, and they pledge themselves to intensive and continued efforts in this direction.

Serious progress toward disarmament is the only way of breaking out of the dangerous contest so sharply renewed by the Soviet Union. The president and the prime minister believe that the plans for disarmament put forward by the United States in the current session of the United Nations General Assembly offer a basis for such progress, along with the treaty for ending nuclear tests that the two nations have so carefully prepared and so earnestly urged upon the Soviet government.

Prime Minister Macmillan later gave permission to use Great Britain's Christmas Island as a place to conduct the atmospheric nuclear weapons testing program of the United States.

The president finally decided to resume atmospheric testing and announced his decision to the nation on Friday, March 2, 1962, following a meeting with congressional leaders:

From 4:00 p.m. to 4:45 p.m. I attended a meeting the president called for the purpose of briefing the bipartisan leaders of Congress on his decision to resume atmospheric testing. Present were: the president, Senators Clinton P. Anderson, Mike Mansfield, Richard B. Russell, John O. Pastore, Everett M. Dirksen, J. William Fulbright, Leverett Saltonstall, Bourke K. Hickenlooper; Congressmen Chet Holifield, Melvin Price, John W. McCormack, James E. Van Zandt, Carl Vinson; Secretary McNamara, John McCone (director, CIA), Harold Brown, Mac Bundy, and I. The president described the general outline of the announcement that he intends to make on nationwide television tonight. Following this, McCone gave a description, using charts, of the results of the Russian tests and their significance. Brown then made a brief statement regarding the four classes of tests in the U.S. program: (1) effects, (2) advanced concepts, (3) verification, and

---

> "The president finally decided to resume atmospheric testing and announced his decision to the nation on Friday, March 2, 1962, following a meeting with congressional leaders."

(4) systems. I mentioned in particular the 2000-kilometer test to develop a capability for testing in outer space and also the role that Johnston Island and Christmas Island will play.

The president said that Prime Minister Macmillan asked him to delay his announcement until today; the original plan had been to make the television announcement last night. The president said that the only dissenting message from a head of state that should be taken seriously was from Japan. Countries such as England, France, and Switzerland have expressed approval. Senator Anderson said he feels the Soviet Union would accept our offer for a treaty but in such a manner that prolonged negotiations would result; but the president assured him he meant it should be a signed treaty before the deadline mentioned. The president said that we won't announce the number of tests but merely give a general description of the types and total fallout. It was the unanimous reaction of the members of Congress present, voiced individually by such people as Vinson, McCormack, Mansfield, Dirksen, and Hickenlooper, that the president has made the proper decision in the proper way.

The president's address to the nation that evening was long and thorough in its coverage of the issues. His intent was to present the entire rationale for his fateful decision to announce resumption of atmospheric testing. It was as though he were addressing the judgment of history.

The U.S. atmospheric test series, designated Operation DOMINIC, began April 25, 1962, with an air drop in the intermediate-yield range (20 kilotons to 1 megaton) off Christmas Island. It was the first U.S. atmospheric test since 1958.

In all, the series comprised 40 tests. It included the firing of 29 nuclear devices dropped from aircraft in the vicinity of Christmas and Johnston islands and 5 detonations of nuclear devices carried to high altitudes by missiles launched from Johnston Island, a dot of land owned by the U.S. about 500 miles southwest of Hawaii. Two nuclear weapons system tests were also involved, one in the Christmas Island area and one in the eastern Pacific. These 36 Pacific tests were conducted by a joint AEC–Defense Department Task force that, at the peak of its activity, numbered more than 19,000 men. In addition to the Pacific tests, four small tests were conducted near the surface at the Nevada Test Site.

Between June 29 and July 1, I visited the Pacific test sites along with Bundy, Arnold Fritsch (my technical assistant at AEC), and Dwight Ink (AEC assistant general manager). On June 30, on Christmas Island:

I arose at 5:15 a.m. and went with others to Observation Point, from which, at 6:20 a.m., we saw an explosion 30 miles south at 5000 feet, the BLUESTONE event, about 1.3 megatons. It was dropped from an airplane. It was necessary to use dark glasses for the first eight seconds. Upon removing them I found the area brighter than full daylight, an awesome sight.

I spent the morning touring the impressive diagnostic facilities that had been set up in the trailers on short notice.

At 6:40 p.m. we flew by military plane to Johnston Island, 1100 miles distant, arriving at 10:40 p.m. We were joined there by Commissioner Leland J. Haworth.

The next morning Gen. Salet (who was in charge of Johnston Island), Alfred D. Starbird (Test Task Force commander), and William Ogle (Los Alamos Laboratory, scientific director of the test series) showed us the facilities on the island, which measures in total approximately one mile by one-quarter mile. We saw the launching pad and the complicated diagnostic facilities operated by the air force and several laboratories.

After lunch on Johnston, we flew on to Hawaii and then home.

In accordance with the restrictions imposed by the president, the total yield of the series was held to

> "It is necessary to recall that almost from its inception, but especially since the Korean War, AEC had maintained a readiness plan for continuity of essential operations in the event of hostilities."

approximately 20 megatons. The Soviet series in fall 1961 had yielded almost 10 times that much.

In fall 1962 came the Cuban Missile Crisis, which played a crucial role in the test ban story. Periodic intelligence reports since late August of 1962 revealed the off-loading of military equipment from Soviet ships and an increase in military construction activity at several locations in Cuba. Although AEC was not a "collector" of intelligence, it did serve as an evaluator and interpreter of nuclear-related intelligence data collected by the CIA, the Department of Defense, and other elements of the intelligence community. I served as a member of the U.S. Intelligence Board, the highest intelligence estimating body in the government. Commencing in October, AEC's director of intelligence, Charles Reichardt, often accompanied by Assistant General Manager for Administration Harry Traynor and General Manager Alvin Luedecke, came to my office in the early morning nearly every day to give me the latest reports and estimates on developments in the Cuban situation. Many of these reports bore classifications above top secret.

The crisis broke on Monday, Oct. 15, when analysis of photographs from reconnaissance overflights by U-2 planes disclosed evidence of a medium-range missile site, though not yet the missiles themselves, in western Cuba. Now a nuclear confrontation with the Soviet Union over Cuba appeared to be a distinct probability.

The president immediately established a top-level group formally named the Executive Committee of the National Security Council (EXCOM) to consider policy alternatives and make recommendations to him. By Wednesday, Oct. 17, launchers and missiles could be seen in U-2 photographs, and it was clear that the missiles could be fired within two weeks.

EXCOM discussions began to focus on two options: (1) a swift air strike to take out the missiles, or (2) a naval blockade while diplomatic pressure was exercised to get the missiles removed.

It is necessary to recall that, almost from its inception, but especially since the Korean War, AEC had maintained a readiness plan for continuity of essential operations in the event of hostilities. Indeed, when the new AEC headquarters was constructed at Germantown, MD, in 1957 (as part of President Eisenhower's plan for the dispersal of critical government functions), a reinforced structure replete with sophisticated emergency communications systems was built into the underground structure of the new complex. It was known as the Emergency Relocation Center (ERC) and was built with compartmentalized sleeping facilities to house 120 people with sufficient water and food to meet their needs for several weeks.

Periodically, mock exercises were held in the ERC during which imaginative efforts were made to write a realistic scenario. For most key officials who participated, these exercises were a bit of a nuisance, interrupting their busy schedule. In mid-October 1962, the exercises commenced to assume a new reality.

ERC was meant to house, in the event of a war emergency, the initial cadre, consisting of the chairman, the commissioners, and those members of the AEC staff essential to operation of the agency in such an emergency situation. It was also contemplated that the members of the initial cadre might be accompanied by their families, although the feasibility of this was in doubt and subject to much debate.

By Friday, Oct. 19, the blockade concept appeared to have won out over the air strike in the deliberations of EXCOM, but with the proviso that an air strike would follow if diplomacy failed. The president's address to the nation on radio and television, which revealed the extent of the crisis to the world for the first time, took place on Monday evening, Oct. 22. This address brought home to the nation the gravity of the situation. AEC employees, who had been enjoined by secrecy, were now for the first time able to discuss and develop with their spouses concrete plans for the safety of their families. This raised serious questions among members of the initial cadres as to whether, if ordered to occupy ERC in the face of impending outbreak of hostilities, they would actually bring their families to take up residence in the underground ERC in Germantown. Helen and I had serious discussions as to our proper course of action should we be faced with such a fateful decision. Fortunately, we never had to make this decision.

The day following the president's address, I informed the commissioners that AEC operations had been placed under Phase I Alert (i.e., instructions to check that communications were in order, 24-hour duty for communications personnel, additional security guards, etc.). It was a tense day that featured a meeting at which the Organization of American States (OAS) endorsed President Kennedy's action, a spirited discussion in the UN Security Council, and reactions of various types from around the world. What the USSR reaction would be was not yet clear.

Fortunately, after an historic exchange of messages between Kennedy and Khrushchev, a message came from the Soviet government on Sunday, Oct. 28, agreeing to remove the missiles under UN inspection.

Although it was not publicly announced at the time, it is now known that, in return, Kennedy conveyed private assurances to Khrushchev: (1) the United States would not attack Cuba, and (2) we would remove Jupiter missiles we had deployed in Turkey.

This brush with disaster brought President Kennedy and Chairman Khrushchev closer together, a prelude to the successful attainment of the Limited Test Ban Treaty less than one year later.

As indicated earlier, I have the feeling that my visit to the Soviet Union during May 1963 and my meeting with Brezhnev may have helped a little to pave the way for the move toward a test ban. A new initiative was already underway. President Kennedy and Prime Minister Macmillan had sent a letter to Khrushchev on April 24, 1963, urging renewal of negotiations on a test ban and offering to send to Moscow a very senior representative empowered to talk directly with him. This, then, was followed by President Kennedy's extraordinary commencement address at American University on June 10, 1963, a speech that has been rated as one of the great state papers of American history. At the end of a speech that described in an eloquent fashion the common interests of the United States and the Soviet Union in peace and the avoidance of war, he concluded with two announcements:

> First: Chairman Khrushchev, Prime Minister Macmillan, and I have agreed that high-level discussions will shortly begin in Moscow looking toward early agreement on a comprehensive test ban treaty. Our hopes must be tempered with the caution of history—but with our hopes go the hopes of all mankind.

> Second: To make clear our good faith and solemn convictions on this matter, I now declare that the United States does not propose to conduct nuclear tests in the atmosphere so long as other states do not do so.

The effect of Kennedy's speech on the primary target audience, the leadership of the Soviet Union, appeared to be profound. Khrushchev later confided to Harriman his view that the speech was the best by any American president since Franklin Roosevelt and that it had taken courage on Kennedy's part to make it. Notwithstanding the fact that it contained some criticisms of Soviet policies and Soviet analyses of history, the address was published in its entirety in the Soviet press and rebroadcast in its entirety in the Voice of America translation.

Khrushchev made a speech on July 2 that all but sealed the agenda for the Moscow talks. He now rejected the concept of on-site inspections, which the United States and the United Kingdom regarded as necessary to insure compliance with a comprehensive test ban treaty. At one time he had seemed willing to accept two or three on-site inspections per year, and the United States seemed willing to go down from 20 to perhaps as few as 8 per year, but it seemed impossible to close this gap. Now he again said the West demanded these on-site inspections for espionage purposes, and he said there would be no bargaining on this issue. He then said that the Soviet Union was ready to conclude an agreement on the cessation of nuclear tests in the atmosphere, in outer space, and underwater.

Kennedy chose W. Averell Harriman, the experienced American diplomat who had the respect of the Soviet leadership, to lead the U.S.–U.K. negotiating team to Moscow. Lord (Quintin Hogg) Hailsham was Harriman's British counterpart, but Harriman was in charge of the U.S.–U.K. delegation. On the specific issue of a test ban, Harriman was told that the achievement of a comprehensive test ban remained the U.S. objective. If that was unobtainable, he was supposed to seek a limited treaty in three environments (atmosphere, water, space) along the lines of an earlier Western draft treaty.

Notwithstanding Khrushchev's July 2 remarks in Berlin about his willingness to accept a three-environment test ban, members of the administration, including Harriman himself, were far from certain that such a treaty on terms acceptable to the United States was in the bag.

As the negotiations opened on July 15, there were several favorable omens. One was the designation of Foreign Minister Andrei A. Gromyko as the chief Soviet negotiator, indicating a welcome seriousness of purpose. Prior Western speculation had been that the Soviet team might have lesser leadership, possibly Vasily V. Kuznetsov or Valerian Zorin, each a deputy foreign minister.

Another good omen was the fact that Khrushchev himself remained throughout the first day's discussion, and his good spirits established an unexpectedly relaxed mood for the start of the negotiations. As Harriman later described to me: Khrushchev was very jovial in our first meeting. He said, "Why don't we have a test ban? Why don't we sign it now and let the experts work out the details?" So I took a blank pad and I said, "Here, Mr. Khrushchev, you sign first and I'll sign underneath." That was the jovial way in which we were talking.

At this first meeting Khrushchev tabled two draft treaties, one for a limited test ban and one for a nonaggression pact. The Soviet test ban draft was simplicity itself. It had only two operative articles. The first said that each party undertook to discontinue test explosions in the prohibited environments: atmosphere, space, and underwater. The second article stated that the agreement would enter into force immediately on signature by the USSR, the United Kingdom, the United States, and France. In response, Harriman gave Khrushchev a copy of the limited test ban treaty the West had introduced in Geneva.

Harriman made an unsuccessful attempt to negotiate a comprehensive test ban treaty, then went on to negotiate the details of the Limited Test Ban Treaty. The negotiations were successfully completed in 12 days of masterful work by Harriman. To achieve agreement with the Soviets, Harriman had to give up the U.S. peaceful uses of nuclear explosives (the Plowshare) provision in exchange for Soviet acceptance of a withdrawal clause.

Although I regard the achievement of a Limited Test Ban Treaty as a great achievement, I also regard the failure to achieve a comprehensive test ban as a world tragedy of the first magnitude. Evidence of the mutual mistrust and suspicion responsible for this unhappy outcome, revolving principally around the issue of on-site inspection, has been repeatedly demonstrated. To put the matter in its baldest form, the Soviets were persuaded that the United States wanted to inspect in order to spy; many on our side were convinced that, without adequate inspection, the Soviets would cheat.

During the 12 days of the negotiations in Moscow, Harriman had a number of opportunities to observe Khrushchev in relatively informal circumstances. As Harriman later told me, toward the end of the final day of a U.S.–Soviet track meet, Khrushchev turned up unexpectedly with a large party that included Janos Kadar (the Hungarian leader), Brezhnev, and their wives. Harriman was convinced that Khrushchev came because he knew Harriman was there. Khrushchev asked Harriman to join him in his box high up in the stadium, along with U.S. Ambassador Foy Kohler and Mrs. Kohler. He was in an ebullient mood and grew especially excited when the famous Soviet high jumper, Valery Brumel, broke his own world record. Harriman particularly remembers Khrushchev's reaction to the closing ceremonies when members of the two teams walked around the field together, arm in arm. Harriman recalls: "He and I stood up, and there was a tremendous ovation from the crowd, which seemed to be watching us as much as the athletes. There were tears in Khrushchev's eyes."

Secretary Rusk invited me to be a member of the U.S. delegation attending the signing ceremony in Moscow. We took off aboard Air Force One at 11 p.m. on Aug. 2. Besides Rusk and myself, others on the flight were Arms Control and Disarmament Agency (ACDA) Director William Foster, Ambassador-at-Large Llewellyn Thompson, UN Ambassador Adlai Stevenson, and six senators. Ambassadors Arthur Dean and Foy Kohler joined the delegation in Moscow. The six senators had been chosen with a view to having the maximum impact on their colleagues. Those selected came from both parties and were leading figures on the key senatorial committees.

From the Democratic side there were John O. Pastore, JCAE chairman, and the three ranking Democrats on the Foreign Relations Committee, which would hold hearings on the treaty: J. William Fulbright, Hubert Humphrey, and John Sparkman. Presaging the hostility of the Armed Services Committee toward the treaty, none of its Democratic members was willing to attend. The Republicans present were Leverett Saltonstall, the ranking Republican on the Armed Services Committee, and George Aiken, the second-ranking Republican on the Foreign Relations Committee.

This historic signing (by Dean Rusk, Soviet Foreign Minister Gromyko, and British Foreign Secretary Lord Home) took place at 4:30 p.m., Aug. 5, 1963, in the Kremlin in Catherine's Hall. A meeting with Premier Khrushchev in his office that morning preceded the signing, and a reception in the

"Although I regard the achievement of a Limited Test Ban Treaty as a great achievement, I also regard the failure to achieve a comprehensive test ban as a world tragedy of the first magnitude."

Georgian Hall followed the signing ceremony.

Thursday, Aug. 2, 1963—Washington, DC (en route to Moscow)

At 11 p.m. I left for Moscow, via Copenhagen, on Air Force One from Andrews Air Force Base with Secretary Rusk; Foster; Thompson; Stevenson; and Senators Aiken, Saltonstall, Pastore, Humphrey, Fulbright, and Sparkman, the delegation in connection with signing the Limited Test Ban Treaty.

Saturday, Aug. 3, 1963—Copenhagen and Moscow

Our special plane arrived in Copenhagen about noon. I toured Copenhagen by car with Mrs. William McC. Blair, Jr. (wife of the U.S. ambassador to Denmark), Adlai Stevenson (U.S. representative to the UN) and Roy O. Carlson (economic officer, U.S. embassy), where we saw Tivoli, the waterfront, and Governmental Square (King's Palace, etc.). Our plane flew from Copenhagen to Moscow (leaving Copenhagen at 1:50 p.m. and arriving at Vnokovo 2 Airport in Moscow at about 6 p.m.). I was driven to the Sovietskaya Hotel with William and Beulah Foster (he is director, U.S. ACDA). I attended a buffet supper given at Spaso House for the Test Ban Treaty delegation. I spent the night at the Sovietskaya Hotel.

Sunday, Aug. 4, 1963—Moscow

I attended meetings in the morning and afternoon at the U.K. embassy (Chancellery) with Secretary of State Rusk, Ambassador Kohler, Stevenson, William Bundy (deputy assistant secretary of defense), Foster, Alexander Akalovsky (ACDA), Thompson (U.S. ambassador at large), Richard Davis (deputy assistant secretary of state), Charles Stelle (U.S. representative to disarmament committee), Lord Home (U.K. foreign secretary of state), Sir Harold Caccia (U.K. permanent undersecretary), P. Thomas (minister of state, U.K.), Duncan Wilson (U.K. assistant undersecretary), Sir Edward Heath (Lord Privy Seal), Trevelyan (U.K. ambassador to the Soviet Union), and others. We discussed the Limited Test Ban Treaty, its effect on West Germany, the French plan for disarmament (delivery vehi-

cles), the Soviet suggestion of observer posts, nonaggression and antisurprise attack agreements, the concept of a NATO steering committee to discuss a nonaggression pact, the possibility that nuclear weapons testing by France will bring an end to the treaty, our suggestions to the Soviet Union of no weapons in outer space, B47-badger exchange, and so on.

I had lunch with the same group at the U.K. embassy. I toured the Kremlin area with Foster, Stelle, and Akalovsky. We then visited Moscow State University, driving by the dachas (residence cottages) of Soviet officials (including Khrushchev's in the Lenin Hills), Gorky (amusement) Park, and along the banks of the Moscow River.

I had a buffet dinner at Spaso House with members of the U.S. and U.K. delegations. I walked in Red Square with Senator and Mrs. Humphrey, Senator and Mrs. Fulbright, Senator Pastore, and William and Beulah Foster. I saw the changing of the guard at Lenin's tomb at midnight (this takes place every hour).

Monday, Aug. 5, 1963—Moscow

Our delegation had an appointment with Andrei A. Gromyko (Soviet foreign minister) in the Foreign Ministry Building from 9 a.m. to 9:30 a.m. Those present included the U.S. delegation plus Gromyko, V.V. Kuznetsov (first deputy minister, Foreign Affairs, USSR), M.N. Smirnovsky (American section of the Soviet Foreign Office), Anatoli F. Dobrynin (Soviet ambassador to the U.S.), and S.K. Tsarapkin (USSR International Organizations Division Ministry of Foreign Affairs). Press and photographers were present at the beginning. Dobrynin and Rusk voiced hopes for the Test Ban Treaty as the first step toward avoidance of nuclear war. Rusk called on Fulbright, who said he was strongly in favor of the treaty.

From 11 a.m. to 12 noon the U.S. delegation visited Nikita Khrushchev in his Kremlin office, which is long and narrow. We sat at a table with a green felt top, like a pool table. There were windows on the west side, pictures of Lenin and Marx, an electric clock on Khrushchev's desk, bookcases, and two telephones at the conference table. (I saw only one telephone at Khrushchev's desk.) Gromyko, Kuznetsov,

Smirnovsky, and Dobrynin were present. After greetings by Khrushchev and Rusk, Khrushchev said the Test Ban Treaty was only a first step and that the main problem is the German problem. He said liquidation of the government of the German Democratic Peoples' Republic would not be a victory for the United States, nor would a Communist win (i.e., liquidation of the Federal Republic) be a victory for the Soviets, and that a common solution was needed. Rusk recognized that the German solution is fundamental, and he will discuss it with members of the Soviet government. Rusk said we understand the historical reasons that it is important to the Soviet Union—we, too, went through two World Wars—but the German people need an opportunity for peace so as not to give them reason to start trouble again. There has been a relaxation of tensions in Eastern Europe in the past year.

Khrushchev observed that Rusk doesn't use the term Socialist country but says "the East." Rusk said that some people in the United States call us Socialist. Khrushchev said, "such a man to say that." Rusk said that the Yugoslav government is less involved in its country's economy than the U.S. government is in the American economy. Khrushchev said, "Capitalism gave birth to communism; let's compete in culture instead of rockets." Rusk said that Glenn Seaborg, Stewart Udall, and Orville Freeman have visited the Soviet Union, and advancement of relations continues; let's cooperate in the peaceful uses of atomic energy, education, and so on.

Rusk then called on Senator Fulbright, as chairman of the Senate Foreign Relations Committee. Fulbright recalled Khrushchev's pleasant meeting with his committee four years ago. He referred to U.S. internal trouble of 100 years ago, which has been overcome; and now the South gets along with "the damn Yankees." Similarly, the United States and the Soviets can get along. The United States is capitalistic, but actually, it has a mixed economy. Someone has said the Soviet Union is promoting capitalism. Also, the U.S. Democratic party has been accused of promoting socialism. The differences are less than we think. He recalled that someone said Khrushchev would be a good member of Congress and that

Khrushchev replied, "You don't get to be chairman of the USSR Council of Ministers by being stupid." Khrushchev agreed on our common goals.

Khrushchev said, "You (U.S.) go forward on private property; we (USSR) on common property. We are for everyone; you are for every man for himself." Udall paid eulogy to the USSR achievement in power plants: "The Soviet Union will solve problems in agriculture drastically in seven years, and completely by the 1980s." Khrushchev said they are putting billions of rubles into the chemical industry, agriculture, and so on. He showed a sample of some plastics and a plastic cup. Khrushchev invited Rusk to come down to the Black Sea for a dip because his vacation begins tomorrow, whereupon Gromyko said Rusk shouldn't swim toward Turkey. On the way out of his office I talked to Khrushchev, and he referred to me as "my old friend." Since we had never met, he must have had reference to my recent visit to the Soviet Union, with which he was, of course, familiar.

Khrushchev impressed me as a very able person. He seemed to be in the best of physical condition, full of bounce, and in unfailing good humor.

The U.S. delegation then toured the Kremlin Arms Chamber Museum and rooms in the large Kremlin Palace. We had lunch in the oldest room of the Kremlin (built 500 years ago), at a large U-shaped table with about 100 guests. Khrushchev, Rusk, Home, and U. Thant spoke—all along lines of a test ban as a symbolic step. We retired to an anteroom for coffee and brandy. At lunch I sat next to Leonid F. Ilychev (secretary for Ideology of the Central Committee, an historian and speech writer) and Ya. V. Peive (chairman, Council of Nationalities of the Supreme Soviet, a biochemist from Riga) and near Petrosyants and K.N. Rudnev (deputy chairman, Council of Ministers, and chairman, State Committee for Coordination of Scientific Research Work, a mechanical engineer).

At 4:30 p.m. we attended the historic signing of the Limited Nuclear Test Ban Treaty, in Catherine's Hall, by Rusk, Gromyko and Home simultaneously, followed by speeches by Gromyko, Rusk, Home, and U. Thant. I stood just behind Khrushchev, and he and I tipped our champagne glasses

---

"Krushchev impressed me as a very able person. He seemed to be in the best of physical condition, full of bounce, and in unfailing good humor."

Signing of the Limited Test Ban Treaty in the Kremlin's Catherine's Hall. Seated, left to right: Dean Rusk (U.S. Secretary of State), Andrei Gromyko (Soviet Foreign Minister), Lord Home (British Foreign Minister). Standing, left to right (partial list): Senator George Aiken (partly out of picture), Senator William Fulbright (leaning forward), Alexander Akalovsky, Senator Hubert H. Humphrey, Senator Leverett Saltonstall (behind Humphrey), Adlai E. Stevenson (U.S. Ambassador to the United Nations), UN Secretary-General U Thant, Soviet Premier Khruschev, (unidentified individual) Soviet Deputy Foreign Minister Valerian Zorin, Anatoly Dobrynin (behind and slightly left of Zorin), William Bundy (behind and slightly left of Dobrynin), British Lord Privy Seal Edward Heath, (unidentified individual), and Seaborg. Moscow. August 5, 1963. Courtesy of The Bettman Archive.

together for toasts at least five times. About 50–60 press representatives and photographers were present.

At about 5:15 p.m. we attended a huge reception in Georgian Hall (magnificent!) where Khrushchev pulled a prepared speech out of his pocket and delivered it. I took a picture of him with my Minox camera (which was perhaps a risky endeavor). I had a chance to talk to Brezhnev, Petrosyants, Gromyko, Kuznetsov, Dobrynin, Zorin, Tsarapkin, and Klementi Voroshilov (grand old military leader). I noticed that Khrushchev mixed freely among all the people present, maintaining his ebullient good humor.

That evening I went to the circus and walked through Gorky Park with Senator Pastore.

A glorious day.

Tuesday, Aug. 6, 1963—Moscow

I paid a short visit to the embassy. I toured the Kremlin, saw Moscow sights, including a collective market and the world's largest swimming pool, with Bill Bundy (assistant secretary of defense, international security affairs) and Mr. and Mrs. Ken Kerst (brother of Don Kerst, professor of physics, University of Wisconsin), who drove us around.

I attended a luncheon given by the Ministry of Foreign Affairs at the Government Reception House, Lenin Hills (near Khrushchev's dacha), with all the members of the U.S. delegation and their wives. Gromyko and Rusk spoke briefly. I sat next to Petrosyants and Tanga Sorotkina (interpreter) and near Nikolai Tikhenov (Leningrad poet, president of the Lenin Prize Committee for Literature, and president of the Soviet Peace Committee.)

After lunch I visited Moscow State University with Bundy and Senator Pastore. I later attended a huge reception (300 guests) at the American embassy. Here I talked to Rudnev and seconded Rusk's invitation that he visit the United States. He implied he might come with Petrosyants. I talked to Petrosyants here as well as at lunch. He asked me to suggest an itinerary for his U.S. visit to places not seen by Soviets and said there will be 12 persons in his party. He said they are rushing construction of the Melekess Boiling Water Reactor in view of my reference to it in my press conference (implying that I didn't think they could meet the schedule). He referred to my National Press Club speech, of which he apparently had read a Russian translation, and which he said he liked. I had dinner at Uzbekistan (Asiatic Republic type) Restaurant (at Neglinuaya) with Bundy and Mr. and Mrs. Kerst.

A Chemist in the White House: From the Manhattan Project to the End of the Cold War

Kennedy threw himself into the ratification process with every resource available to him. He did so out of a sense of conviction that he probably felt for no other measure sponsored by his administration. Indeed, he confided to his associates that he "would gladly forfeit his re-election, if necessary, for the sake of the Test Ban Treaty."

The treaty was referred for study to the committee on foreign relations, which began hearings on Aug. 12, four days after the Senate received the president's message.

The first three witnesses before the Foreign Relations Committee—Secretary Rusk, Secretary

McNamara, and I—were each questioned for an entire day.

Without doubt, the most important aspect of my testimony on Aug. 14 had to do with the effect of the treaty on AEC's Plowshare program for peaceful nuclear explosions. Several senators, perhaps most notably Clinton Anderson, an ex-chairman and a highly influential JCAE member, had let it be known that their support of the treaty had been weakened because its ban on atmospheric testing seemed to exclude the possibility of Plowshare explosions. My optimistic testimony had a strong influence in moving Senator Anderson, and perhaps others, from a doubtful to

Soviet Premier Nikita Khrushchev delivers speech at the reception in the Kremlin's Georgian Hall following the signing of Limited Test Ban Treaty. Photo was stealthily taken by Seaborg with his tiny Minox camera. Adlai E. Stevenson (U.S. Ambassador to the United Nations) is visible on the right. Aug. 5, 1963.

a favorable position on the treaty's ratification.

Gen. Maxwell Taylor (chairman, Joint Chiefs of Staff) then recommended four safeguards that the chiefs thought necessary: (1) the conduct of comprehensive, aggressive, and continuing underground nuclear test programs; (2) the maintenance of modern nuclear laboratory facilities; (3) the maintenance of the facilities and resources necessary to institute promptly nuclear tests in the atmosphere, should they be deemed essential to our national security; and (4) the improvement of our capability, within feasible and practical limits, to monitor the terms of the treaty.

On Sept. 24, 1963, the momentous vote on the treaty was taken. Every able-bodied senator was present. The treaty was approved by a vote of 80 to 19. This was 14 votes more than the required two-thirds majority, a margin that satisfied the president's desire for a strong endorsement. Senator Clair Engle, too ill to be present, announced that he also would have voted aye.

The 80 senators who voted for the ratification included 55 Democrats and 25 Republicans. Of the 19 opposed, 11 were Democrats and 8 were Republicans. All the Democrats who opposed the treaty were Southerners except Senator Frank J. Lausche of Ohio. All the opposed Republicans were from west of the Mississippi except Margaret Chase Smith of Maine.

I have described in my book, *Kennedy, Khrushchev, and the Test Ban* (written with my colleague Benjamin S. Loeb), Kennedy's central role in achieving the Limited Test Ban Treaty. (Visits to the John F. Kennedy Library in Boston were helpful.) On April 11, 1988, accompanied by Helen, I participated, along with numerous other members of the Kennedy administration, in the commemoration (which was videotaped) of the 25th anniversary of the Limited Test Ban

Treaty at the John F. Kennedy Library:

Helen and I rode in the special van for the symposium participants (who are staying at the Boston Harbor Hotel) to the Kennedy Library, where we had a continental breakfast. Helen and I sat at a table with Solly Zuckerman, Adam Yarmolinsky, Norris Bradbury, and Herb York.

After breakfast we rode with John Stewart (director of education, John F. Kennedy Library) to Newton North High School (36 Lowell Ave., Newton-ville), where I was scheduled to talk to two special classes. The classes ran for 45 minutes with a 5-minute interval between classes; and both met in the same room, one after the other. The first class consisted mainly of students interested in history. The second class included, besides history students, some students interested in science. (Some students stayed for both of my lectures.) My lecture, which was similar for both groups, was a discourse on how the Limited Test Ban Treaty was achieved, and it included an expression of disappointment that a comprehensive test ban has not been achieved. Each lecture was illustrated by 19 slides. The first lecture included a 10-minute question-and-answer period, and the second, a 15-minute question-and-answer period. In both instances the questions were very good and included such topics as the Strategic Defense Initiative, the future of nuclear electric power, the role of the Intermediate-range Nuclear Forces Treaty, my participation in the "Franck Report," and so forth.

Helen and I then rode with John Stewart back to the Kennedy Library, where I joined a press conference already in progress in the theatre lobby. Senator Ted Kennedy introduced the symposium participants and made a short statement that was followed by a statement from Soviet Ambassador Anatoliy Dobrynin concerning the importance of the Limited Test Ban Treaty. We then went up to the seventh floor for a buffet lunch. Helen and I sat at a table with Roswell Gilpatric, Carl Kaysen, Walt Rostow, and Solly Zuckerman.

Following lunch we went to the theater to participate in the conference "The First Step: The Negotiation, Signing, and Ratification of the Limited Test

"On Sept. 24, 1963, the momentous vote on the treaty was taken. Every able-bodied senator was present. The treaty was approved by a vote of 80-19."

Ban Treaty." The conference was opened with some welcoming remarks by Charles Daly (director of the Kennedy Library) and Senator Kennedy. Paul Doty (director emeritus of the Center for Science and International Affairs, Harvard University) served as moderator. After some introductory remarks, he introduced me as the lead-off speaker (for a 10-minute talk), followed by McGeorge Bundy (professor of history at New York University and former special assistant for national security affairs for President Kennedy) and Theodore Sorensen (former senior counsel to President Kennedy from 1961 to 1963 and currently senior partner with the New York law firm of Paul, Weiss, Rifkind, Wharton, and Garrison), each of whom also talked for 10 minutes. The entire proceeding was videotaped and audiotaped.

I began with a description of the call from John F. Kennedy inviting me to be AEC chairman and the vote by my family that evening. I described my first meeting with President Kennedy and the meetings that followed immediately of the Committee of Principals (first alone and then with the president and key members of Congress). I explained how this led to a draft test ban treaty, described its provisions for verification (including the requirement of 20 on-site inspections), and said it was tabled at Geneva on April 18, 1961. I went on to describe how the number of inspections acceptable to Khrushchev was a much lower number (namely, three), how we came down to eight inspections but never close enough so that there was agreement. (The problem was that the Soviets thought we wanted these for espionage purposes.) I said that it was unfortunate and emphasized that we would be much better off today had a comprehensive test ban treaty been negotiated then. I ended by describing the successful negotiation of the Limited Test Ban Treaty by Averell Harriman in Moscow and emphasizing the general disappointment that a comprehensive test ban treaty was not attained.

Following our 10-minute talks were 5-minute talks, not necessarily in this order, by the following: Norris Bradbury (director of Los Alamos Scientific Laboratory from 1945 to 1970), Philip Farley (special assistant for Atomic Energy and Outer Space to the secretary of state from 1960 to 1962), Roswell Gilpatric (deputy secretary of

defense from 1961 to 1964), James Goodby (professor of the School of Foreign Service at Georgetown University; officer-in-charge of the nuclear test ban negotiations for the U.S. Arms Control and Disarmament Agency from 1960 to 1963), Betty Lall (senior fellow and director of arms control verification studies of the Council on Economic Priorities), Franklin Long (professor emeritus of science and society in Cornell University's program on science, technology and society; assistant director of the U.S. Arms Control and Disarmament Agency from 1962 to 1963), Benjamin Read (executive secretary of the State Department from 1963 to 1969), Walt W. Rostow (Rex G. Baker, Jr. professor of political economy at the University of Texas, Austin; President Kennedy's deputy special assistant for national security affairs), Jack Ruina (director of the defense and arms control studies program at the Massachusetts Institute of Technology; director of the Advanced Research Projects Agency from 1961 to 1964), Adam Yarmolinsky (provost and vice chancellor for academic affairs at the University of Maryland; special assistant to the secretary of defense from 1961 to 1964), Herbert York (director of the Institute on Global Conflict and Cooperation at the University of California, San Diego; with the U.S. Arms Control and Disarmament Agency from 1962 to 1969), and Lord Solly Zuckerman (professor emeritus at the University of East Anglia, Norwich; deputy chairman of the U.K. Advisory Council on Science Policy from 1948 to 1964).

Carl Kaysen (who served as deputy special assistant to President Kennedy for national security affairs) told an interesting story about a meeting with Richard Garwin (of IBM, who served on PSAC), Carl Kaysen, and Mac Bundy, in which the idea of order of magnitude was sprung on President Kennedy, and the president turned to me and said that he now understood how to deal with me when I make predictions that might be off by an order of magnitude. The session started a little after 2:30 p.m. and lasted until precisely 4:30 p.m., as scheduled.

I then rode with Richard Wilson (physicist, Harvard University) to the Democratic headquarters (105 Chauncey St.), where we met Christopher Edley (political aide to Governor

Michael Dukakis). He took us to a nearby hotel for tea and refreshments so we could talk without being interrupted. I told him why I think Dukakis cannot be elected president on his present vociferous antinuclear power platform. I gave him poll material (furnished to me by Harry Finger) indicating that the public favors an adequate energy supply in the future, including nuclear power. I also told Edley the reasons for favoring a comprehensive test ban, gave him the reasons I favor funding of the Superconducting Super Collider (this and the human genome can be funded even on a level budget if military spending is decreased in a compensating manner). I suggested that on nuclear power Dukakis could criticize the present way that it is being handled by the Nuclear Regulatory Commission and suggested he could even criticize other aspects of the U.S. nuclear power program, but that he should do this in a positive manner, saying that he will reform and change the way the program is conducted so that, in the future, we can have the nuclear power that we need for the welfare of our country. I emphasized the success of the French and Japanese programs to indicate that it is possible to have successful programs. With respect to waste disposal, a subject brought up several times by Edley, I said that this can be technically solved and that Dukakis is wrong in saying he is going to wait until there is a political solution (this could be 10 or 15 years); he should say that he is going to insist on a proper solution of this so that we can get on immediately with nuclear power development in the United States. I think Edley was somewhat impressed by my argument. He is going to pass on my arguments to Dukakis and suggest that maybe Dukakis meet with me when he comes to campaign in California for the primary.

Richard Wilson then drove me back to the Kennedy Library (Helen had gone back to the hotel with the others in the van and then returned to the Kennedy Library to meet me).

At the reception preceding the dinner, Helen and I met and talked to a number of people, including Senator Kennedy, his sister Pat Kennedy Lawford and her son Chris, Joseph and Sheila Kennedy (he is the son of Bobby Kennedy and a representative to the U.S. Congress from Massachusetts), Alan Goodrich (archivist of the Kennedy Library who had also been at Newton North High School and taken pictures of me there), David Powers (Kennedy Library official, former aide to President Kennedy), Charles Daly (director of the Kennedy Library), Soviet Ambassador Dobrynin, and Soviet UN Ambassador Aleksey Roshchin (Armand Hammer wasn't there because he was visiting the Soviet Union). In talking to Ted Kennedy I told him that I think that the Reagan administration is just using their program of work on verification of the Threshold Test Ban Treaty and so forth as a ploy to delay action on the comprehensive test ban treaty for years, and he agreed. I also complained about this to Soviet Ambassador Dobrynin, and he said that the Soviets did this as a last resort because it was their only way to get the Reagan administration to talk about the testing problem at all; I told him that I think this was a mistake.

At the dinner (which was on the ground floor in a sort of rotunda section where the ceiling goes all the way to the top of the seven or more stories and is surrounded by windows), Helen and I sat at a table with Teddy Kennedy, Jr. (26 years old, he is the campaign manager of Ted Kennedy and the son who lost his leg to cancer some years ago), Kara Kennedy (Ted Kennedy's oldest child, 28 years old), Benjamin Binswanger (Ted Kennedy's deputy campaign manager), and Teddy Harkwell (she also works in the Ted Kennedy campaign headquarters, as does Kara). Helen and I had an excellent conversation with this group and were impressed by Teddy Kennedy, Jr., who conversed with us on a number of subjects (I also tried to convince him of the value of nuclear power and the danger of the Democrats taking an adamant adverse stance on this, the value of a comprehensive test ban, and so forth); he indicated that he would like to keep in touch with us.

The program of speeches preceded the dinner and was presided over by John Calinane, chairman of the Kennedy Library. He introduced the following (in order); all made very impressive speeches: Ted Sorensen, Solly Zuckerman, Soviet Ambassador Dobrynin, and Ted Kennedy. There was a convocation by someone whose name I didn't catch (Cardinal Law, who *was scheduled to give this, did not attend).*

After the dinner there was a program introduced by Charles Daly. He gave a toast to a singing group from Middlebury School of Children, introduced by their teacher, and they performed for a while. Toward the end of the dinner Senator Kennedy came around and introduced John F. Kennedy's son John to us. Ted Kennedy, at the end of the evening, suggested that we keep in touch on such things as the comprehensive test ban treaty. I also talked with Jerry Wiesner, whom I'd seen throughout the day but who didn't participate in the program (I learned that he had an interesting op ed. piece on the comprehensive test ban treaty published in today's New York Times).

Following the program, Helen and I rode in a van with our group back to the Boston Harbor Hotel.

As a comment on my meeting with Christopher Edley (political aide to Dukakis), I can add that Dukakis did not meet with me. As is well known, he was soundly beaten by George Bush in the presidential election of 1988. I believe that a more moderate view on a number of issues, such as nuclear power, might have helped him, although perhaps not enough to win the election.

One day in June 1963, I received a telegram at home from President Kennedy expressing delight that I had consented to re-appointment as AEC chairman. This was later confirmed by a letter on June 27, 1963. Although I couldn't remember ever discussing this with him, it seemed to me that I couldn't turn down the president of the United States, so I accepted the re-appointment.

As I have indicated, I accompanied Kennedy on a remarkable number of visits to AEC installations: AEC headquarters at Germantown, MD; Lawrence Radiation Laboratory at Berkeley; Los Alamos Scientific Laboratory and Sandia Laboratory in New Mexico; and the Nuclear Weapons Test Site in

Nevada. He also visited the Hanford plant in Washington, accompanied by Commissioner Gerald Tape, on Sept. 26, 1963 (while I was in Europe attending the IAEA general conference). Even before he became president, while he was a U.S. senator from Massachusetts, he and his wife Jacqueline visited Oak Ridge National Laboratory.

I last saw President Kennedy on Nov. 8, 1963, when we discussed the production of fissionable materials. At that time, he was, as always,

President John F. Kennedy's letter reappointing Seaborg as chairman of the Atomic Energy Commission. June 27, 1963.

ORDER

Pursuant to the provisions of the

Atomic Energy Act of 1954, I hereby designate

Glenn T. Seaborg as Chairman of the Atomic

Energy Commission.

[signature]

THE WHITE HOUSE,

June 27, 1963

Alvin M. Weinberg (right), Director of the Oak Ridge National Laboratory, describes the Oak Ridge Research Reactor to Massachusetts Senator John F. Kennedy and Jacqueline Kennedy during a 1960 presidential campaign trip. In the background: L. P. Riordan (ORNL) and Edward E. Stokely (Atomic Energy Commission, Oak Ridge Office). Oak Ridge, TN. Feb. 24, 1960. Courtesy of the Oak Ridge National Laboratory, Oak Ridge, TN. Photo by Ed Westcott.

cordial and attentive and quickly grasped the problem. He displayed his usual quick wit and friendly humor:

From 12 noon to 12:15 p.m. I met alone with President Kennedy to discuss with him a question in weapons technology and how this might affect the time scale in which the French might be able to produce and successfully test thermonuclear weapons in the range of energy of 1/2 to 1, 2, 3 or 4 megatons. I explained to him means by which the French might be able to attain this objective sooner than most people think but I said I doubted the French have these particular methods in mind. We agreed this is something to remember in connection with this general problem with the French, which includes their attitude toward the limited test ban treaty, but that no particular steps are called for at present. We also briefly discussed the matter of curtailment of fissionable materials production and

possible timing with respect to such reduced production.

There was a tragic irony in the circumstances in which the news of President Kennedy's assassination reached me. As on Jan. 9, 1961, when he called to invite me to assume the AEC chairmanship, I was once again in the Lawrence Radiation Laboratory of the University of California and in the HILAC building, this time participating in the reciprocal visit of Soviet scientists. I was called aside because of some very important news. I stepped away and was told that the president and Texas Governor John Connally had been shot:

Friday, Nov. 22, 1963

The group returned to the Radiation Laboratory, where we visited the Nuclear Chemistry Division. We viewed fission work (Stanley Thompson), ultramicrochemistry (Burris Cunningham), the hot lab, and so on.

While at the HILAC, I received the news over the radio that President Kennedy (and Governor Connally) had been shot in Dallas at 10:30 a.m. Harold Fidler (associate director, Lawrence Radiation Laboratory) called me aside to give me the sad news. Ironically, it was during a visit to the HILAC building while I was chancellor at Berkeley that I had received the telephone call from President-elect Kennedy inviting me to come to Washington to serve as AEC chairman.

President Kennedy died at 11 a.m. (PST). After informing the Russian delegation and making alternate plans for them to go directly to Yosemite, I flew back to Washington on United flight 868, which left at 3:30 p.m. and arrived at Friendship Airport at 11 p.m.

The personal shock of President Kennedy's death is tremendous. How unnecessary it is! This raises some questions in my mind as to how much longer I want to stay on as chairman.

I was devastated to learn of the death of this vital and vibrant young man, John Kennedy, who had brought such a new air of optimism to Washington with his exciting "New Frontier." He had an almost unique appreciation and understanding of the important role of science and scientists in modern society. I particularly appreciated his complete dedication to the attainment of a treaty to end the testing of nuclear weapons. I believe that if Kennedy had lived and served out a second term, and if Nikita Khrushchev had survived in office, significant further steps in arms control, including a comprehensive test ban, would have ensued. The resolve of both Kennedy and Khrushchev to make progress on arms control was strengthened greatly by the searing experience of the Cuban Missile Crisis. Khrushchev in particular seems to have been persuaded by that experience to recognize that the enormous power at his disposal gave him responsibilities not only to the Soviet Union but to all of mankind. He became, increasingly, a responsible world leader.

The Missile Crisis also had a dramatic effect on the relationship of the two men. Their attitudes toward each other showed evidence of having been somewhat ambivalent. The brush with calamity seems to have forged a bond between them. They began to consult each other more frequently and work together on problems of common interest. This was done in large part through their private correspondence, the pace of which quickened after the Missile Crisis.

Some 16 years later, on Oct. 20, 1979, I was pleased to attend the dedication ceremony for the John F. Kennedy Library (as described in Chapter 8.) Helen and I have been pleased to be readers of *The John F. Kennedy Library Newsletter*.

chapter 5 LYNDON BAINES JOHNSON

1963–1969

AN OVERWHELMING PERSONALITY SUPPORTS THE NONPROLIFERATION TREATY

The presidency of Lyndon Baines Johnson is perhaps

remembered chiefly for his extraordinarily successful promotion of his concept of The Great Society and, unfortunately, his role as a leader in the Vietnam War. As chairman of the Atomic Energy Commission (AEC), I didn't play any important role in the achievement of the former, but I did watch (as a result of my attendance at numerous meetings), with awe and admiration, his complete dedication to and success in initiating this ambitious social program. I was also in a position to see how the overoptimistic advice and urgings of his military advisers led inevitably to our ensnarement in the Vietnam War. I recall that I was visiting the San Francisco Bay Area with my administrative assistant, Arnold Fritsch, when we saw and heard on March 31, 1968, the televised speech by President Johnson, in which he announced the de-escalation of the Vietnam War and the cessation of bombing on North Vietnam and, surprisingly, his decision not to seek or accept the Democratic nomination for president.

The Johnson presidency was a period of great arms control activity, and it produced two significant results: the Nonproliferation Treaty (NPT), signed in Johnson's last year, although not ratified until Nixon's first year; and the intellectual groundbreaking for the Strategic Arms Limitation Talks (SALT), the long campaign by which Johnson and Robert McNamara finally persuaded the Soviet leadership to embrace the concept of a mutual limitation on strategic weapons, both offensive and defensive. Indeed, but for the Soviet invasion of Czechoslovakia, SALT would have begun on Johnson's watch. His administration deserves much more credit for this accomplishment than it is generally given.

There were other events, as well. The Outer Space and Seabed treaties reserved those two environments for peaceful uses. Major cutbacks were made in our capacity to produce fissionable materials, which had grown far beyond our foreseeable needs. For example, a program was begun to stop operation of all but 4 of the nation's 14 plutonium production reactors. There was also discussion, without agreement, of adding further restrictions on nuclear testing and more preliminary consideration of a number of other arms control initiatives, including a freeze on strategic missiles.

And there were, of course, the other activities attendant with my continuing AEC chairmanship. Of special interest and great satisfaction to me were my visits to the LBJ ranch in Texas each December, when I met with President Johnson to successfully promote the AEC budget.

I first met Lyndon Johnson during 1947–48 when, as a Texas congressman, he served on the Joint Committee on Atomic Energy (JCAE), while I was serving on AEC's General Advisory Committee (GAC). Our next contact came during my chancellorship at Berkeley, when I invited him to come and speak to our students (an invitation he didn't find possible to accept).

As I have recounted in the previous chapter, I became very well acquainted with him while he was serving as vice president in the John F. Kennedy administration.

Many remember the remarkable courage Jacqueline Kennedy exhibited after her husband's assassination. But I was also impressed with the performance of Lyndon Johnson, the new president. He acted resolutely in several matters I brought to his attention, some of which were so sticky that President Kennedy had asked that decisions be postponed until after the 1964 presidential election. When I briefed Johnson on these issues and pointed out the political implications, his bold answers surprised me. "Glenn," he stared me in the eyes and said, "if it's the right thing to do, let's do it. The hell with the election!"

I described in detail in my book, *Stemming the Tide—Arms Control in the Johnson Years* (written with my colleague Benjamin S. Loeb), Johnson's important role in the final achievement of the NPT. (Visits to the Lyndon B. Johnson Library in Austin, TX, were helpful for the writing of this book.) Success came only as he turned his attention seriously to the attainment of this objective.

By the time Johnson became president, the Arms Control and Disarmament Agency (ACDA) had adopted nonproliferation as its number one objective and attempted to enlist the new president in that quest. Johnson went along by making nonproliferation the centerpiece of both his address to the General Assembly of the UN within the first month (Dec. 17, 1963) of his presidency and his first message to the Eighteen Nation Disarmament Committee, sent early in 1964. His latter message essentially proposed that all inter-

United States Senate
Office of the Democratic Leader
Washington, D. C.

February 8, 1960

Dear Chancellor Seaborg:

Thank you very much for your invitation to
speak to a meeting of your student body
during the Spring semester. I had hoped
to be able to come to California this spring
but Senate activities are becoming so pressing that I don't see how I can get away. If
something does work out, however, I will
get in touch with you immediately.

I appreciate your thoughtfulness in asking
me to come.

Sincerely,

Lyndon B. Johnson

Chancellor Glenn T. Seaborg
Chancellor at Berkeley
The University of California
Berkeley 4, California

Texas Senator Lyndon B. Johnson's letter, declining with regrets, Seaborg's invitation to speak before the University of California–Berkeley student body. Feb. 8, 1960.

national transfer of nuclear materials take place under effective international controls to prevent their use for weapons purposes. But the United States had no detailed proposal ready at that time.

At first ACDA's enthusiasm for an NPT was not shared throughout the government. It conflicted with another objective, which had strong support in the State Department: the establishment of a North Atlantic Treaty Organization (NATO) naval force, manned by personnel from several nations and equipped with U.S. nuclear weap-

ons, the so-called Multilateral Force or MLF. This idea, introduced in 1960, had several purposes: to give NATO countries, particularly Germany, a greater role in planning their own defense, thereby helping to dissuade them from wanting to be independent nuclear powers; to preserve allied cohesion in the face of the Soviet threat; and to encourage the budding movement toward a unified Europe. It was enthusiasm for the unification of Europe that made George Ball (undersecretary of state) and others in the State Department very enthusiastic advocates. Although it was argued that the MLF and the NPT were not inconsistent, the former tended to exclude the latter because of the Soviet Union's attitude. The Soviets were fiercely hostile to a scheme that seemed to place a West German finger on the nuclear trigger. It began to seem clear that the Soviets would agree to an NPT only if we first gave up on the MLF.

West Germany was at the center of most of the tugs of war about the NPT. I think it is fair to say that the Soviets' main interest in an NPT was to achieve thereby a German renunciation of nuclear weapons. Remembering two German attacks in a generation, the Soviets were haunted by the prospect of a vengeful Germany with nuclear weapons. They repeatedly declared or implied that they would ratify the NPT only when West Germany did. The Germans, feeling threatened by the Soviets, lacking confidence that the United States would always be there to protect them, and having unquestioned technical ability to make nuclear weapons, were very hesitant to forswear this possibility forever. The United States, and the State Department in particular, wanted to appease the Germans, in part because we wanted to retain the cohesion of the alliance and also because it was always possible that, after we and the Soviets agreed on an NPT, a disgruntled Germany might refuse to sign it.

Germany, and to a lesser extent Italy, seemed interested in the MLF from the start. The British were opposed–they didn't think this was any way to run a navy. Other NATO allies were indifferent at best. President Kennedy was rather cool toward the idea, although he was willing to go forward if the allies showed a clear desire to do so. Later, after France began to distance itself from NATO, Kennedy showed more interest because of a desire to give the Germans an alternative to nuclear cooperation with France. But there was strong opposition in Congress to sharing U.S. weapons with anybody, and to do so would have required amending the Atomic Energy Act.

Despite the political problems, technical work on the MLF went forward, and when Johnson became president he was immediately subjected to strong pressure from MLF advocates in the State Department. Following some intense discussion within the administration, he authorized a campaign to sell the idea to our allies, hoping to reach agreement by the end of 1964.

But then, on Oct. 16, 1964, I made the following entry in my diary:

The big news today is that at 3 a.m. Washington time, the Red Chinese exploded an atomic bomb in the atmosphere. Our electromagnetic and acoustic detection devices picked it up, and the Chinese announced it. President Johnson made the U.S. announcement of the test (after the Chinese announcement). The yield appears to be in the range of tens of kilotons (as expected). Intelligence sources had picked up indications that this test would come soon.

Our analysis of the debris convinced us, to our surprise, that the Chinese had detonated a U-235 device of sophisticated design, not a plutonium bomb such as the one the other four nuclear powers had used for their first tests. I reported these findings to a cabinet meeting on Oct. 20. The main topics of discussion were three unsettling events that had occurred in distant places during the preceding few days: the first nuclear weapons test in the People's Republic of China; the unexpected ouster of Soviet Chairman Nikita Khrushchev, who was replaced by Leonid I. Brezhnev (and A.N. Kosygin); and the electoral loss of the incumbent Conservatives in Great Britain to the Labour Party, led by Harold Wilson (thought to be less friendly to U.S. policy).

Tuesday, Oct. 20, 1964

At 1:10 p.m. I attended the cabinet meeting held in the Cabinet Room of the White House. Present at the meeting were President Johnson, Dean Rusk (secretary of state), Douglas Dillon (secretary of the treasury), Robert McNamara (secretary of defense), The Postmaster General John Gronouski, Willard Wirtz (secretary of labor), Orville Freeman (secretary of agriculture), Anthony Celebreeze (secretary, Department of Health, Education, and Welfare [HEW]), Adlai Stevenson (U.S. representative to the UN), Luther Hodges (secretary of commerce), Kermit Gordon (director of the Bureau of the Budget [BOB]), Bill Moyers (special assistant to President Johnson), McGeorge Bundy, George Reedy (presidential press secretary), James E. Webb (administrator of the National Aeronautics and Space Administration [NASA]), Sargent Shriver (director of the Peace Corps), S. Douglass Cater (special assistant to President Johnson), Donald Hornig (director of the Office of Science and Technology), Walter W. Heller (chairman of the Council of Economic Advisors), and Jack Valenti (special assistant to President Johnson).

Again, the Russian changeover, the Chinese test, and the U.K. election were discussed as well as individual reports by cabinet members, the economic and political situation, and the Jenkins situation. (Presidential assistant Walter Jenkins had recently been arrested on a morals charge.) I gave a report on the Chinese test saying that analyses of air debris now indicate, to our surprise, that the device apparently contained uranium-235. I explained the Chinese production capability and the use of plutonium and uranium-235 to make fission and thermonuclear weapons. I admonished them to be very conservative in estimating Chinese weapons capabilities at this time.

The Chinese test had long been expected, but the actual occurrence nevertheless shook up the whole international equation. Powerful forces in India immediately began agitating for an Indian

Lyndon B. Johnson Cabinet meeting, White House Cabinet Room. Left to right: John Gronouski (Postmaster General), Adlai E. Stevenson (U.S. Ambassador to the UN), Seaborg, Robert S. McNamara (Secretary of Defense), Orville Freeman (Secretary of Agriculture), Willard Wirtz (Secretary of Labor), Donald Hornig (Presidential Science Advisor), Walter Heller (Chairman, Council of Economic Advisers), Luther Hodges (Secretary of Commerce), Dean Rusk (Secretary of State), Johnson, Douglas Dillon (Secretary of the Treasury), Kermit Gordon (Director, Bureau of the Budget, James E. Webb (Administrator, NASA), and Sargent Shriver (Director, Peace Corps). Washington, D.C. Oct. 20, 1964. Courtesy of the Lyndon Baines Johnson Library, Austin, TX. Photo by Cecil Stoughton.

bomb to match China's. This made the Pakistanis edgy. The Australians began to stir. Proliferation seemed to be in the air. The need for an NPT seemed more urgent.

President Johnson had to confront the MLF issue seriously in December 1964. The occasion was a visit by British Prime Minister Harold Wilson. The principal item on the agenda was the MLF, and the British had made no secret of their opposition. However, it was probably the run up to the meeting rather than the meeting itself that had the biggest effect on the president's decision. In five days of intensive meetings with his principal advisers, Johnson grappled with the MLF question, seeking a policy position of his own. Following the summit with Wilson, he leaked to the *New York Times* the essence of a new position. Essentially, he determined that the United States, while not opposing the MLF, would no longer actively try to bring it about.

The president's new position really energized the diplomatic process. In August 1965 the United States unfurled a complete NPT draft at the Eighteen Nation Disarmament Conference (ENDC). The draft did not fully rule out a future MLF, however—die-hards in the State Department had managed to keep it alive—so the Soviets promptly rejected the draft. The Soviets wanted to outlaw any transfer of nuclear weapons whatever—their position seemed to bar even existing NATO arrangements by which U.S. weapons were stationed in Europe. Then McNamara devised a substitute for the MLF—the idea of a consultative committee to determine NATO nuclear strategy. This seemed to satisfy the motive of giving Germany and other NATO allies a voice in their own nuclear defense. The situation now seemed ready for forward movement on an NPT.

The missing ingredient was lack of presidential involvement. President Johnson had become somewhat disengaged from arms control matters because of his preoccupation with the Vietnam War following the major escalation there early in 1965. Pressures to get him to focus again on the NPT came from a number of directions. One was a resolution sponsored by Senator John O. Pastore and passed by an 84–0 Senate vote in May 1966. The resolution began, diplomatically, by praising what the president had already done and then urged additional efforts by the president. . . for the solution of nuclear proliferation problems. Next, certain individuals inside the administration managed, through Bill Moyers, to reach the president and advocate the urgency of getting an NPT. The break seemed to come on July 5, 1966, when, in answer to a question at a news conference, the president stated: "We are going to do everything within the power of our most imaginative people to

find language which will bring the nuclear powers together in a treaty which will provide nonproliferation." I was told by someone who was at Geneva that this press conference statement electrified the process there. It gave the U.S. delegation a signal it had been waiting for—a signal to put on a full-court press for an NPT. In Washington, too, the situation changed. Rusk, previously quite removed from the issue, now became an active and very effective NPT advocate. His participation from this time forward made a great difference.

Just to allay any doubts there might have been about where he stood, President Johnson stepped up the pressure in a speech at the National Reactor Testing Station (NRTS) on Aug. 26, 1966. I report in more detail about this visit below. Speaking of the NPT negotiations, the president said, "I believe that we can find acceptable language on which reasonable men can agree." The search for such language was under way in hard, intense, *private* negotiation between the United States and the Soviet Union.

On Oct. 10, 1966, Soviet Foreign Minister Gromyko visited the White House full of smiles, indicating that the process had borne fruit. On Dec. 5, 1966, the two sides unveiled the text of the first two articles of an NPT. Article I forbade states having nuclear weapons from transferring them "to any recipient whatsoever." Article II forbade states not having nuclear weapons from accepting their transfer or manufacturing them. Article I essentially ruled out the MLF. The United States, however, prepared a series of interpretations that we told the Soviets would be submitted to the Senate with the treaty. The most important of these was that the treaty would not prevent a federated European state, if one ever developed, from inheriting the nuclear weapons of Britain or

"We are going to do everything within the power of our most imaginative people to find language which will bring the nuclear powers together in a treaty which will provide nonproliferation."

A Chemist in the White House: From the Manhattan Project to the End of the Cold War

France, or both. Apparently, the Soviets considered this eventuality sufficiently remote that they were willing to take a chance on it. In the end, although we gave up on the MLF, we did not give up entirely on two of its purposes. The allies gained a voice in planning their own defense through the McNamara committee. And we provided for the bare possibility of eventual European unity.

After the breakthrough on the core provisions of the treaty, embodied in Articles I and II, there was still one other important matter to clear up. This concerned so-called safeguards, meaning inspections and other mechanisms for detecting on a timely basis any diversion of nuclear materials from peaceful to weapons uses. In this matter AEC became embroiled in a dispute with other parts of the U.S. government. AEC wanted safeguards, preferably administered by the International Atomic Energy Agency (IAEA), to be made mandatory. Our European allies resisted mandatory safeguards, ostensibly because they did not like the idea of inspectors from other countries roaming around in their nuclear power plants. They were supported in this attitude by some members in our State Department. ACDA, bowing to allied and State Department pressure, at first introduced in Geneva a miserably weak treaty provision specifying merely that the parties to the treaty would "cooperate in facilitating the application of safeguards." AEC bitterly protested the weakness of this provision, and our position won support from the then-powerful JCAE. In fact, JCAE implied that any treaty lacking mandatory safeguards would be in trouble in the Senate. This helped tilt the balance; mandatory safeguards for all non-nuclear weapon countries soon became the U.S. position.

It did not, however, settle the question of who would administer the safeguards. In deference to our European allies, the United States argued in Geneva for a formula specifying "International Atomic Energy Agency or equivalent" safeguards. "Or equivalent" was a reference to safeguards already being applied to its members by the European Atomic Energy Community (EURATOM). Several allied countries very much preferred EURATOM to IAEA safeguards. Their argument was that IAEA inspectors might make off with industrial secrets about their growing nuclear businesses. They seemed to be concerned about losing secrets not only to the Russians, but also to the Americans and the British. My own feeling was that these arguments about commercial disadvantage were somewhat contrived—they may have masked a resistance to the main provisions of the NPT. I traveled to West Germany in March 1966 to try to convince the German Atom Forum of the value of IAEA safeguards. I doubt that I made many converts.

But AEC had had experience with EURATOM safeguards and thought they were equal in effectiveness to IAEA safeguards. We were willing to accept them, as long as they were made mandatory. The Soviets, on the other hand, said that self-inspection by EURATOM of its own members was unacceptable. Various compromise proposals were then thrown into the mix, all seeking some way to keep EURATOM safeguards, at least for a while, subject to some verification of their adequacy by IAEA. The U.S. proposal was that all non-weapon states be required to accept IAEA or IAEA-verified safeguards within three years.

In August 1967 the United States and the Soviet Union formally recognized the impasse on safeguards—they presented identical treaty drafts in Geneva in which they reserved space for an Article III about safeguards and then left it blank. Soon thereafter, however, informal talks among negotiators from the two sides produced basic agreement on a compromise solution: Each non-nuclear party to the treaty would, within a specified time, reach a safeguards agreement with IAEA. This formula allowed for the possibility of continued EURATOM safeguards in that the agreements could be negotiated either individually or together with other countries. In time there was an agreement between IAEA and EURATOM. It still exists, and it provides for continued application of EURATOM safeguards under IAEA supervision.

Having informally reached this compromise formula, the Soviet and American delegations faced a problem: They could not submit the compromise solution to their governments because both delegations were under instructions not to budge from previous positions. Eventually it was decided that each would submit the compromise to its government as a suggestion emanating from the other side. Once our government had decided to accept the compromise, it submitted it to the North Atlantic Council for allied consideration, still maintaining the fiction that it was a Soviet suggestion. There would have been difficulties had it been submitted as a U.S. idea, because Rusk had promised German Foreign Minister Willy Brandt that we would not change the existing U.S. proposal without prior consultation.

A key step to soften allied opposition to the proposed safeguards article was taken on Dec. 2, 1967, when President Johnson announced that the United States would accept the application of IAEA safeguards to all its own peaceful nuclear activities at the time that such safeguards were generally applied to other nations under the NPT (as described below). This announcement was the culmination of a series of prior suggestions and events in which I

had played a key role. The British immediately followed our example. These actions tended to cut the ground from under previous allied objections based on presumed commercial disadvantage. The allies then agreed to the text of the safeguards article and, after some last-minute haggling with the Soviets over wording, the agreement was announced in Johnson's State of the Union message in January 1968.

The first three articles of the NPT (Articles I and II setting out the basic obligations of nuclear-weapon states not to transfer and nonweapon states not to acquire nuclear weapons, and Article III prescribing safeguards) pretty well encompassed what the superpowers hoped the final treaty would be. Not so for the non-nuclear countries that were the main object of the treaty. They greatly resented what they considered to be the discriminatory nature of the draft NPT. They were being asked to renounce a future means of defense and without any compensation. One delegate characterized the draft treaty vividly as an attempt to castrate the impotent. The proceedings in Geneva and at the UN in New York teemed with such bitter complaints, particularly from nonaligned countries.

The United States and the Soviet Union found themselves in unaccustomed alliance as they sought to fend off or moderate the demands of the non-nuclear nations. Ultimately three articles were added to the treaty in an effort to appease them.

Article IV stated the right of all countries to pursue the peaceful atom without discrimination. It also announced the obligation of more advanced countries to provide technical assistance on peaceful uses to others, particularly to those in "the developing areas of the world."

Article V referred to a technology that has since declined in importance: the use of nuclear explosions for peaceful purposes such as excavation, mining, and research. Both Brazil and India objected to the draft NPT on the grounds that it would preclude their independent development of such explosives. In a trip to Brazil in 1967 I spoke to Brazilian officials at length about this. I pointed out that the U.S. AEC stood ready under an NPT to provide a peaceful nuclear explosives service at a fraction of what it would cost them to provide it for themselves. I found that they were generally not well informed about the issues and that their arguments did not hold up. I became convinced that their avowed interest in peaceful nuclear explosions was mainly a cover to keep alive a nuclear weapons option. Nevertheless, to meet such objections as the Brazilians advanced, an Article V was added to the NPT providing for such a nuclear weapons service as I had described to them.

Also in 1967 I traveled to India, Pakistan, Australia, and Argentina to try and convince the leaders of those countries to support and join an NPT. Australia, India, and Pakistan would support an NPT only if it contained adequate guarantees against nuclear aggression. Australia and India were most seriously disturbed by China's growing nuclear prowess, and India and Pakistan were afraid of each other. Earlier I had visited Egypt and Israel to discuss these and other matters, but I did not have the opportunity to visit the Dimona facility, which included a reactor that was believed to produce sufficient plutonium to give Israel a nuclear weapons capability.

The most clamorous demand of the non-nuclear countries was that, in exchange for their abjuring nuclear weapons, the superpowers must do something to halt their bilateral arms race, which was regarded as a threat to everybody. The tide of revolt on this issue ran very strongly—so much so that the superpowers felt that if they did not give ground they might lose the treaty. They therefore added an Article VI, pledging "to pursue negotiations in good faith on effective measures regarding cessation of the nuclear arms race and disarmament. . . ." Later they were forced by the efforts of Sweden's Alva Myrdal to agree to an amendment requiring that these negotiations take place "at an early date."

Formal UN debate on the NPT began in the General Assembly on April 24, 1968. It was approved on June 12 by a vote of 95 to 4, with 21 abstentions.

The treaty was opened for signature on July 1, 1968, in Washington, London, and Moscow. It was signed on that day by the Big Three (the United States, Britain, and the Soviet Union) and more than 50 other countries. Senate hearings began on July 10. There was little opposition, but the Foreign Relations Committee did not vote out the treaty until Sept. 17. On Oct. 11, with the presidential election campaign in full swing, the full Senate voted to postpone action. After Nixon's election, he made it clear that he wanted action still further deferred, until after his inauguration. On Feb. 5, 1969, President Nixon recommended ratification in a special message to the Senate. The Senate gave its consent on March 13; and two days later, having been ratified by the requisite number of countries (the Big Three plus 40), the Treaty on the Nonproliferation of Nuclear Weapons entered into force.

Only one additional country is known to have exploded a nuclear device since that time. On May 17, 1974, India conducted a nuclear explosion underground in a desert area near Pakistan, using plutonium produced in their CIRUS (Canadian-Indian Reactor Uranium System) reactor and extracted by the processing plant I inspected during a visit to their Trombay Nuclear Establishment in

January 1967. To this date, India has not proceeded beyond the one explosion. At the same time, however, India has not shown any interest in signing the NPT. Neither, for that matter, have the nuclear-capable nations of Brazil, Israel, or Pakistan signed. Nor can we any longer confidently assert that a nation that has not tested doesn't already have nuclear weapons.

Although the NPT has not ended the proliferation problem, it still is an impressive accomplishment. The treaty's 178 parties account for 98% of the world's installed nuclear power capacity, 95% of the nuclear power capacity under construction, and all of the world's exporters of enriched uranium.

The NPT provides that review conferences be held at five-year intervals to assure that the purposes and provisions of the treaty are being realized. At conferences held in 1975, 1980, 1985, and 1990, there was vigorous complaint that the superpowers were not living up to Article VI—that they were not negotiating in good faith an end to their arms race. There was even concern that this dissatisfaction would threaten the renewal of the treaty when it came up for a vote in 1995.

Nevertheless, at the review conference on May 11, 1995, the member nations agreed by acclamation to extend in perpetuity the Treaty on the Nonproliferation of Nuclear Weapons. Coming out of this meeting, the nuclear weapons nations also committed to concluding a comprehensive ban on nuclear testing in 1996.

On July 1, 1968, the very day they signed the NPT, President Johnson and Soviet Premier Kosygin announced their intention to enter into talks on the limitation and reduction of offensive and defensive nuclear weapons.

The formal arms limitation talks thus announced were to be the first of their kind, but there had

been several approaches to this subject matter, dating back a number of years. President Johnson had made several arms limitation proposals in his first message to ENDC on Jan. 21, 1964. The proposal that attracted the most attention was that the "United States and the Soviet Union and their respective allies. . . agree to explore a verified freeze on the number and characteristics of strategic nuclear offensive and defensive missiles." The president expressed his conviction that "this initial measure preventing the further expansion of the deadly and costly arms race [would] open the path to reductions in all types of forces from present levels."

After many exchanges of messages with Premier Kosygin, and pushing by McNamara to seek a treaty placing limitation on antiballistic missile (ABM) systems, arrangements were completed for a summit meeting to be held at Glassboro, NJ, on June 23, 1967.

The approach of the Glassboro get-together, President Johnson's only summit meeting with a Soviet leader, produced great excitement. For its part, the White House was determined to capture for history every detail of the proceedings; and a presidential secretary, Marie Fehmer, was assigned to take detailed notes. These make fascinating reading, and I have reproduced them in their entirety in my book, *Stemming the Tide—Arms Control in the Johnson Years*. McNamara took the lead on our side and pleaded for limitations on both offensive and defensive weapons.

On Aug. 19, 1968, the Soviet Union finally agreed to schedule a summit conference that would initiate the talks on missile and launcher limitations. The date was to be in the first 10 days of October, the site probably Moscow. A joint announcement was agreed upon; it was to be released on the morning of Aug. 21. The White House press corps was alerted.

However, on the night of Aug. 20 (Washington time), armored units of the Soviet Union and several other Eastern bloc nations rumbled into Czechoslovakia to bring to a halt the liberalization initiatives of the Dubcek regime. Ambassador Dobrynin personally delivered information about the invasion to President Johnson at 8 p.m. His message concluded with the hope that "the current events should not harm Soviet–American relations, to the development of which the Soviet government as before attaches great importance." Later that night, following an urgent meeting of his top advisers, the president instructed Rusk to call Ambassador Dobrynin and insist that he call Moscow immediately to tell them not to issue the summit conference announcement the following morning. Johnson realized that, in the general state of outrage over the invasion, it was impossible to proceed as scheduled.

And so the task of carrying forward the missile talks, which Johnson and McNamara had conceived and ardently sought, passed to a new administration. Of all the tragedies that befell Lyndon Johnson, this must rank among the most grievous.

Johnson had perhaps the strongest and most overwhelming personality of any person I have ever known. His ability to bring key members of Congress around to his way of thinking and to support his legislative program was something to behold. I watched in amazement as he pursued his successful methods of persuasion. My wife Helen characterized his presence, on the basis of her personal experience, as literally surrounding the person to whom he was talking. However, during such conversations, Helen also felt that he had a genuine interest in her activities.

When Johnson wanted information or advice, he wanted it right away. For this reason, one had to be prepared to present the full story on the spot. If you did not manage to make all your points the first time around, it was unlikely that you would be allowed a second opportunity. I remember one time when I was swimming at the University Club after work, as was my custom. The University Club is a men's club (or was, at that time), and so

Seaborg gets the word from President Johnson. Jan. 14, 1965. Courtesy of the Lyndon Baines Johnson Library.

Seaborg and President Johnson "chewing the fat" in the White House Oval Office. Washington, D.C. Jan. 17, 1964. Courtesy of the Lyndon Baines Johnson Library, Austin, TX. Photo by Y. R. Okamoto.

members generally swam in the nude. As I completed a lap, an excited pool attendant informed me that the president of the United States wished to speak with me on the phone. Dripping wet, I took the call and marshaled all of my arguments against a proposition I did not support. I remember feeling rather foolish, debating an issue with the president in that pose; however, I knew that the president might not wait until I grabbed a towel to bolster my sense of dignity.

It has been said that Johnson tended to place people in two categories—those he liked and trusted, and those he did not. Clearly, on the basis of his relations with me, he placed me in the favorable category. He apparently was convinced that I was straightforward, without a hidden agenda, in my dealings with him, although this journal entry (April 27, 1968) would seem to indicate that he realized I am not without wily ways. I reported the following remarks made to me by Johnson at a reception for King Olaf of Norway at the Norwegian embassy:

The president told me he was pleased with the way the large nuclear test had gone this morning. He said he had confidence in me that it would go all right, and he again said he regarded me as one of his most able public servants. I expressed appreciation for his backing so as to allow the test to continue under the circumstances. The president said he had received a message from Howard Hughes, which required an answer, and he would like me to draft a reply for him in my own special way. He referred, in this connection, to my ability to handle people, getting them frothing at the mouth and then quietly devastating them in my inimitable way, and he wanted a reply in this vein.

Illustrative of his appreciation for the performance of AEC are his letters to me of April 17, 1965, and April 6, 1966:

April 17, 1965
Honorable Glenn T. Seaborg
Chairman, Atomic Energy Commission
Washington, D.C. 20545
Dear Dr. Seaborg:
I wish to thank you for the two very informative reports describing the Atomic Energy Commission's activities during 1964.

Since my association with our atomic energy programs began in the House of Representatives nearly 20 years ago as a member of the Joint Committee on Atomic Energy, I have followed the program closely.

I want you and your fellow Commissioners to know that your reports impress me from a number of points of view.

First, they present solid evidence that the Commission is pursuing a vigorous program of nuclear weapons research and development;

Second, they make it clear that a steadily increasing proportion of the Commission's budget is being devoted to the peaceful applications of the atom, a matter which is particularly gratifying to me; and

Third, they clearly reflect that the Nation is being well served through the healthy partnership of our Government with our industries and universities.

On these accomplishments, I congratulate you and thank you especially for the personal service you are rendering the Nation by your distinguished chairmanship. As you and I have often discussed, it is essential at all times that we look far ahead in our planning for this vital activity. I would, therefore,

President Lyndon B. Johnson's letter of appreciation for Seaborg's work as chairman of the Atomic Energy Commission. April 6, 1966.

THE WHITE HOUSE
WASHINGTON

April 6, 1966

Dear Doctor:

In reviewing your Annual Report to Congress, I am struck once more by the perfect marriage of opportunity and talent it represents.

The Commission's task is a formidable and constantly expanding one. You have more than kept up with the challenges of this venturesome age. You have harnessed them to our nation's purpose, and in so doing you have charged that purpose with new hope and energy.

On behalf of every American, I extend to you and your colleagues my gratitude and respect. Tomorrow, I know, will give me fresh cause for pride and satisfaction.

Sincerely,

Honorable Glenn T. Seaborg
Atomic Energy Commission
Washington, D. C. 20545

like to convey to you some of my views and hopes in relation to the program.

We have been able to maintain our clear superiority in nuclear weapons, while at the same time we have been responsible and realistic about our needs. The orderly cutback in the production of fissionable materials is a significant example of this realism.

I appreciate the Commission's cooperation in the advancement of measures for effective arms control. I look forward hopefully—and confidently—to the day when our national security and the security of the human race can be further increased through agreements and actions among nations which build upon the important first step of the limited test ban treaty.

I look for the continuation of the important progress that is being made in the peaceful uses of nuclear energy. For example, in the field of civilian nuclear power, I look forward to the development of the advanced converter and breeder reactors, which will be required for the more efficient and economical use of our Nation's nuclear fuel resources. Nuclear energy will fill an important role in partnership with fossil fuels in meeting the growing energy requirements of our Nation. As you know, I also anticipate that nuclear power will play a significant role in the desalting of sea water.

It is characteristic of nuclear energy that its great potential is continually expanding. The full range of its ultimate contributions cannot be foreseen. We must continually press toward the discovery of areas and applications of which we have not yet dreamed, even as we strive to realize the full potential of the areas already defined.

Basic to all of the applications of nuclear energy is the conduct of fundamental research in the physical and biomedical sciences, and I favor the vigorous pursuit of these activities.

On the other hand, we must also remember—keeping in mind always the essentiality of Government control of the uses of nuclear energy in the interest of the national security and public safety—that nuclear energy, after a period of intensive development, is now an integral part of the American industrial scene. It should not be regarded as a Government preserve. I look forward to the assumption by the private sector of our economy of a steadily increasing share of the responsibility for the development of the applications of nuclear energy.

In the field of the application of radioactive isotopes, I would like to see continued emphasis on the development of this humanitarian tool for the diagnosis and treatment of disease. I believe that we have only begun to realize the potential of these remarkable substances for the alleviation of human suffering. I also want to encourage continued development of their application to industrial and other processes.

In the field of space, we should continue the development of isotopic and reactor SNAP devices to enable us to take advantage of their unique application to the generation of electric power for our spacecraft. The recent successes of the nuclear rocket reactor tests indicate that nuclear rockets can be ready for the long-range space missions of the future.

In the field of education, the contributions made by the Commission are many and appreciated. I believe we can achieve even closer cooperation between the many Government laboratories and the universities throughout this country. The national resources in these laboratories can benefit the research and education processes in the universities. The laboratories will, in turn, greatly profit from their association with the universities.

I wish to commend particularly a use of advanced planning by the AEC which is being carried out without much fanfare, but so very effectively. Thus, for example, the cutbacks in special nuclear materials production were planned sufficiently in advance so that the Commission, in cooperation with the local officials and business and labor people, could take appropriate actions, such as diversification programs, to minimize any significant economic impacts.

Our capacity for achievement in atomic energy development never has been greater. The Commission has achieved a high degree of cooperation with private industry and the universities. The Congress, especially the Joint Committee on Atomic Energy, has effectively supported our nuclear program. This team in being—of government, industry, and the educational community—constitutes an unparalleled force for accomplishment. I look to the Commission to continue and further enhance these effective and harmonious relationships.

On this course, I believe we shall ultimately achieve a society in which man can live in peace, enjoy the freedom and personal security to shape his destiny according to his individual beliefs, and have the leisure to contribute to the culture of his civilization. I recognize that our goals will not be easily reached. There will be disappointments and hard choices in priorities to adjust to continually changing requirements and circumstances. We have the will and the capacity. We also clearly have the duty. For if man would inherit from the generations that have preceded him, he must bequeath something of value to the generations that succeed him.

Sincerely,
Lyndon B. Johnson

Another example of his appreciation came in a phone call on Wednesday, Sept. 13, 1967:

At 11:30 a.m. Walt Rostow (White House) called and said that the president has approved my trip to Europe. He went on to say that the president had appended a note to the paper: "Delighted. Would be glad to see Dr. Seaborg before he leaves on or after his trip. I have deep affection and admiration for this man whom I consider tops in the administration."

He also had words of praise for Helen:

Friday, Jan. 27, 1967
Helen attended the signing ceremony at the White House for the treaty outlawing nuclear weapons in outer space. President Johnson, in commenting on our recent world trip, told her that she is a credit to her country.

During the Johnson administration I continued the practice of submitting biweekly reports to the president. These are illustrated by quoting the first and the last of these reports to Johnson.

November 27, 1963
Personal and Confidential
The President
The White House
Dear Mr. President:

I have the honor of submitting to you our bi-weekly report on significant activities in the atomic energy program. Since this is our first report to you, I should like to identify, in addition to a few current events, some of the areas of the atomic energy program which involve national policy considerations.

1. Nuclear Testing—The Test Ban Treaty (Unclassified)

Obviously, the most important activity of the Commission relates to the development and production of nuclear weapons. On October 7, 1963, the underground nuclear test program was re-oriented to comply with the provisions of the Limited Nuclear Test Ban Treaty. The Treaty, as you know, permits the testing of nuclear devices underground but does prohibit the delivery of radioactive debris (fallout) outside the territorial limits of the nation under whose control the detonation was conducted.

Since the Test Ban Treaty, the Commission has conducted all underground nuclear detonations at depths to provide reasonable assurance of containing radioactive debris. Based on past experience and performance, there is high assurance that the past five events scheduled for December in the Niblick XX series will not vent nor produce measurable radioactivity. Our inclination is to go ahead with these events as now scheduled. We would issue the usual low-key, post-shot announcements as in the past.

In addition to the underground test series mentioned above, the Commission is maintaining a sense of readiness to conduct atmospheric tests should the Treaty be abrogated; and readiness posture was stated as national policy by President Kennedy.

The Joint Committee on Atomic Energy continues its active interest in safeguards connected with the nuclear Test Ban Treaty. Senator Henry Jackson will visit with AEC technical experts in Albuquerque on Friday, November 20, to discuss the subject. Representative Craig Hosmer has given evidence of continued criticism of the adequacy of readiness plans to permit resumption of atmospheric testing if necessary.

> "For if man would inherit from the generations that have preceded him, he must bequeath something of value to the generations that succeed him."

2. Plowshare Program (Unclassified)

Our Plowshare program encompasses development of the tremendous potential of nuclear explosives for excavating earth, mining, developing water resources, and conducting scientific research. However, the Test Ban Treaty has caused us to defer planned large-scale excavation experiments in favor of concentrating on the development of nuclear explosives with less radioactivity, smaller-scale excavation experiments, and experiments in scientific and other engineering applications. The first of these small-scale excavations experiments is now under consideration with the Commission. Ultimately, it will be necessary to make arrangements under the Treaty if we are to utilize this promising technology.

3. Private Ownership of Special Nuclear Material

Proposed legislation was submitted by the Commission in March 1963 which would eliminate mandatory government ownership of special nuclear material. Hearings were held July 30–31 and August 1. The Joint Committee on Atomic Energy has indicated that further consideration of the legislation may be given priority in the next session of Congress.

The legislation would permit and ultimately require the nuclear fuels involved in the nuclear power industry to be privately owned and financed. This legislation has been characterized by the Joint Committee on Atomic Energy as "the most sweeping amendment to the Atomic Energy Act since 1954" which "will vitally affect the future legal and economic structure of the entire atomic energy industry." Industry generally supports this legislation.

4. Special Materials Production Cutback

The Commission is reviewing several alternate plans for meeting the projected military and civilian requirements for special materials (enriched uranium, plutonium, and tritium). The military requirements are based on the weapons stockpile objective recommended on August 15, 1963, to President Kennedy by the Secretary of Defense. The long-range nuclear schedule proposed by the Secretary would entail a major reduction in AEC production of special nuclear material (enriched uranium, plutonium and tri-

tium). The major considerations in the development of a detailed program to meet reduced weapons requirements, as well as to provide for contingencies and other requirements, were submitted to the president on August 21, 1963. The reduction in production of special materials, if adopted, would result in a significant economic impact on the areas immediately surrounding the AEC's material production plants.

5. Civilian Nuclear Power Program (Unclassified)

A year ago, the Commission submitted to the President of the United States a Report on Civilian Nuclear Power. That report deals with the need for nuclear power as an energy resource, the role of the Government, and the future program. We believe that in the long term nuclear fuels represent an energy resource equivalent to many times our fossil fuel reserves.

Six civilian nuclear power projects are scheduled to become operable this year. This will result in a total of 14 operable plants at year's end with a generating capacity in excess of one million kilowatts of electric energy.

The low fuel cost of projected nuclear power plants appears to offer over-all economic advantages when applied to large dual-purpose plants for generation of electricity and desalinization of water. The vital importance of water to man's continued expansion on earth gives this posture special appeal. Studies related to this application currently are under way. The findings will be of great significance in considering alternative proposals involving long distance transport of fresh water to areas with current or projected water shortages.

6. Naval Reactors

New types of naval reactor cores are scheduled to be "placed on test" next year which will last approximately half the life of a ship. This represents doubling the life of cores presently being installed in ships and a six-fold improvement over the first core installed in *Nautilus*. Emphasis is being placed on design and development of still longer life cores, with the ultimate objective of developing cores that will last a ship's lifetime.

A total of 87 nuclear submarines has been authorized through Fiscal Year 1964. Thirty-four are now in operation, including 16 *Polaris* missile-launching submarines and 18 attack type. Two additional submarines are expected to be in operation by the end of calendar year 1963. In the surface ship program, four nuclear powered ships have been authorized for construction. The nuclear powered carrier *Enterprise*, cruiser *Long Beach*, and destroyer leader *Bainbridge* are in operation, and a second destroyer leader, the *Truxtin*, is under construction.

A prototype of a new reactor plant for submarine propulsion is under construction and will be completed in calendar year 1984. This reactor plant uses the natural circulation principle, which eliminates the need for large pumps, and is expected to result in significant noise reduction and plant simplification.

7. Nuclear Carrier (Unclassified)

You are of course aware of the recent decision regarding a second nuclear aircraft carrier. Secretary McNamara has subsequently called me and we propose to discuss plans for nuclear-powered surface warships at an early date.

8. Maritime Nuclear Propulsion (Unclassified)

The N.S. *Savannah*'s nuclear reactor will be brought to criticality on Thursday, November 28, 1963, for use in further training by the Maritime Administration of new crew members. The N.S. *Savannah* has been docked at Galveston, Texas, since labor difficulty forced cancellation of its scheduled 1963 voyages.

Sea trials with the new crew are scheduled for mid-February 1964. It is expected that the *Savannah* and crew will be ready for resumption of foreign and domestic voyages by mid-1964. However, before final approval is given to resume vessel operation, the Commission will review safety of port visits.

The Commission shares the Maritime Administration's views as to the national significance and importance of nuclear propulsion for the U.S. Merchant Fleet. The development of new compact maritime reactors and a land-based prototype of a compact gas-cooled reactor for maritime applications are the next steps in the maritime nuclear propulsion program.

9. Nuclear Rocket Program (Rover) (Unclassified)

You are already familiar with the Commission's activities in developing nuclear power for space (Rover). Accordingly, I will not at this time pro-vide detailed information on this program. However, you may recall the briefing given you and President Kennedy at Los Alamos on the technological difficulties being encountered in developing the experimental reactors for the Rover Program. Results of recent laboratory and "cold flow" reactor tests indicate that the design changes being made to the reactor appear to be effective in eliminating the vibrations encountered in the November 1962 power test, at least under the cold flow conditions.

10. Nuclear Electric Power for Space (SNAP)

Another program with which I am sure you are familiar is the Commission's so-called SNAP program, which has as its major purpose the development of a family of compact systems to provide nuclear auxiliary power for spacecraft and satellites, and for other specialized uses. We are currently assessing the needs and timing of this program and plan to have a report of the findings of this study available for issuance next month.

Incidentally, the Navy navigational satellite (TRANSIT), launched in September and powered solely by SNAP radioisotope power source, continues to operate as designed. The next launch for this navigational system is scheduled for December 5 from the Pacific Missile Range.

11. MURA (Unclassified)

The AEC has under consideration a proposal from the Midwestern Universities Research Association (MURA) for the construction at Madison, Wisconsin, of a 10-billion-volt, $125 million accelerator, to be used in studying the fundamental particles of matter. The proposal involves difficult budgetary, technical, and political questions.

12. International Affairs (Unclassified)

a. U.S.–Soviet Relations

A Soviet technical delegation headed by A.M. Petrosyants, Chairman of the State Committee for the Utilization of Atomic Energy, arrived in the United States on November 16 for a tour of AEC and commercial nuclear energy facilities. The tour, which will terminate in New York on December 3, is a part of the atomic energy exchange under the 1962–63 U.S.–USSR Cultural Exchange Agreement. You may recall that I headed a U.S. delegation on a similar visit to the USSR last May.

b. Multi-National Accelerator

There have been general discussions among scientists throughout the world on the desirability of a multi-national cooperative effort, including the U.S. and the USSR, to construct a much larger particle accelerator (about 1000 billion electron-volts) than any now in existence or planned. The Director General of the International Atomic Energy Agency, Dr. Eklund, recently suggested that such an effort be coordinated under the auspices of the IAEA; the AEC has agreed that this possibility should be given consideration.

13. AEC Regulatory Activities (Unclassified)

There are three applications for licenses for the construction of large nuclear power plants currently before the Commission which are of considerable significance. The applications by the City of Los Angeles for a plant at Malibu Beach and by Pacific Gas & Electric Company at Bodega Head raise the question of safety of power reactors in earthquake areas. Consolidated Edison Company has applied for a construction permit to build a large power reactor in New York City. Under Commission site criteria, practice has been to locate power reactors some distance from large cities. However, substantial improvements have been and are being made in the design of safety features of nuclear reactors and our regulatory procedures provide for a most exhaustive and careful review of safety considerations. Strong local opposition to these projects has developed. White House, Congressional and AEC mail has been considerable. Public hearings will be held under the Administrative Procedure Act before Commission decisions on these license applications.

The purpose of the foregoing report has been to identify some of the programs and problems in the atomic energy program that involve national policy considerations. It by no means represents a complete listing. I should be pleased to have the opportunity to brief you more fully on the foregoing items as well as others which would be of interest to you should your time permit.

Respectfully submitted,
Glenn T. Seaborg

January 14, 1969
The President
The White House
Dear Mr. President:

This is my last biweekly report to you during your term of office. I am deeply grateful for the confidence and support which you have so generously extended during the past five years. In nuclear energy matters, as in so many other governmental areas, your administration has made enormous accomplishments for the benefit of the United States and for the people of the world.

Details of my report follow:

1. In response to the Australian Atomic Energy Commission's informal expression of interest in the possibility of excavating a harbor in Australia by using nuclear explosives, we have suggested to the Australian AEC that the matter should be pursued initially through a technical feasibility study in which the U.S. would participate and that the international aspects of the proposed project should be dealt with through State Department channels.

2. A tour of the Nevada Test Site was conducted on January 7 for newsmen, officials of the state of Nevada, Las Vegas area local officials, members of the Howard Hughes organization, and assistants to the U.S. Senators from Nevada. The tour, which followed a short briefing on the public interest aspects of the recent BENHAM test, included visits to the sites of BENHAM and other previous nuclear tests.

3. The U.S. will supply Japan's second nuclear power plant, the largest ever exported from the U.S. (about 775 megawatts). GE will supply the reactor, and AEC and Allied Chemical Company will provide the enriched uranium fuel. The plant will be financed in part by a $69 million loan by the Export-Import Bank.

4. A summary of commercial nuclear power plant activities in the U.S. in 1968 indicates that orders declined to 61 percent of the net capacity ordered in 1967. The number of reactors ordered fell from 31 in 1967 to 17 in 1968. The average capacity of the reactors ordered climbed to 915 megawatts in 1968, an increase of 100 megawatts over last year. The decline in orders placed is attributed largely to the

historically cyclical nature of power plant construction in the United States.

5. The first nuclear power plant in Argentina will be built by a West German company. The Bonn Government has granted Argentina a $25 million credit to help finance the $70 million plant. The AEC will sell Argentina the heavy water for the reactor.

6. Greece recently stated its intent to acquire a nuclear power plant during the next four years. In 1968 the Greek Government heard presentations on nuclear technology by potential suppliers from the U.S., United Kingdom, Germany, and Italy.

7. The Soviet Government announced that it recently began to operate an experimental fast breeder reactor.

8. The AEC, the Department of Interior, and the El Paso Natural Gas Company have agreed to study a proposed experiment using nuclear explosives to stimulate the release of natural gas in the Pinedale, Wyoming, area. The study, which will define the bounds of the proposed experiment, will be completed in 1970.

9. Four possible geographical areas have been identified for the proposed Gondola Project, an experiment using nuclear explosives for large scale excavations. Preliminary studies indicate that two areas in Carter County, Montana, one in Butte County, South Dakota, and one on the Humboldt-Pershing County line in Nevada may have the geological characteristics required for this experiment. The results of Gondola would be relevant to the study of a possible sea-level canal in Central America being conducted by the Atlantic-Pacific Interoceanic Canal Study Commission.

10. France will participate in Euratom's research program during 1969 but not in the technology program, which the Community is trying to reshape.

11. The European project to build a 300 BeV accelerator now has 85 percent of the required financial support. France, Germany, Italy, Belgium, and Austria plan to participate in this project, which is being developed by the European Organization for Nuclear Research (CERN).

12. The U.S. firm, Western Nuclear, Inc., has received an exclusive concession to explore for uranium over a 445,000 acre tract in Somali.

13. The Republic of Korea expressed its appreciation for our Atoms-In-Action Demonstration Center that recently closed in Seoul by presenting a scroll letter to me from the ROK Ministry of Science and Technology.

Respectfully,
Glenn T. Seaborg
cc: Mr. Bromley Smith,
Executive Secretary
National Security Council

Vice President Hubert Humphrey and I visited the Laboratory of Nuclear Medicine and Radiation Biology at UCLA on April 2, 1965. On April 13, 1965, he chaired his first meeting of the National Aeronautics and Space Council:

At 3:30 p.m. I met with the national space council to discuss international cooperation in space. Present were Vice President Humphrey, Rusk, Webb, Harold Brown, Ed Welch (exec-

Vice President Hubert H. Humphrey and Seaborg visit the Laboratory of Nuclear Medicine and Radiation Biology, UCLA. Los Angeles. April 2, 1965. Courtesy of the Laboratory of Nuclear Medicine and Radiation Biology, University of California–Los Angeles.

utive secretary of the council), Robert C. Seamans (associate administrator of NASA), Arnold Frutkin (NASA), Al Little (my AEC aide), L. M. Hale, and Norman Sherman (Yale University).

The vice president opened the meeting by indicating that this was his first meeting of the space council. He said he hoped his association with the council would be noted for progress in aeronautical and space developments and that he personally intended to be a strong advocate for progress. The vice president also emphasized that many international aspects of space development may result in considerable mutual benefits to the United States and cooperating foreign countries. He sees his role with the council as one specifically involving the following: (1) strengthening national defense; (2) furthering international cooperation; (3) maintaining and increasing cooperation among federal agencies; (4) examining all aspects of the various space and aeronautical programs, including all expenditures in these areas and all recommendations to Congress; (5) examining Soviet programs, including progress and failures; and (6) examining the status of man-in-space programs, space laboratory programs, and nuclear power-in-space programs.

Rusk stated that about 49 foreign ministers had expressed the view the United States must spend $20 billion or $30 billion, or any amount necessary to send a man to the moon, simply as a means of counteracting Russian progress in space activities. Rusk indicated that the State Department's support of progressive space programs is based largely on ideological grounds that, as far as his department is concerned, override scientific and technical considerations. The secretary also said that our space programs provide the opportunity to enlist a large measure of international cooperation by providing foreign nations with the types of programs that allow them to participate. U.S. cooperation in this area has a beneficial effect in another area: It strengthens the development of international law, principally through UN activities.

Rusk then expressed his concern about the posture of the United States in the next decade. He believes that inactivity now would give other nations a distinct advantage. He recalled that our inactivity in the 1940s in planning and developing space programs is now handicapping our efforts. We should not be placed in the same position in the 1970s. He recommended that the council undertake to develop a decade-long planning program.

The vice president asked Rusk about the number of U.S. scientific attaches now stationed in embassies abroad. Rusk answered that there are about 16 at present but that the department hopes to improve this situation.

Webb introduced NASA's international cooperative programs in space activities by referring the council's attention to two huge scrapbooks containing newspaper clippings from around the world. Webb said that the scrapbooks would indicate the extensive attention being given to U.S. cooperative efforts in space programs and would also reflect the extent to which this cooperation has resulted in favorable publicity to the United States. Webb then introduced Seamans, who presented five charts showing (1) NASA's international cooperative flight projects involving satellites, experiments, and sounding rockets; (2) NASA's international cooperative ground-based projects; (3) NASA's international station agreements; (4) NASA's international personnel exchange statistics; and (5) distinguished foreign visitors to NASA in 1964. Seamans also spoke of NASA activities that involve telegraphic transmissions between the United States and the Soviet Union.

The vice president suggested that NASA should concentrate on developing a cooperative program specifically for Latin American countries. He stated that, in his opinion, the most important location for any international space exhibit would be in South America. The vice president expressed the feeling that the space exhibit would provide the Latin Americans an excellent opportunity to demonstrate their warmth and friendship for the United States. I advised the council of the favorable and impressive reception the AEC exhibits have received in Latin America. Webb stated that NASA's foreign exhibits are limited to about 10 space mobiles operating throughout the world. Webb also said that he sees a great opportunity for expanding NASA's international exhibits programs.

Rusk said that the United States is prepared to cooperate with any foreign country to the full extent of that country's willingness; if a country wants to cooperate with us in a lunar landing program, we stand ready to undertake such a joint project.

I stated that AEC activities at present are devoted principally to interagency cooperation. Prospects for important international cooperation exist in nuclear power-in-space activities. I briefly summarized the scope of AEC participation in international bilateral arrangements that provide, among other things, extensive information exchange programs, training programs, and technical and financial assistance in a wide range of nuclear projects.

The vice president, apparently getting an idea from my statement, strongly urged that the council members make announcements about the willingness of the United States to cooperate peacefully with foreign nations in joint space programs. He emphasized that there is a great need for constantly reminding foreign nations and U.S. citizens alike of this willingness to cooperate.

Webb cautioned against oversimplification of the problem, stating that there now exists a genuine feeling of cooperation between the United States and technical personnel of foreign countries in space and other scientific programs. Webb stated that emphasis on international cooperation, as a means of enhancing propaganda objectives, might tend to delimit rather than advance our cooperative efforts. I corroborated Webb's statement concerning the close cooperation that now exists between the United States and various foreign nations in scientific and technical programs but is lacking in the field of space nuclear energy projects. I said, however, that during the last Geneva convention, the Soviets exhibited a small fast neutron space nuclear reactor. (We have since learned that this device does not compare very favorably with our Snap-10A reactor system.)

The vice president asked that the council be informed of the aeronautical activities taking place in the respective agencies and departments that may be considered for international cooperative programs. The vice president further stated that many senators and congressmen are strong supporters of international space programs. Accordingly, information should be made available

The members of the National Aeronautics and Space Council pour over newspaper clippings from around world about NASA's international cooperative programs on space activities. Seated, from left to right: Seaborg, Vice President Humphrey, and Dean Rusk (Secretary of State). Standing, from left to right: Harold Brown (Director of Defense Research and Engineering, Department of Defense), Edward C. Welsh (Executive Secretary, Space Council), and James E. Webb (Administrator, NASA). Washington, D.C. April 13, 1965. Courtesy of the National Aeronautics and Space Council, Washington, D.C.

to them to be used in speeches or inserted in the Congressional Record. (The vice president said that the Congressional Record is a vital source of ideas for small-town newspaper editorial writers.)

The vice president said that the space council would circulate to council members a listing of questions to which he wants answers.

The first meeting of the National Council of Marine Resources and Engineering Development took place on Wednesday, Aug. 17, 1966:

At 10 a.m. Arnold Fritsch and I attended the meeting of the National Council on Marine Resources and Engineering Development, held in Room 474 of the Executive Office Building. Participants included Vice President Humphrey, Edward Wenk (executive secretary of the council), Don Hornig (director, Office of Science and Tech-

nology), Secretary Eugene Fowler (Treasury), Secretary Paul Nitze and Assistant Secretary Robert Frosch (U.S. Navy), a representative of the Department of the Interior, Secretary John Connor (Commerce), Lee Haworth (director, National Science Foundation), Undersecretary George Ball (State Department), Charles Schultze (director, BOB), Bill Carey (BOB), Herman Pollack (State), Herbert Hollomon (Commerce), Dave Robinson (Office of Science and Technology), and George A. Silver (a representative from HEW).

Before the meeting pictures were taken of the vice president with all statutory members of the council.

The vice president then called the meeting to order, noting that although this was the first session of the National Council on Marine Resources and Engineering, it would have to be expedited because of the presidential commemoration of the Alliance for Progress scheduled for 11 a.m. The vice president devoted most of the 45 minutes of the meeting to reading a statement, with numerous interjections on the purpose, plans, and programs of the marine council. He said that the purpose of the council should be to coordinate the federal program in marine sciences, with emphasis on international cooperation. At present, more than 20 agencies are involved in oceanographic work with the coordination being supplied by the Inter-Agency Committee on Oceanography (ICO), a standing committee of the Federal Council on Science and Technology.

The vice president noted that the ICO route has been successful and that its mechanism should be kept intact. However, Congress felt that this coordination should be carried out at the highest levels of government. Congress in a sense wanted to formalize by statute these coordination activities in oceanography. ICO had been established only by executive order. In setting up the mechanism of the council, the vice president had extensive discussions with Hornig and Schultze to ensure that there would be coordination and not competition. The council will undertake the preparation of the Annual Report to Congress on oceanography advising the president on the subject. The council will further have roles in long-range planning, evaluation and interpretation of the citizens' commission report on oceanography, and coordinating an international program of cooperation.

The vice president recommended that this council proceed with a small but high-level staff. He announced that they had obtained Wenk's services as executive secretary of the council. He gave many accolades to Wenk. He also noted that Civil Service Commission Chairman John Macy is gathering a list of individuals for possible nomination to the Citizens Commission on Marine Science, Engineering, and Resources and asked council members to give suggestions to either Macy or Wenk. The vice president noted that the citizens' commission has 18 months to recommend a long-range plan and program. After council consideration, this report should be useful for preparation of the FY 1969 budget.

The vice president also pointed out that the marine council differs from the space council in one important aspect: There is no similar single mission agency, such as NASA, in oceanography. This again emphasizes the need for careful coordination among the many agencies. The vice president also mentioned his many meetings with Congress on this subject and noted Congress' intent that the marine council proceed expeditiously. The main parallel studies in the oceanography area were discussed. The vice president called particular attention to the President's Science Advisory Committee report, which he said will be the subject of the next meeting. In keeping with Congress' intent, the vice president expressed a wish to schedule regular monthly meetings between now and January and encouraged good attendance.

Wenk was then sworn in as executive secretary by the vice president. The vice president then discussed the continuing role of ICO. He noted it is a good base on which to proceed; he wants to see ICO not disband, but rather integrate its activities carefully with those of the marine council. Again, the key words are coordination and communication. The vice president mentioned the Pell Bill now pending in Congress for sea grant colleges. If it passes, the marine council will set the criteria for sea grants and the National

Science Foundation (NSF) will administer them.

The vice president then opened the meeting to discussion, of which there was little, because of the minimum amount of time left. Nitze, Ball, Silver, Hornig, and Schultze made no comments. Fowler noted the ongoing Coast Guard program on the use of buoys for marine research undertaken at the request of ICO. I noted the wide variety of programs AEC has in oceanography and associated myself with the vice president's remarks. Haworth called attention to the fact that NSF has endorsed the Pell Bill and is ready to cooperate with the council and Congress. Connor strongly associated himself with the vice president's remarks, particularly those that called for not dismantling ICO, and noted that the marine council's statute calls for disbanding it at the end of two years and that the ICO should remain as a fallback resource. Frosch (ICO chairman) mentioned that he and ICO members will be delighted to cooperate with the council. The vice president closed the meeting by circulating a list of policy questions that will serve as the basis for further discussion.

The Committee of Principals continued to meet regularly during the Johnson administration. As an example, I quote from my diary for Tuesday, March 14, 1967:

At 4 p.m. I attended a meeting of the principals in Rusk's conference room. Present were: Rusk, McNamara, General Wheeler, Hornig, Richard Helms, Walter Rostow, Adrian Sanford Fisher (deputy director, Arms Control and Disarmament Agency), Webb, Foy D. Kohler (deputy undersecretary of state and former ambassador to the Soviet Union), Nicholas Katzenbach (undersecretary of state), John T. McNaughton (deputy secretary of defense), Leonard H. Marks (director of the U.S. Information Agency [USIA]), Herbert Scoville (ACDA assistant director for science and technology), Allan Labowitz (my assistant for disarmament), Arthur W. Barber (deputy assistant secretary of Defense Department Arms and Trade Control), Raymond C. Garthoff (deputy director of the State Department Bureau of Political Affairs), et al.

Rusk opened the meeting by calling on Fisher, who referred to the State/ACDA paper, "Proposal on Strategic Offensive and Defensive Missile Systems," which had been circulated before the meeting. Fisher said this is a simple plan that will not fully satisfy Soviets' wishes for complete disarmament. Kohler said that this plan has been essentially agreed to by all the deputies and that its potential success is linked to the dependability of the unilateral detection system. Rusk mentioned that in the paper there are a couple of suggestions that the U.S. make concessions, even before the Soviets have said no, and he thought we should at least feel them out before establishing these less desirable positions as the U.S. position.

Rusk then called on Wheeler for his reactions. Wheeler said that the Joint Chiefs of Staff have studied the paper and came to the following five or six conclusions. They want a plan that would maintain U.S. strategic superiority at all times. They would insist on on-site inspection or, if that was impossible, they would limit the plan to those systems that could be verified unilaterally, which would mean only the fixed land-based systems. They do not wish to forego the possibility of the United States deploying an ABM in the future. They believe that the plan should be formalized as a treaty. They think there should be withdrawal provisions to come into effect in the event of hostile actions interfering with our information-collection systems, attempts to deceive, or deterioration of our ability to detect.

Rusk then called on McNamara, who said he does not think the United States is ready to put forward such a plan in written form. He also said that he thinks we should not agree to anything we can't check ourselves. Rusk then asked me for my opinion. I said I agree with McNamara. Rusk found it a frustrating situation, because he felt we should do something along these lines, but we shouldn't do anything that we can't check, and that doesn't leave very much. Rusk called on Webb, who said that he could live with the plan. Rusk called on Helms, who said he is queasy about our ability to carry out the unilateral verification under the plan. Rusk

"Rusk found it a frustrating situation, because he felt we should do something along these lines, but we shouldn't do anything that we can't check, and that doesn't leave very much."

called on Hornig, who said he agrees with McNamara but thinks some steps should be taken, and the question is how to begin.

Rostow said he is especially interested in the nature of the Soviet Talinn system (a defense system using an advanced surface-to-air missile), and he thinks we should somehow learn more about it. McNamara then suggested that our first objective should be to try to establish the nature of the Talinn system. He thinks we might use this as a basis for opening a dialogue with the Soviets, and perhaps we might be successful in 6–12 months. If we aren't successful, this particular plan is no good; and if we are successful, we might be in a position to proceed. He said he thinks we should start with some exchanges on our capabilities and maybe even show each other reconnaissance photographs. Rostow said that when we reach such a point, that might be the proper time to table a paper.

Fisher pointed out that we have been talking a long time and that we must talk in a context of holding down offensive missiles. Kohler agreed with Fisher and expressed doubt that the dialogue approach would be possible with the Soviets. Katzenbach also indicated that he doubts such discussions would be successful, but that there might be a byproduct: We might end up in deciding on a mutually advantageous unilateral showdown. McNamara said he doesn't think we are very close (within a year or two) to an agreement with the Soviets on any of these matters. (He left the meeting about 4:45 p.m.) Rusk made the important point that he regards detection by national means as limited to means within the power of the United States itself. He said he would have another go at this with Kohler and Fisher in order to prepare a cable for Thompson, to give him (Thompson) guidance to begin the dialogue with the Soviets. These instructions would perhaps be along the lines of probing further on three or four points.

Rusk indicated that verification is a key issue, and we are already having difficulty with it. He referred in this regard to the Space Treaty and to "Dr. Seaborg's treaty" (the NPT). I said I bought the idea of this being my treaty provided there were proper safeguards in Article III. Rostow expressed optimism that a dialogue with the Soviets would be successful. Rusk expressed the interesting thought that, as time goes on, we will find ourselves in the situation that we won't be able to make progress in these matters in bits and pieces; we will need something dramatic to take a big step, and this within a few years. Rusk ended the meeting by saying he would circulate to the principals the instructions that will be sent to Ambassador Thompson.

I continued to meet with the AEC's GAC, as exemplified by their 100th meeting on Monday, May 1, 1967:

At 11 a.m. I met with GAC (all present). I brought them up to date on the 200-BeV accelerator, the NPT, the uranium miners' exposure problem, the FY 1968 budget, Plowshare, our decision to cut off private work on the gas centrifuge, and the price set by AEC on uranium separative isotope work.

I met with them again on Wednesday, May 3, 1967:

At 2 p.m. I met with GAC to hear a report on their 100th meeting. I decided to discontinue the present practice of destroying the GAC letters summarizing their meetings; past practice has been to destroy all copies after they have served their initial purpose.

As usual, the GAC chairman sent me a letter summarizing the meeting:

General Advisory Committee to the U.S. Atomic Energy Commission
P.O. BOX 19029
Washington, D.C. 20036

May 3, 1967
Dr. Glenn:

The meeting of the General Advisory Committee just adjourned represented a milestone in the operations of the Committee—it was the 100th formal meeting of the group. The meeting was held at the AEC offices in Washington, D.C., on May 1, 2, and 3, 1967. All

Committee members were in attendance for the entire period except for Mr. William Webster who was unable to attend on May 3rd. The members of the Committee are Manson Benedict, J. C. Bugher, E. L. Goldwasser, J. H. Hall, Stephen Lawroski, N. F. Ramsey, H. G. Vesper, William Webster, and L. R. Hafstad, chairman. D. C. Sewell, Scientific Officer, and A. A. Tomei, Secretary, were also present.

The Committee transmits the following comments on subjects discussed at this particular meeting.

1) Weapons Development Program

The Committee received a concise and informative briefing from Gen. Delmar L. Crowson (director, Division of Military Application) and some members of his staff. While we note that this year's weapons testing program is somewhat smaller than that of last year, it appears to us to be an imaginative balanced program and much should be learned from it. Next year's testing, we observe, continues at the slower pace and, in addition, contains plans for only one advanced development experi-

ment. We recognize that this is brought about by the large increase in Phase 3 weaponization programs, by budget restrictions, and by the need to invest considerable funds in a supplemental test site. However, we are concerned to see the number of advanced development shots shrink so much, for it is only from these experiments that new ideas can be developed and evaluated.

We are pleased to see the increase in the Phase 3 weaponization program. Although this is one of the important objectives of the weapon development program, an equally important objective is the development and testing of new ideas. It is only from this latter source that the AEC can generate a body of tested ideas upon which future stockpiles will be based. We understand that the Commission shares our concern about this important aspect of the development program, and we add our support to the effort to correct this deficiency in Fiscal Year 1969.

Two items of specific interest to the Committee were:

1. The use of the Sandia seismic detection system in Vietnam.
2. The new scaling principle that may allow a reduction in the scaled depth for large underground nuclear explosions.

The GAC was pleased to hear that the readiness program is being reviewed. The Committee was interested to learn of the use of the readiness teams for scientific studies in conjunction with a recent eclipse and believes that such activities not only are of direct value, but also are excellent means for maintaining the morale and the training of the readiness teams.

Periodic reviews of this program are necessary if it is to be kept up to date and effective. The Committee looks forward to receiving a report on the conclusions of the review.

2) Meeting with MLC Chairman

The Committee received an informative report from Dr. M. Carl Walske, the new MLC chairman, and his associates. The discussion was of sufficient interest that we feel consideration might be given to a joint meeting of the MLC and the GAC during the next year.

The Committee fully agrees with the need to increase the reliability and lifetime of the stockpile weapons, par-

Centennial meeting of the General Advisory Committee of the Atomic Energy Commission. Seated around table, left to right: Howard G. Vesper, Duane C. Sewell, Lawrence R. Hafstad, Samuel M. Nabrit, James T. Ramey, Seaborg, Wilfrid E. Johnson, Gerald F. Tape, Robert E. Hollingsworth, Anthony A. Tomei, Edwin L. Goldwasser, Manson Benedict, and William Webster. Standing from left to right: Stephen Lawroski, Jane H. Hall, Norman R. Ramsey, and John C. Bugher. Washington, D.C. May 1, 1967. Courtesy of the U.S. Department of Energy, Germantown, MD.

> "The Committee was impressed by the large numbers of nuclear weapons now under U.S. custody in foreign countries, and was concerned about the problem of maintaining adequate control."

ticularly with regard to Zippers, and would welcome a technical discussion of this program with Los Alamos staff at its next meeting.

The Committee was impressed by the large numbers of nuclear weapons now under U.S. custody in foreign countries, and was concerned about the problem of maintaining adequate control. A continuing and vigorous development program should be maintained to improve the safety, command and control methods.

The Committee was interested in the weapons effects studies, particularly as concerns anti-ballistic missiles, and hopes to learn more about this from Los Alamos staff at its forthcoming meeting.

3) Reactor Safety Programs

The Committee appreciates the invitation it received from Mr. Milton Shaw (director, Division of Reactor Development and Technology) to comment on the Commission's reactor safety program as detailed in the two reports entitled "Water Reactor Safety Program—Summary Description—Jan. 1967," and "Liquid Metal Fast Breeder Reactor—Program Plan—Section 10, Safety, Feb. 1967." We first offer some general comments on the Commission's reactor safety program applicable to both reports, and then make some specific comments on each report.

These reports provide a complete classification of the kinds of accidents that might conceivably occur in these two types of reactors. The reports then outline, in great detail, the analytical and experimental research programs which are being conducted or are to be carried out to provide an understanding of these accidents and to limit their consequences. For each safety research project the reports spell out what is to be done, how it is to be done, and who it is to be done by.

The Committee notes the extensive analysis of reactor safety made in these reports and recognizes the necessity for a considerable degree of centralized direction of the reactor safety program. We wish to emphasize, however, that too rigid central direction will discourage initiative and original thinking on the part of organizations potentially capable of contributing to a broader understanding of reactor safety and, therefore, we urge that the groups working with the Commission in this area be encouraged to follow certain of the

more promising leads of their own toward the general goals set by the Commission.

In conducting research on reactor safety we believe that it would be a mistake to seek complete understanding of the course of events in every possible reactor accident. Instead, reactor safety research should be aimed at providing general information on the phenomena accompanying accidents, so that the main features of an accident can be anticipated without the expectation of being able to predict every detail of it. By their very nature, accidents contain many random elements which make complete predictability unattainable.

In formulating the reactor safety program we recommend that more attention be given to assessing the probability of the different types of accident, so that principal effort and expenditures may be concentrated on those accidents which are most likely to occur. It would seem desirable, for instance, to assemble information on the frequency of failure of conventional high-pressure water or steam piping systems, in order to estimate how likely pressurized or boiling water reactor systems are to experience this type of accident.

The Committee would like to stress the importance of having the fullest possible participation by industry in the reactor safety program. It is disappointing, for example, to note so little industry involvement in the 107 series of projects dealing with development of design methods and standards for water reactors.

In the water reactor safety program, the Committee feels that too little attention is being given to devising means for preventing fuel melting after a break in the primary system, with a disproportionate amount of attention being given to understanding the spread of fission products after fuel melts. If a reliable emergency core cooling system could be devised and proved to be effective, research on the spread of fission products and on containment integrity could be substantially curtailed. In our opinion, development and demonstration of a reliable emergency cooling system is the most important single goal for a research program on water reactor safety.

In the liquid metal fast breeder safety program, we feel that too little emphasis is being given to building a complete reactor system, so that one can begin to learn through the design of a complete system and its subsequent operation what the real safety problems of a new type of reactor are. It is true that knowledge of the behavior of individual components of a fast reactor system is also important but it is only by gaining experience with the complete system that the relative importance of different possible events and phenomena can be assessed. We were disappointed in Section 10, therefore, to find only vague mention of the schedule for construction of complete LMFBR systems and no inter-relation between the schedule proposed for fast reactor safety research and the schedule for reactor construction. In developing a new type of reactor it is just as important to take prudent risks in building one or more complete systems which may experience unexpected failures as it is to conduct research on individual components and specific accident phenomena.

Except for these general criticisms, we believe that the research programs on water reactors and on liquid metal systems should yield information vital to the prevention of reactor accidents and assessment of reactor safety.

4) Uranium Enrichment

Mr. George F. Quinn (assistant general manager for Plans and Production) gave the Committee an informative summary of the studies the Commission has made of alternative plans for operating the gaseous diffusion plants between now and 1980, and described the factors which were taken into account in arriving at the specific terms to be recommended to the Bureau of the Budget for toll enrichment contracts. The Committee concurs with the Commission's choice of criteria leading to about 0.2 weight percent U-235 for tails assay. The choice of $26 per kg for separative work seems prudent as an initial value, but we would hope that this might be reduced if operating experience proves the contingency allowances to be excessive. Moreover, when this value is published we recommend that the Commission state its estimate that the government's costs till 1975 could be recovered by a charge of $22.50 per kg of separative work and should disclose that this was arbitrarily increased by 15% to cover initial uncertainties and business risks in providing toll enrichment services.

5) Gas Centrifuge

Mr. R.E. Hollingsworth (general manager) gave the Committee a summary of the deliberations and conferences the Commission held before deciding to issue its March 21 regulation discontinuing privately-sponsored research on uranium isotope separation by the gas centrifuge. The Committee commends the Commission for the careful consideration which was given alternative courses of action, for the thorough preparations made in advance of the final action, and for making the difficult decision to discontinue private work in this field. We are glad to learn of the Commission's determination to maintain the U.S. lead in gas centrifuge technology through classified, government-sponsored work, and of the high priority to be given programs for checking the reliability and life-time of production centrifuges. In this connection, it is important that the Commission succeed in obtaining funds in the FY 1969 budget for the planned construction of the multimachine pilot plant for reliability testing at Oak Ridge.

6) Exposure of Uranium Miners

In the course of his program review, Dr. Charles L. Dunham (director, Division of Biology and Medicine) gave a brief summary of the studies in Colorado which are giving indications of the degree of occupational hazard to uranium miners. The GAC learned that a study, initiated almost 15 years ago by the Division of Biology and Medicine, the U.S. Public Health Service, and the Colorado State Board of Health, has involved the continuing health review of over 1,000 uranium miners. The incidence of lung cancer appears to be about 10 times higher than normal expectancy in the age group. Further study will be required to identify the contributions to this increase by other factors, such as smoking. In recent years, radon concentrations have been lowered by improved ventilation in the more highly radioactive mines. We hope the AEC will continue its active interest in this serious health problem.

7) R & D Briefing

Dr. S.G. English (assistant general manager for Research and Development) and his division directors reported to us on the status of the research and development program under his supervision. Informative presentations were given on the five subdivisions of this work.

Dr. English presented a general discussion of the potential of the AEC laboratories for contributing to the solution of new national scientific problems. Environmental pollution is a current example of one of the burgeoning problems of our society. He pointed out that at least one small project along this line already has been undertaken by an AEC laboratory.

We believe that applications of the special existing expertise of the laboratories for the solution of new problems of this kind are entirely appropriate and should be pursued. The mobilization of the more general scientific competence of the existing laboratories to find new missions and to solve new problems is much broader than this in its scope. During the coming months, the GAC intends to devote attention to the long-range future of the AEC laboratories.

Dr. Paul McDaniel (director, Division of Research) expressed general satisfaction with progress in the physical research program. On the other hand he properly expressed concern about the funding of existing programs, in the face of increasing commitments of operating funds brought about by authorized new construction.

The Committee is of the opinion that serious thought must be given to the procedures which govern the planning of the physical research program. In particular, new construction should be undertaken only when a clear understanding of operating costs exists among the persons responsible for funding such work.

We strongly support the continuing studies of the Atomic Bomb Casualty Commission. The Committee was interested in the work that is in progress in the development of an artificial heart.

We were interested to learn that because of increasing restrictions imposed by the Food & Drug Administration, the basic work on food preservation is still continuing. We believe that if radiation processing is to become a viable industrial procedure, the time for a transfer of responsibility to the industry is rapidly approaching.

We were disappointed by the postponement of the Cabriolet shot of the Plowshare Program and strongly favor its firing on the new schedule set for the coming autumn.

A report was given on the organization of the Environmental Science Services Administration. The Committee has been concerned with possible conflict between the work of this new agency and AEC programs, and was pleased to note that such does not appear to be the case.

8) Intelligence Briefing

The GAC was given a review by Dr. C. H. Reichardt (intelligence officer) of the nuclear weapons capabilities and potentials of foreign powers. We were impressed by the progress in collating highly significant information and by the political and military implications for the U.S. of the rapid progress toward sophisticated weapons systems being made by Communist China.

9) AEC Personnel Changes—DMA and DBM

The GAC members wish to make note of the fact that both Dr. C. L. Dunham and Gen. D. L. Crowson are leaving their present positions in the immediate future. Their effective and dedicated work in their respective fields and their helpful cooperation with the Committee has been much appreciated. The Committee members wish them continued success and look forward to further interaction with them in their new activities.

10) Weapons Subcommittee Meeting

The Weapons Subcommittee will meet at the Lawrence Radiation Laboratory at Livermore on Friday, July 21, 1967, to discuss weapons matters in general and the Lawrence Laboratory weapons program in particular. The members of the Subcommittee are Dr. Ramsey, chairman, Dr. Hafstad and Mr. Vesper. They expect to be joined by other members of the GAC.

11) Research Subcommittee Meeting

The Research Subcommittee plans to meet in San Francisco on July 20, 1967, to discuss the missions of the AEC major laboratories in relation to the national scientific resources and priorities. This is a prelude to the more extended consideration by the GAC of these matters. The Subcommittee may be joined by other members of the GAC.

12) Reactors Subcommittee

The Reactors Subcommittee expects to attend meetings of some of the Reactor Evaluation Task Forces during the summer in order to become acquainted with the current development status of the principal power reactor types and their place in the Commission's reactor development program.

13) Visit of GAC to the Nevada Test Site

The GAC plans to tour the weapons test areas at NTS on August 1, 1967, and view various AEC and DOD experiments being prepared for detonation.

14) 101st GAC Meeting

The 101st GAC Meeting will be held at the Los Alamos Scientific Laboratory on August 2, 3, and 4, 1967.

Following are some agenda items for the 101st Meeting:

a. Discussion of test results from the DOD weapons effects program with representatives of the Defense Atomic Support Agency.

b. Discussion of the Los Alamos program and tour of selected facilities. We propose that the entire second day be devoted to this topic, and we would like to leave to the Laboratory Director the matter of arranging the schedule for August 3rd.

15) 102nd GAC Meeting

The 102nd GAC Meeting is tentatively scheduled to be held in Washington, D.C. on November 13, 14, and 15, 1967. Among other matters, the GAC would like a report on the up-to-date

plans and status of the weapons test readiness program.

Sincerely yours,
L. R. Hafstad Chairman

Soon thereafter, on May 8, 1967, I visited the Lawrence Livermore Radiation Laboratory for a briefing on the nuclear weapons program.

Johnson was not one to worry much about appearances. In fact, he cultivated his folksy image and was in his element at his ranch. Both LBJ and Lady Bird took obvious pleasure in showing guests their ranch. On one occasion, Johnson presented Schultze (director, BOB) and me (see below) with the challenge of continuing the debate on major items in the AEC budget while he drove us around the ranch in his white Chrysler station wagon. In addition to other guests, passengers in the car included his six-month-old grandson Patrick Nugent and his dog Yuki, who both competed with us for his attention.

LBJ was very proud of his ranch, particularly his collection of wild animals. On tours I saw lots of deer, including the native white-

AEC Chairman Seaborg on a visit to the Lawrence Radiation Laboratory at Livermore. Michael May (Director, LRL–Livermore), John W. Gofman (LRL–Livermore), Seaborg, and Roger E. Batzel (Director for Chemistry, LRL–Livermore). Livermore, CA. May 8, 1967. Courtesy of the Lawrence Livermore National Laboratory, Livermore, CA, and the U.S. Department of Energy.

tailed Texas deer, English deer, Japanese deer, Axis deer, nilgai (a species of Cambodian antelope), and other antelope. On one tour, the president stopped his car, got out and, with considerable difficulty, chased some quail back into an enclosure from which they had escaped. I commented that this was probably "as high-priced help as had ever chased quail"—which seemed to tickle him no end.

As I have indicated earlier, my meetings with President Johnson, at the LBJ ranch in Texas each December, to debate with the BOB director disputed items in the AEC budget, were uniformly successful. Following are some excerpts from my diary for Friday, Dec. 10, 1965:

I flew to Texas on a Jet Star, leaving National Airport at 7:30 a.m., with Connor, Webb, Hornig, Mr. and Mrs. Arthur Krim (friends of President Johnson), Gen. John W. Green (commandant of the Marine Corps), and Gen. H. K. Johnson (chief of staff of the army). We arrived at the LBJ ranch landing strip about 10 a.m. and were met by President and Mrs. Johnson. The president drove some of the group in one car; Mrs. Johnson drove another part of the group, including me and Webb, in another car to the ranch, which is just a few hundred yards from the landing strip. The president then went into a meeting with the Joint Chiefs of Staff, McNamara, and Deputy Secretary Cyrus Vance to discuss budget matters and decisions concerning the Department of Defense.

Later in the afternoon, after the president had seen Webb and before he had seen Connor and Freeman, I met with the president, along with Schultze, Hornig, and Joe Califano (presidential special assistant) to discuss AEC budget issues.

We then discussed the four "A" items at issue in the FY 67 budget. The first item was the BOB suggestion that one of the 10 presently operating plutonium production reactors be closed down. I resisted this strongly, emphasizing that a study is under way for the whole requirements picture, that this will be due in May, and that it would

be unfortunate to prejudge the eventual outcome. I pointed out the nonweapons needs for reactor products, such as radioisotopes, and emphasized the need for isotopes such as Pu-238 to power an artificial heart. This seemed to impress the president, despite the fact (as pointed out by Hornig) that the cost of such a power source would be about $10,000 for each artificial heart. However, the president replied that such a cost would not be prohibitive and that the project might be worthwhile even if only one such artificial heart were available. I also mentioned the need for continued high-neutron flux reactor operation and our suggestion that we should stockpile tritium. I noted the political problems with JCAE and other congressional committees attendant with shutting down another reactor. The president ruled in my favor and said that we shouldn't shut down a reactor.

The second "A" item issue involved the BOB recommendation to cut $4 million from the weapons research and development budget, which would mean a substantial reduction in weapons laboratory personnel. I argued that this would be contrary to the safeguards endorsed by the president and would have an adverse effect on the morale of the laboratory people and on the ability of the laboratories to recruit. The president ruled in my favor that this cut should not be made.

We then discussed the item of $4 million for A&E (Architect and Engineer, i.e., design funds) on the 200-BeV accelerator. I pointed out the need to have such a line item in the budget in order to be consistent with our actions to date, to assure the scientific community that we are serious, and to keep the design scientists on the job. Schultze pointed out the high cost of building the accelerator and the high operating cost. He said the construction cost would be $308 million, and by the time you add equipment needed at the accelerator site and at the universities for the users groups, the total cost would be about $400 million. He also indicated that the operating cost would be about $60 million per year at the site, and by the time you add the operating costs for the users at the universities, this would amount to $90 million to $100 million per year. He said he wanted to emphasize the tremendous cost entailed by any commitment to build.

I described the status of the process for selecting the site for the 200-BeV accelerator and mentioned some of the beneficial side effects of this site selection process. For example, people in various parts of the country would be awakened to the value of educational institutions, scientific research, and cultural activities. I said that, as a result of the competition for the 200-BeV accelerator, some universities had obtained from their state legislatures money earmarked for research for the first time. I also mentioned the time scale for the selection. The president ruled in favor of including the A&E money but with the understanding that there would be no commitment at this time for building this very expensive facility.

The final "A" item was the $2.9 million for research and development and the $3 million for A&E for the Los Alamos Meson Physics Facility (LAMPF). Schultze pointed out that we do have in the "A" budget a $45 million project to convert the AGS (Alternating Gradient Synchrotron) to higher intensities. He said that this AGS conversion, plus the 200-BeV accelerator, is a very expensive program, and he doubted that we could add LAMPF to it. I emphasized the great scientific need for this project. I said it is a matter of making up our mind one way or the other; that is, either kill the project or add the requested funds to the budget, because the work has gone on in a state of indecision so long that it would not be possible to retain the scientists for another year without a decision. I also pointed out the strong support from Senator Clinton Anderson and the strong (essentially unanimous) support of JCAE. The president said he would like to think about this further. (Just before I left the ranch, at about 5 p.m., I learned that the president had decided to include this in the FY 67 budget.)

During these discussions, the president sat in a reclining chair in a very relaxed fashion and at times closed his eyes. The room in which we met was rather dimly lit, so I couldn't use the extensive backup material I had brought along; however, this didn't turn out to be necessary because I had studied the material thoroughly the night before as well as earlier in the week and on the plane coming down to the ranch. At one point I mentioned that I thought McNamara is making a mistake by not

including nuclear power for an aircraft carrier and a frigate in his budget.

Connor, Freeman, Webb, Hornig, Califano, and I then went by helicopter to Austin, landing at the Civic Center, and on to the Driskill Hotel, where we participated in a press conference conducted by Moyers. I spoke about the president's interest in the peaceful uses of atomic energy, particularly desalting, advanced reactors and breeders, and the use of radioisotopes in medicine, such as to power an artificial heart. One question put to me was about the time scale for choosing a site for the 200-BeV accelerator. I replied that the National Academy of Sciences (NAS) committee will give us their recommendations in a few weeks, and the commission will make the choice of a site no sooner that a few months after that.

On this occasion I was able to rescue the 200-BeV accelerator and the Los Alamos Meson Facility, and similar success in December meetings in subsequent years made possible the construction of these important accelerators. Many other AEC projects, nearly all that I presented for review to President Johnson at these yearly December meetings, were similarly approved. He rescued many smaller projects from the understandable money-saving efforts of BOB and supported the development of civilian nuclear power. In the overall budgetary process, however, he was very conscious of the need for prudence and had no qualms about eliminating some projects from the scientific arena, such as the Midwest Universities Research Association (MURA) accelerator, nor about curtailing unneeded production facilities. In the latter category, when he assumed office in 1963, he was completely unafraid to cut back drastically on the production of fissionable material despite the anticipated vigorous opposition from powerful elements in Congress.

Mary I. Bunting (president of Radcliffe College) was sworn in as an AEC commissioner on June 29,

"On this occasion I was able to rescue the 200-BeV accelerator and the Los Alamos Meson Facility, and similar success in December meetings in subsequent years made possible the construction of these important accelerators."

President Lyndon B. Johnson's support for Civilian Nuclear Power is exemplified on this occasion when he presents the pen he uses to sign the Private Ownership Bill (for nuclear fuel) to Seaborg during a ceremony in the White House Cabinet Room. Aug. 26, 1964. Left to right: Representative Chet Holifield (Member, Joint Committee on Atomic Energy), Representative Melvin Price (Member, JCAE), Representative Jack Westland (Member, JCAE), Johnson, Gerald F. Tape (AEC Commissioner), James T. Ramey (AEC Commissioner), Seaborg, Mary L. Bunting (AEC Commissioner), Ernest R. Tremmel (Director, AEC Division of Industrial Participation). Courtesy of the White House.

1964, replacing Robert E. Wilson.

As an example of our interactions, I recall a meeting I had with the president in the Oval Office, lasting nearly all afternoon (2:30 to 6 p.m. on Thursday, June 17, 1965). I had the impression that here was a man who suffered periods of lonesomeness in his position of exalted isolation and thus longed to spend some time in conversation with a friend. The flavor of this meeting is captured in the following extract from my diary:

In connection with the following topics I discussed with the president, he had the pertinent passages from my letter of June 16, 1965, which I had furnished as a basis for this appointment, before him for ready reference. I first took up the matter of the IAEA safeguards system and its contributions to nonproliferation. I pointed out that

much progress has been made, and I think he should refer to this in his June 26 speech in San Francisco to the UN on the occasion of its 20th anniversary. He said that he would be glad to do this and asked that I furnish his office with a possible draft of material to be used, and later he mentioned Dick Goodwin (presidential assistant) in this connection. (Later I also told presidential assistants Valenti, Moyers, and Cater that I was going to furnish this material, and the material referred to below, for the president's San Francisco speech.)

I then made a short progress report on the status of civilian nuclear power in the United States. The president seemed to be quite impressed by this and suggested that I arrange a meeting between him and a group of utilities interested in nuclear power. He wanted to speak to them about the importance of starting construction projects as soon as possible to help alleviate the economic situation.

The president raised the question of the U.S.–USSR exchanges in atomic energy, referring to the discussion of this in the memorandum. I mentioned this was a positive contribution to improving relations with the Soviet Union because science could act as a bridge and a common language between the two countries. He immediately took this up as an idea for his San Francisco UN speech and asked that I furnish some text on the theme of science acting as a bridge and a common language between the East and West. He suggested mentioning that AEC and his science adviser would be asked to help work out, through the UN, a means of implementing the exchange of 1000 scientists a year.

This also brought to the president's mind other possibilities for the San Francisco speech. He asked me to try to find some dramatic suggestions for the speech—two or three real proposals, if possible. The president suggested that I develop for him some text about a worldwide desalting program, perhaps suggesting that the United States would be willing to spend $100 million for this purpose if this amount were matched by the other countries involved. He said the speech might suggest that, with cooperation from the UN, a real breakthrough in the saline water problem might be achieved, and we might express the hope of bringing the price of water down to some designated cost. He asked me to see Hornig to help work out these suggestions.

Referring to the idea of exchanges between East and West, the president said that the San Francisco speech might also suggest the expenditure of some $25 million to $50 million to bring different types of people from Africa, East Europe, and other areas to the United States. He asked me to contact Harry McPherson (special counsel to the president) and Senator William Fulbright (Arkansas) to see whether I could work out some more ideas on such an exchange program.

I invited the president to visit either the Livermore Laboratory or the Lawrence Radiation Laboratory in Berkeley. After a little discussion with Moyers, who was present during this part of the appointment, it was decided that a visit to Livermore might not be consistent with his mission to the UN. The president did indicate that he would be glad to visit the Lawrence Radiation Laboratory, provided it was put on the basis of peaceful uses. I told him that this would be possible and that I would arrange for such a visit and for some text that he might use in a talk to scientists at the laboratory.

The president had his lunch—a bowl of soup—during this conference because he was scheduled to greet astronauts James A. McDivitt and Edward H. White and was running late. He also accepted and placed numerous phone calls during the conference. At the end of the appointment, the president asked me to stay on for the press conference later in the afternoon. In the intervening period I talked to Valenti and others at the White House about the president's visit to the Lawrence Radiation Laboratory and about his request for help with his San Francisco speech. I participated in the press conference, sitting behind the president's desk with him and the vice president. The president read almost verbatim from the material furnished with my letter of June 16, 1965, the sections on the IAEA safeguards system, nonproliferation, the progress report on nuclear power, and the U.S.–USSR exchanges in atomic energy. The press conference ran until almost 6 p.m.

The exalted isolation of the presidency. Seaborg and President Johnson converse about national policies and friendships in the White House Oval Office. Washington, D.C. June 17, 1965. Courtesy of the Lyndon Baines Johnson Library, Austin, TX. Photo by Y. R. Okamoto.

Such brainstorming sessions were not uncommon with LBJ. Fortified with root beer, he jumped rapidly from one idea to the next, never limiting his notion of what one had to contribute to a particular area of expertise and always keenly attentive to the potential public relations impact.

As already indicated in Chapter 2, I accompanied the president on *Air Force One* to San Francisco on Friday, June 25, 1965:

Air Force One left Kansas City at 8:15 a.m. and arrived at 10:30 a.m. (This was the same plane and the same pilot who flew us to Moscow in 1963.) Aboard, the president and I discussed a number of matters, including the situation in Vietnam. The president said he is under pressure from the Joint Chiefs of Staff to bomb the Russian-supplied jet planes (RL-28s) and the surface-to-air missile bases in North Vietnam. He said that he is using his own less expert judgment to overrule them and is not bombing these objectives, and I urged him to continue that policy. He feels very strongly about the aggressive actions from the North and by the Viet Cong and said he intended to resist them and to protect Americans in South Vietnam as well as continue to bomb inanimate objects, such as bridges, in North Vietnam.

We flew over the flood area of the Arkansas River in Kansas, including such towns as Hutchinson, Great Bend, and Dodge City. Senator Frank Carlson (Kansas) and Senator Frank Church (Idaho) were aboard and participated in viewing the flood area and also in the conversations regarding Vietnam. Photographs were taken during the trip by White House photographer Yoichi Okamoto. Movies were made by USIA people—Ray Long (cameraman) and John Peckham (director). Jake Jacobson (legislative counsel), Horace Busby (presidential assistant), and secretaries Marie Fehmer and Vicki McCammon were also aboard.

After having read the president's speech, Busby told me in confidence that the reason my IAEA, scientist common ground between nations, and desalting proposal were missing from the speech is that the president was annoyed by Senator Robert Kennedy's speech on nonproliferation yesterday. Busby thinks this material, including Gilpatric's report, was leaked to Kennedy through State Department people and Bill Foster, because he used some of the same language as the State Department and his draft speech material. (Roswell Gilpatric, deputy secretary of defense during the Kennedy administration, served as chairman of a presidential panel to advise on the nuclear weapons proliferation problem.) The president then asked me how I liked it, and I was rather noncommittal, saying I liked the section where he referred to the future through science. Somewhat annoyed, he asked, "Is that the only part you liked?"

Upon arriving in San Francisco I rode in a motorcade with Church and Carlson and Representative Phillip Burton (San Francisco) to the Opera House where we met Governor Pat Brown, Mayor John Francis Shelley, and others. The president gave his short speech, and the audience seemed to receive it with some disappointment because they had been led to expect more. The president apparently was also annoyed by this unjustified expectation, which is another reason he changed to a less substantive speech. Also, he hadn't been able to clear some sections, such as that on the UN financial crisis, with members of Congress.

The president's changed plans also included cancellation of the planned visit to the Lawrence Radiation Laboratory in Berkeley. However, I visited Berkeley the next day:

Saturday, June 26, 1965

I attended the convocation, "Honoring the UN on the Occasion of the Twentieth Anniversary," at the Greek Theatre, where UN officials U. Thant (secretary general), Carlos P. Romulo (former member of the Security Council), and Alex Quaison-Sackey (president of the General Assembly) spoke. I escorted Adlai Stevenson in the Academic Procession. Clark Kerr presided and introduced Stevenson, Ralph Bunche, and me. (Bunche received the 1950 Nobel Prize for his work on the UN Palestine Commission.) I sat on the stage between Bunche and Stevenson. Stevenson unburdened his soul to me in

On stage at the Greek Theatre, Adlai E. Stevenson (second from left), Seaborg, and Ralph Bunche attend the University of California's convocation honoring the 20th anniversary of the United Nations. The hole in Stevenson's shoe received national press coverage. Berkeley, CA. June 26, 1965. Courtesy of Wide World Photos, Inc., New York.

a long recitation of all the trouble he is having with President Johnson, who, he says, is not very knowledgeable or sympathetic about international affairs.

The hole in Stevenson's shoe sole received national attention. Even though Johnson did not worry much about how he appeared to others, he was very concerned with the image projected by his administration. A meeting he held at the White House on Wednesday, July 20, 1966, with a number of subcabinet-level employees, makes this concern with image vivid:

At 11:45 a.m. Commissioners Gerald F. Tape, James T. Ramey, and I attended a meeting with President Johnson at the White House in the Cabinet Room. Present were William Foster and colleagues (ACDA), William Gaud and colleagues (Agency for International Development), Leonard Marks and colleagues (USIA), Harold Linder and colleagues (Export-Import Bank), Robert Seamans, and others. The president opened the meeting by saying that this was one of a series of meetings he is having with subcabinet-level presidential appointees to chat with them about a number of important items.

The president reviewed the accomplishments of his administration, referring especially to the progress made in education, health, and poverty. He described the numerous reverses that communism has suffered throughout the world within the last year or two and mentioned the antagonism between the Soviets and the Chinese. He described the situation in Vietnam, pointing out that the North Vietnamese were showing some signs of losing strength in their prolonged struggle. He spoke of the need to strengthen NATO as well as all our agreements with our allies. He said he asked Rusk yesterday whether there is some more acceptable language that could be used in a possible NPT. He mentioned the progress that has been made in establishing the Asian Development Bank.

He reiterated that the big problem facing this country is Vietnam. The key items are food, health, and education, and he wants to spread these to the 120 nations of the world to the extent that our resources permit. He also mentioned that he wants to continue to foster our exchanges with the Communist world.

The president said that, like President Eisenhower, President Kennedy, and other presidents, he is suffering from "mid-termitis"—a sag in popular-

ity; he suggested that this is somewhat normal and will be overcome. The president emphasized that he wants every agency and department to assume responsibility for their legislative programs and to give attention to their information programs. He said, "Get the senators and congressmen off my neck; and, if you do your job, I'll never hear from you." But he did indicate that if we got bogged down we should come see him.

On the subject of information, he said there must be something of value coming out of our agencies that he might use in his press conferences. He asked us to have adequate contact with the news media and to make the truth available so that they can make judgments based on facts. He said he would like to have something every week and that we should get in touch with Moyers.

The president admonished us to be frugal in our expenditures; he said we should handle the government dollar as if it were our own. He said, "I can change the light bill in the White House from $6000 to $3000 per month," as an example of the impact of small savings. He asked us to check our own telephones, for example, to be sure we don't have more than we actually need. He admonished us to be efficient and to get a dollar's worth of value for every dollar. He asked us to bear in mind that we are acting for the president of the United States and that we are judged as his personal representatives.

In conclusion, the president referred to the importance of the three "I's"—ideas, imagination, and innovation—and asked us to bear these in mind.

The stress on frugality was typical of President Johnson. His well-known habit of turning off the lights in the Oval Office, as an

Ceremony at the White House East Room to swear-in AEC commissioners Samuel M. Nabrit and Wilfred E. Johnson, and to commemorate the 20th anniversary of the signing the Atomic Energy Act. Left to right: Commissioners W. E. Johnson, Nabrit, and Seaborg, President Lyndon B. Johnson, and Commissioners James T. Ramey, and Gerald F. Tape. Washington, D.C. Aug. 1, 1966. Courtesy of the U.S. Department of Energy, Germantown, MD.

A Chemist in the White House: From the Manhattan Project to the End of the Cold War

economy measure, approached the point of idiosyncrasy. I remember when we received instructions to give up our chauffeur-driven Cadillacs and ride in smaller, less ostentatious cars like Chevys. In the midst of an important meeting, LBJ interrupted the conversation to glare at me and demand to know why I was still riding in a Cadillac. I hastened to assure him that he had been misinformed; I had traded in my Cadillac as soon as I received his instructions and was already modestly riding in a more compact car.

A heartwarming event took place on Aug. 1, 1966:

At 4:30 p.m. Helen, Pete, Lynne, Steve, Dave, Eric, Dianne, and I attended the ceremony commemorating the 20th anniversary of the Atomic Energy Act, which was signed in 1946, and also the swearing-in ceremony of AEC commissioners Samuel Nabrit (president, Texas Southern University), and Wilfrid Johnson (general manager, Hanford Atomic Works). The ceremony was held in the East Room of the White House. Helen and the kids met President Johnson in the receiving line following the ceremony. The president said, "Hi, Honey Bun!" to Dianne, which thrilled her very much. The commissioners had their picture taken with the president. The White House photographer took individual pictures of the kids and Helen with the president. A large crowd of atomic energy, industry, JCAE, and academic people was present at the ceremony.

A picture of our family had been taken earlier (April 1964) to use in the Blair House gallery of government employees.

A notable event was President Johnson's visit to the NRTS in Idaho. Quoting from my diary of Aug. 26, 1966:

At 9 a.m. I left Andrews Air Force Base on Air Force One (the president's plane) for Idaho Falls (via Pocatello) with President and Mrs. Johnson; we arrived at Pocatello at 9:45 a.m. I had

Helen and Dianne "Honey Bun" Seaborg (age 6) greet President Lyndon B. Johnson in the White House East Room. Washington, D.C. Aug. 1, 1966. Courtesy of the Lyndon Baines Johnson Library.

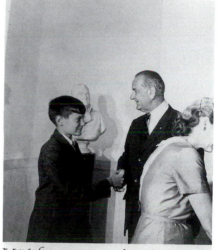

Eric Seaborg (age 11) greets President Johnson in the White House East Room. Washington, D.C. Aug. 1, 1966. Courtesy of the Lyndon Baines Johnson Library.

Stephen Seaborg (age 15) greets President Johnson in the White House East Room. Washington, D.C. Aug. 1, 1966. Courtesy of the Lyndon Baines Johnson Library.

David Seaborg (age 17) greets President Johnson in the White House East Room. Washington, D.C. Aug. 1, 1966. Courtesy of the Lyndon Baines Johnson Library.

Lynne Seaborg (age 18) greets President Johnson in the White House East Room. David Seaborg (center) with back to camera. Washington, D.C. Aug. 1, 1966. Courtesy of the Lyndon Baines Johnson Library.

Peter Seaborg (age 20) greets, President Johnson in the White House East Room. Washington, D.C. Aug. 1, 1966. Courtesy of the Lyndon Baines Johnson Library.

an opportunity to talk with the president on the way and to brief him on AEC's NRTS and the Idaho Operations Office.

Upon arrival at Pocatello, the president was introduced by Congressman Compton I. White, Jr., of the Idaho congressional delegation. The president then introduced a number of people on the platform, including me, and spoke generally on the U.S. domestic scene to a large crowd at the airport.

Among those present at the airport were AEC commissioners Ramey, Tape, and Nabrit; Fritsch; Rosel Hyde (Federal Communications Commission); John A. Carver (undersecretary, Department of the Interior); Representative Chet Holifield; Representative Wayne Aspinall; John Conway (JCAE); Idaho Senators Frank Church and Len B. Jordan; Idaho Representatives George V. Hansen, Compton I. White, Jr., and Ralph Harding; Idaho Governor Robert E. Smylie; Marvin Watson; William L. Ginkel (manager, Idaho AEC Operations Office); Walter H. Zinn (vice president, Combustion Engineering); Stephen Lawroski (chemical engineer, Argonne National Laboratory); A.V. Crewe (director, Argonne National Laboratory); newspaperman Robb Brady (of the Post Register); and Meyer Novick (Argonne National Laboratory).

After the brief ceremony, the president, Church, Holifield, Watson (special assistant to the president), and I boarded a helicopter [to travel] to the Central Facilities area of the NRTS.

On the way, the president berated Watson for omitting the names of some of the people that he (the president) should have introduced at Idaho Falls. During the ride I had a chance to brief the president further on the NRTS.

At 10:40 a.m. we arrived at the Central Facilities and were met by William L. Ginkel (manager, Idaho Operations Office). At 10:45 a.m. we attended the ceremony dedicating the Experimental Breeder Reactor No. 1 (EBR 1) of the NRTS as a national historic landmark. Ramey, Tape, and Nabrit, representatives from JCAE, the Idaho congressional delegation, Idaho state officials, and members of the Idaho legislature were included on the platform. Ginkel made the introductory remarks, and Smylie introduced the president. The president made an historic speech,

The family Seaborg photo for the Blair House Gallery. Washington, D.C. April 25, 1964. Photo taken at the Seaborg home, 3825 Harrison Street, Washington, D.C. Left to right: Lynne, Dianne, Peter, David, Stephen (in front of David), Eric, Glenn, and Helen. Photo by Fabian Bachrach, New York.

using material prepared by AEC on the growth and prospects of civilian nuclear power, and going into consideration of the nuclear threat and ways of alleviating it through U.S.–USSR actions, such as an NPT. The first part of his speech was marred by a faulty public address system (set up by White House staff, not AEC). The television and radio reception was not impaired.

Following the president's speech, Ginkel introduced the people on the platform. He then introduced John A. Carver (undersecretary, Department of the Interior), who made the presentation of a plaque designating the EBR 1 as a

registered national historic landmark. I responded, first making reference to the difficulties with the public address system, alluding to the other technological accomplishments of the NRTS and thus the irony of such an occurrence, then praising Argonne National Laboratory for the advancement of peace and the peaceful uses of atomic energy. (I had some trouble with a slight lip bleeding at the end of my remarks, but I doubt it was noticeable.)

After the ceremony President and Mrs. Johnson and Lawroski in one car, with the Idaho congressional delegation and me in the following cars, drove to *the site of the EBR 1. President Johnson and I used screwdrivers to finish putting the plaque on the wall of the reactor. President and Mrs. Johnson toured the area, and the original EBR 1 group met the president and had their pictures taken with him. The president and his party then drove back to the Central Facilities and took a helicopter to Idaho Falls, where the entire party, including Holifield, proceeded to Denver and Tulsa. The president went on to Texas.*

As indicated earlier (in Chapter 4), Vice President Johnson had visited AEC's Los Alamos Laboratory and Sandia Laboratory in New Mexico with President Kennedy on Dec. 7, 1962.

When the occasion called for it, Johnson withheld nothing in support of accomplishing his objective. At the Dec. 27, 1966, meeting at the LBJ ranch, I told him about a trip I was making to atomic energy installations in India and Pakistan. I mentioned that the State Department had been urging me to visit Bangkok, but I felt that I had to be back in Washington by Jan. 16. The following excerpt from my diary describes how LBJ changed my plans:

The president asked why I felt the need to be back then, and I said that, among other things, I feel I should be

At the dedication ceremony of Experimental Breeder Reactor No. 1 (EBR-1) as a National Historical Landmark, President Lyndon B. Johnson and Seaborg are "putting on the screws" to the plaque marking the occasion. National Reactor Testing Station, Idaho Falls, ID. Aug. 26, 1966. Courtesy of the Argonne National Laboratory, Argonne, IL.

here when Congress is back. He said he does not think this is necessary and that, because I make such a good envoy, it is much more important that I take this trip and stay as long as necessary, even if it takes all of January. The president then turned to Califano and told him to arrange to place a converted 707 (i.e., a KC 135) at my disposal so that I could take some of my staff and colleagues. This would be McNamara's plane. The president suggested that I be given McNamara's pilot or, if he is not available, the president's own pilot. He suggested that I give speeches on the peaceful uses of atomic energy, desalting, and so forth, and that I speak at universities, if possible. He suggested the inclusion of a press relations man, a steward, our wives, and a doctor, if that would be helpful.

As a result, we had a very successful trip around the world—visiting Australia, Thailand, India, and Pakistan—with my wife, Helen, and Tape and his wife Jo as members of our group.

Also at the Dec. 27 meeting:

In the course of the hour-long discussion, the president made a number of interesting observations. He was repeatedly very laudatory of me, saying that he wanted me to know how much he appreciates the way I am running AEC so as not to give him problems with either the public or with Congress. He feels it is a very efficient job and that I am one of the best and more effective men working for him and a source of pride and joy to him. In connection with my handling of the site search for the 200-BeV accelerator, he said it was an example of how a professor, who is not supposed to understand such things, handled the politics of the situation in a very knowledgeable and effective manner.

A noteworthy occasion involving President Johnson was the 25th anniversary observance of the first nuclear chain reaction at the University of Chicago on Dec. 2, 1967:

Following the warm-up session, I opened the program carried on satellite television between Italy and the United

Prime Minister Indira Gandhi and Seaborg at the dedication of Bhabha Atomic Research Center in Trombay, India. January 12, 1967. Left to right: Vikram Sarabhai (Chairman, India Atomic Energy Commission), Homi Sethna (Director, Bhabha Center), Gandhi, and Seaborg. Courtesy of the U.S. Information Service, Bombay, India.

States by sending greetings to some of the group at the capitol in Rome (President Guiseppi Saragat, Minister of Industry and Commerce Giulio Andreotti, and Professor Carlo Salvetti). I made a few remarks about the significance of the Fermi experiment. I then introduced to the television audience Mrs. Laura Fermi, the Fermi team as a group, Walter Zinn, Herbert Anderson,

Seaborg at the lectern commemorating the 25th anniversary of the first sustained nuclear chain reaction at the University of Chicago. Front row, left to right: Betty (Mrs. Arthur H.) Compton, Laura (Mrs. Enrico) Fermi, Egidio Ortona (Italian Ambassador to the U.S.), Richard Daley (Mayor of Chicago), and George Beadle (President, University of Chicago). Middle row, left to right: Alice and Paul Wiener (Fermi grandchildren), General Leslie R. Groves, Gerald F. Tape, Emilio Segrè, and Seaborg (at lectern). Back row, left to right: Herbert Anderson, Robert R. Wilson, Robert Duffield, and Kenneth Dunbar. Chicago. Dec. 2, 1967. Courtesy of the United States Department of Energy, Germantown, MD. Photo by Ed Westcott.

Mrs. Arthur Compton, Emilio Segré, Gen. Leslie R. Groves, Italian Ambassador Ortona, Chicago Mayor Richard Joseph Daley, and University of Chicago President George Beadle. I then introduced President Johnson, in Washington, DC, who spoke for 10–12 minutes. He discussed the significance of the Fermi experiment and the peaceful uses of nuclear energy, suggested that the 200-BeV accelerator be dedicated to Enrico Fermi, and said we would place our U.S. peaceful nuclear program under IAEA safeguards as soon as an NPT was in effect. I then switched the program to Rome, where President Saragat spoke to an auditorium full of people. I brought the program to a close by thanking President Saragat, President Johnson, and the Fermi team. Although the program was not broadcast domestically, the major television networks taped it for later broadcast.

I had convinced the president to make the offer to place our U.S. peaceful nuclear facilities program (i.e., our civilian nuclear power reactors) under AEC safeguards. This was the step that broke the logjam in obtaining the NPT the following year (1968).

My last trip to the LBJ ranch to discuss budget items took place on Wednesday, Dec. 27, 1967. Here I emphasize some human interest sidelights:

I flew in President Johnson's Jet Star from Andrews Air Force Base to the president's Perdenales Ranch in Texas. On board were Schultze, Mrs. and Mrs. Califano, Freeman, and Robert Wood (associate administrator, Department of Housing and Urban Development). We arrived at the landing strip (the so-called Johnson City Airport) at the president's ranch on the Perdenales River, about 11:20 a.m. President Johnson had started to come out to meet us, but we actually rode to the ranch house in two other cars because his car had arrived only about halfway to the plane by the time we started, so the president turned around and met us at the ranch house.

The president then invited all of us to ride in his white Chrysler station wagon. The president was at the wheel and had young Patrick Lyndon Nugent (age six months) on his lap as he drove. Also in the car was the president's dog, Yuki. We were followed by two station wagons with Secret Service people in them. The president drove us around the part of his ranch that is at the opposite end from the airfield, where we saw a number of deer. We then returned to the ranch house, and the president conferred with Schultze in the living room about the broad outline of the budget for FY 1969, while the rest of us waited in the study just off the living room. Mrs. Johnson joined us there, and the conversation concerned their arrival at the ranch the previous night and getting settled for their two-week stay.

When the president finished with Schultze, we all went to lunch in the dining room, about 12:30 p.m. The president sat at the end of the table in front of the window, with Mrs. Califano at his right and Schultze at his left, while Mrs. Johnson sat at the opposite end of the table, with Freeman at her right and me at her left. Also present were Califano, Wood, and Jim Jones (presidential assistant). Hot tamales and enchiladas were served, and they were very good. The conversation at lunch concerned mainly the president's recent four-and-one-half-day trip around the world. He described the circumstances concerning the drowning of Prime Minister Harold Holt of Australia; apparently, Holt had been swimming at high tide, in the company of a younger government official, who turned around when the going got tough and just barely made it back to shore. It isn't clear what happened to Holt, but at one point a huge 12-foot wave engulfed him. He may have been seized by a back cramp or possibly was overcome by the ferocity of the sea.

The president also described his visit to Rome and his discussion with Pope Paul VI. The circumstances of entry into Rome and the visit with the Pope presented some problems with respect to the possibility of picketing, but these were circumvented by maintaining secrecy as to the time of the visit and the very question of whether the visit would take place at all. He discussed with the Pope stopping the bombing in Vietnam and maintained that more lives would be lost in the long run by stopping bombing because the North Vietnamese would take advantage of this and built up concentrations of troops for immediate entry into South Vietnam. He said that the roads were lined with trucks and conveyances for miles, ready to proceed south at any signal for stopping of the bombing. The president did indicate that the Pope differed with him on this and felt that a halt in bombing would save lives.

I reminded the president that his appearance on satellite television in connection with the Dec. 2 observance of the 25th anniversary of the Fermi chain reaction in Chicago was seen on national television in Italy and, from our indications, was well received. I told him that the scientists at the gathering in Chicago appreciated very much his participation in this program. The president gave me an autographed copy of his recent book, *No Retreat from Tomorrow*, which is a compilation of his 1967 messages to the 90th Congress.

After lunch, the president started his conferences with us on the FY 1969 budget. These meetings were held in a yellow room just behind the room where we met last year, which is just off the living room. I waited in the living room while the president met first with Freeman; toward the end of this meeting, I entered the room to begin my session. I had the impression that Freeman failed to convince the president on a number of his appeal items.

During our meeting, the president read the section in the working paper on the 200-BeV proton accelerator, which contains the BOB recommendation of $25 million for the start of construction. The president made no comment on this, and Schultze told me later that he regards such silence as an indication of the president's concurrence.

President Johnson continued President Kennedy's custom of presenting AEC's Fermi Award to the recipients. After some hesitation attributable to the anticipated strenuous disapproval by some members of Congress, Kennedy had approved presentation of the $50,000 award to J. Robert Oppenheimer. Johnson unhesitatingly agreed to go ahead with the presentation on Dec. 2, 1963:

The presentation of the seventh Fermi Award to Oppenheimer took place at the White House at 5 p.m. I presented Oppenheimer to President Johnson, who made a very moving presentation. Oppenheimer made an excellent response. Just before the ceremony Helen and I, Robert and Kitty Oppenheimer, their children Peter and Toni, and Mrs. William S. Parsons (widow of Oppenheimer's wartime colleague) met with President and Mrs. Johnson in his office. The ceremony, held in the Cabinet Room, was attended by a group larger than that of either of the two previous ceremonies held in the White House, and the arrangements for this ceremony were better than for previous ones. After the ceremony Helen and I hosted a reception at NAS that was attended by about 150 people.

Seaborg and President Lyndon B. Johnson present the Fermi award to J. Robert Oppenheimer in the White House. Washington, D.C. Dec. 2, 1963. From left to right: Seaborg, Johnson, Mrs. William S. Parsons, Oppenheimer, Peter Oppenheimer, Kitty Oppenheimer, Lady Bird Johnson, Robert E. Wilson, James T. Ramey, and Helen Seaborg. Courtesy of the U.S. Department of Energy, Germantown, MD.

He presented the Fermi Award (which had been reduced to $25,000 following the adverse reaction to Oppenheimer's receipt of the award) to Adm. Hyman G. Rickover on Jan. 14, 1965, in a White House ceremony.

On Dec. 2, 1968, he made his last presentation of the Fermi Award—to John A. Wheeler:

At 5:30 p.m. Helen and I attended the Fermi Award ceremony in the East Room of the White House. Some 50–100 people were present. President Johnson presented the 10th Fermi Award to John A. Wheeler in the presence of Mrs. John (Janette) Wheeler, his father, son, and grandson. I opened the ceremony with a few remarks; then President Johnson made his remarks; I read the citation; the president presented the citation, a $25,000 check, and gold medal to Wheeler; and Wheeler made his response. The president was in a very good humor and made flattering remarks about me during his talk and as I prepared to read the citation. He said that I have been consistently successful in getting what I want in my budget, and I responded by suggesting that this might be an invitation with regard to this year's budget.

Following the ceremony we went to the Blue Room, where pictures were taken of the Wheelers with the president and the five commissioners with the president.

The White House dinners were similar to those that Helen and I had attended during the Eisenhower and Kennedy administrations. Following initial greetings in the East Room and dinner in the State Dining Room, the guests gathered in the intervening rooms (the Red, Blue, and Green Rooms) for conversation before going back to the East Room for the entertainment. For example, on Nov. 14, 1967, Helen and I attended a White House dinner given in honor of Prime Minister and Mrs. Eisaku Sato of Japan. President and Mrs. Johnson welcomed their guests at the North Portico entrance. After the usual predinner reception in the East Room, the guests proceeded to the State Dining Room and the Blue Room. I sat at a table with Mrs. Kenneth H. Burns (Washington, DC), Congressman Charles N. Youngblood (Michigan), Mrs.

Leo Jaffe (Leo Jaffe was president of Columbia Pictures), Arnold Olsen (Montana), and Mrs. Rosel Hyde (Hyde was chairman of the Federal Communications Commission).

Preceding his toast to Japan and its emperor, President Johnson referred to Abraham Lincoln: "Lincoln gave us a faith that no time or crisis can kill. It took time and patience, but Lincoln won peace at home. It is taking time and patience to win peace in the war." I believe that he had in mind the war in Vietnam. Then Johnson quoted these words of Lincoln: "I am here. I must do the best I can and bear the responsibility of taking the course which I feel I ought to take." Pursuing this theme in his response, Prime Minister Sato praised the president for his war efforts. President Johnson seemed to be in unusually good spirits and spent time with nearly every one of the approximately 185 guests, paying particular attention to the leaders of the new DC government.

Besides President and Lady Bird Johnson, Helen and I talked to Vice President and Mrs. Hubert Humphrey; Ambassador and Mrs. Takeso Shimoda of Japan; Florida Representative and Mrs. Claude Pepper; Tennessee Representative and Mrs. Joe Evins; Texas Representative and Mrs. Jim Wright; General and Mrs. Maxwell D. Taylor; DC Mayor and Mrs. Walter E. Washington; Dr. and Mrs. Leland J. Haworth (director, NSF); Richard Adler (composer); Mr. and Mrs. Kirk Douglas (actor); Mr. and Mrs. Robert Gibson (baseball pitcher); Mr. and Mrs. Frank Stanton (president, CBS); Mr. and Mrs. Werner P. Gallander (president, National Association of Manufacturers); Illinois Governor and Mrs. Otto Kerner; Tony Bennett (popular singer); Mr. Howard Duff (actor) and his wife, Ida Lupino (actress); Gen. and Mrs. William D. Eckert (commissioner of baseball, New York); West Virginia Representative and Mrs. John D. Rockefeller, IV; and others.

After dinner we all went to the East Room, where Tony Bennett, accompanied by a five-piece orchestra, entertained everyone by sing-

Seaborg and President Lyndon B. Johnson present the Fermi award to John A. Wheeler (center) in the White House East Room. Washington, D.C. Dec. 2, 1968. Courtesy of the U.S. Department of Energy, Germantown, MD.

ing his famous "I Left My Heart in San Francisco" and many other songs, including "Country Girl," which he dedicated to our great President Johnson. It was a lively and very interesting evening.

Lyndon Johnson was dedicated to the goal of equal opportunity for all Americans. As I can testify from first-hand experience, Johnson was very determined that women and minorities be represented on the five-member AEC. In effect, more than 30 years ago, Lyndon Johnson anticipated and promulgated the importance of diversity to our nation, which we are only beginning to recognize and advance. His strong commitment to the Civil Rights Bill is clear in the following description of a cabinet meeting held on Thursday, July 2, 1964, the day the bill was signed:

> At 1:50 p.m. I attended the cabinet meeting held in the White House. Those present included President Johnson, Rusk, Robert Kennedy, Celebreeze, Wirtz, Freeman, McNamara, Gronouski, Udall, Dillon, Hodges, Shriver, Haworth, Weaver, Webb, Hornig, Gleason, Bundy, Ackley, Gordon, Bell, Dutton, Moyers, Heller, Reedy, Rowan, and Valenti. The president spoke first about the Civil Rights Bill and said it will be signed at about 7 p.m. in the East Room. He said it is important to impart to our top staff the sense of urgency in complying with this bill. There are long, tough, difficult days ahead; however, we should be inspired by how far we have already come. The Attorney General, the Justice Department, and Lee White will administer the implementation. He asked us to become familiar with the sections on withholding funds in connection with noncompliance. He suggested that we make calls for compliance in speeches, department meetings, and so on. He asked us to submit within a week a memorandum, through White, on the steps we are taking to implement the bill (i.e., our formalized plan). This is to be followed by periodic progress reports.
>
> The president then called upon Attorney General Kennedy, who said

> that in many areas the difficulties are just beginning (e.g., in the area of application to hotels, restaurants, etc.), and trouble may break out this weekend. We are approaching a crisis involving the whole country. He said there will be a big problem in September: Negroes and whites will be mixing socially for the first time in about 2000 schools that are still not desegregated. He said this is the law of the land and therefore must be observed.

The story of the site selection for the 200-BeV accelerator (the National Accelerator Laboratory) illustrates how effective a threat to withhold funds could be in the civil rights arena. At the beginning of the site contest, we made clear that we would demand from all the finalists appropriate local commitments and measures to prevent discrimination in community facilities and services. In February and March 1966, the commission added civil rights as one factor in evaluating site proposals. Throughout the evaluation process, we kept in contact with the Commission on Civil Rights (CCR), the Equal Employment Opportunity Commission, and the Department of Housing and Urban Development, and we requested that the six governors comply with steps recommended by CCR.

Having selected Weston as the site for the 200-BeV accelerator, we moved rapidly to build the machine and the laboratory. Illinois Governor Kerner and Chicago Mayor Daley cooperated enthusiastically with the commission. Months before, the Universities Research Association had appointed as laboratory director physicist Robert R. Wilson, who quickly assembled an excellent staff for the design and construction of the laboratory. There was, however, one problem: Commitments on fair housing that we sought from communities in the Weston area had not materialized. In January and February 1967, JCAE, along with many of its wit-

---

"Lyndon Johnson was dedicated to the goal of equal opportunity for all Americans. As I can testify from first-hand experience, Johnson was very determined that women and minorities be represented on the five member AEC."

nesses, showed strong interest in using the 200-BeV accelerator project to ensure local civil rights for minority groups. In particular, the committee chairman, Senator John Pastore (D-Rhode Island), threatened to stall authorization for the project if fair housing was not enacted at Weston.

On April 12, 1967, at the invitation of Governor Kerner, the commission met with mayors of the local communities in the site area to discuss civil rights. I implored them to act quickly to eliminate discrimination; otherwise, it certainly would be feasible to move the site if discrimination persisted and if Congress refused to authorize the project in Illinois. The commission had already stated any of the six finalists was suitable for the project, I reminded the Illinois group. My statement was perceived by the Chicago area media as an "ultimatum."

After this meeting, there followed a very agonizing period as Senator Pastore moved to cut funds for the project and the Illinois leadership struggled to get some fair housing measures passed. Indeed, many civil rights activists urged us to withdraw the 200-BeV accelerator from Weston as a symbolic protest. Sympathetic as I was to their motivation, and at times adopting a hard-line position myself, as at the meeting with the Illinois group, I did not want to pull the project out of Weston. As I stated before JCAE, we thought it would be better to stay and fight for human rights there:

> A satisfactory solution to the human rights problem is more important than this accelerator. However, they are not in conflict here at the Weston site. We believe that construction of the accelerator at Weston and advancement of human rights can complement one another.

Fortunately, we convinced Illinois to move forward in civil rights

and persuaded JCAE not to cut our funds or reduce the machine's proton beam intensity, as requested by BOB. Within months, no less than 14 communities within the commuting distance of Weston announced fair housing measures. The National Accelerator Laboratory actually became more than a catalyst for change. From the beginning, the lab's leadership, like the commission, made strong commitments on civil rights:

> In any conflict between technical expediency and human rights we shall stand firmly on the side of human rights. . . . Our support of the rights of members of minority groups in our laboratory and in its environs is inextricably intertwined with our goal of creating a new center of technical and scientific excellence. The latter cannot be achieved unless we are successful in the former.

In practice, the laboratory cooperated with local unions and other groups in recruiting and training minorities in the construction work. This success story demonstrated that, with enlightened federal, state, and industry cooperation, scientific advance and human progress indeed can go hand in hand.

Lyndon Johnson's interest in increasing the role of women in government was emphasized at a meeting early in his administration with the heads of regulatory and independent agencies, as indicated in my diary on Jan. 17, 1964:

> *The president also emphasized the value of women in government. He wants to see more of them in high places, including subcabinet positions. He wants them in all kinds of places because we have many qualified women in this country, and he mentioned the possibility of their serving on commissions such as the Civil Aeronautics Board and the Federal Power Commission and serving as examiners. He read a list of women with high qualifi-*

cations as an example. He would like us to make a report within a month (he suggested by Feb. 1) on what we have been able to do about bringing women into our agencies. Mrs. Esther Peterson was present, and he called on her to expound upon this subject. She said women constitute a great skill bank in the United States, and this information needs to trickle down to the lowest level in government. The use of women must be implemented, and she said that women can help in our agencies. The president said he would like to have a woman like Mrs. Peterson in our agencies and, if we have such a woman, the president would like to know her name.

My fellow commissioners and I were ordered to help him find a woman and an African American (then called Negro) for appointment to the commission. Thus Mary (Polly) Bunting (president of Radcliffe College) and Samuel Nabrit (president of Texas Southern University) were appointed, and both served with distinction.

A dramatic example of changing times is a journal note I made of a telephone conversation I had with a prominent person in the nuclear power industry. He called to inquire about Polly Bunting's qualifications for appointment to the commission. He voiced one particular concern:

He said that, via the grapevine, you hear that some of the commission meetings can get a little heated—although this was probably more true under Strauss and McCone—and that the same is true of meetings between AEC and JCAE. He asked about the presence of a woman at these meetings. I said, actually, there has not been much heat generated at meetings, and there really isn't any loss of tempers.

The implication, of course, was that ladies should not be exposed to temper and that gentlemen could not conduct business in their customary manner when they were present. How times have changed—as, no doubt, such prominent women as Secretary of State Madeleine Albright, Attorney Gen-

eral Janet Reno and recent Secretary of Energy Hazel O'Leary could testify!

In some ways LBJ was larger than life, towering over others both physically (although I had the advantage of exceeding his height) and emotionally—in the sense of his exuberant personality filling the room—but he was, nonetheless, very human. In the midst of the Vietnam War, it became more and more clear how much the war weighed on him, and he showed this to others.

I was in a position to watch the process of his gradual ensnarement in the Vietnam War as he followed the insistence of his military advisers. U.S. involvement in Vietnam was discussed in numerous cabinet and National Security Council meetings, of which I quote a few excerpts from my diary:

Thursday, April 8, 1965

From 1:40 to 3:50 p.m. I attended a cabinet meeting in the Cabinet Room of the White House. Present were the president, Vice President Humphrey, Ball, Connor, Celebreeze, Wirtz, McNamara, Freeman, Udall, Fowler, Gronouski, Katzenbach, Stevenson, Shriver, Buford Ellington (director, Office of Emergency Planning), Carl Rowan (director, USIA), Wilber J. Cohen (assistant secretary, HEW), Webb, Haworth, Fritsch, Hornig, Gardner Ackley (chairman, Council of Economic Advisors), M. Bundy, Busby, Macy, Moyers, Cater, White, Valenti, Watson, Richard Goodwin (presidential special assistant), and others.

The president opened the meeting by pointing out that this was the fifth cabinet meeting in 1965.

The president discussed Vietnam. He doubted that any president had communicated as much as he had with so many people in 100 days. He said that surveys showed that 96.4% of the listening television audience had heard his April 7 speech on Vietnam. He also said that the ratings on his State of the Union message and his Civil Rights speech were high.

> "In the midst of the Vietnam War, it became more and more clear how much the war weighed on him, and he showed this to others."

He then called on McNamara to discuss the situation regarding Vietnam. McNamara said that the North was increasing its strength and preparing for a massive attack. He said that the ratio of South forces to North was only 4 or 5 to 1, whereas 10 to 1 is needed to counter guerrilla operations. The South forces were placing delayed fuse bombs on the roads and elsewhere to hamper operations from the North.

The president said that our forces have struck 17 times at the North, mostly against radar stations. He spoke emotionally about the fact that not a single word seems to be spoken against those who killed at Pleiku, or those who killed the girl at her typewriter in the embassy, but much criticism is directed against our bombing of concrete bridges in North Vietnam. He referred to unjustified and damaging reports that McNamara was using crates for airplanes, second quality ammunition, and poisonous gas. He pointed out that the British have used the same type of riot gas 125 times in the past five years.

The president reiterated what he had said in his speech the previous evening: (1) we are firm, and we are going to keep our commitments; (2) we will discuss the situation anytime and anywhere; and (3) we have a developmental program to aid Southeast Asia.

The president then called on Ball, who said that the speech last night was the most important diplomatic move that had been made for some time. He said that the meaning of "unconditional discussions" was that it had to be applicable to both sides and that there must not be any set preconditions. This applies to any kind of diplomatic exchange or communication.

Friday, June 18, 1965

I attended a cabinet meeting from 11 a.m. to 12:30 p.m. at which Rusk and McNamara discussed Vietnam. Present were the president, Rusk, Katzenbach, Connor, Celebreeze, William F. Raborn (director, CIA), Freeman, McNamara, Shriver, Robert C. Weaver (administrator, Housing and Home Finance Agency), Gronouski, Schultze, Ellington, Macy, Fowler, Goodwin, Bundy, Valenti, Ackley, O'Brien, Haworth, Reedy, White, Cater, Busby, Moyers, Watson, and others.

The president opened the meeting by saying this was the eighth meeting of the cabinet this year. He welcomed Schultze to the group.

The president then spoke about the international situation. He said that this, of course, is not rosy but it could be worse. He is occupied with it every day, and he spends much time briefing other nations to convey our picture to other people. He wants people to know that ours is a government of peace. The worst problem in his 18 months in office is Vietnam. He said he would have Rusk describe the efforts we have made to reason out a solution. He has made it clear that he is willing to go anywhere, anytime to discuss the situation. He wanted to negotiate, but our opponents have refused because our bombs drown out the negotiations and what is needed is a pause. They had a pause, but still they didn't negotiate. They certainly won't stop now until the monsoon season is over. They think they are winning, and we think we are winning. He said he would call on Rusk and McNamara to make reports and said we might want to take notes for a later discussion. Members of the administration need to have a united position, understand the international situation, perhaps criticize it in places like the cabinet meeting, and then support it publicly. He then called on Rusk.

Rusk said that Vietnam offers a good focal point for comment on the total world situation. He said that Vietnam is of concern to 42 of our allies as well as our country, and that the integrity of these allies depends on the United States. Many of these nations wish us well, mostly privately. He said the Vietnam situation must be understood in the context of tensions in the Communist world, and especially the tensions between Moscow and Peking. If the South Vietnam problem could be solved peacefully it would be a great step toward world peace, but this is not an isolated chapter in the struggle; it is part of the struggle in a choice between a UN world or a Communist world. If the United States hadn't held its ground in the UN, then Cambodia, Thailand, Malaysia, the Philippines, and other nations would have been taken over by Communists long ago. He recalled our patience in taking over 100,000 casualties in Korea to avoid nuclear escala-

> "This is all part of an old problem—we have had differences with the Communist world since 1945, but there are limits to what we can do without cooperation from the other side and, of course, the free world has the problem of deciding what to do."

tion. He said that the United States doesn't have a belligerent attitude and wants a peaceful settlement; we don't want this to grow into a major engagement. He described the efforts over many years to explore a settlement with Moscow, and indirectly with Peking and Hanoi, as well as efforts through such intermediary countries as Poland.

Rusk said it was thought the Agreement of 1962 did form the basis for peace, but the North Viet forces didn't withdraw or reduce their numbers as they were supposed to. Following this, Poland suggested a conference and Hanoi and Peking refused. Cambodia and other countries attempted to take the lead in attempts at negotiation, and these were rejected. There were attempts to bring the UN into the picture in a constructive way, in motions supported by the United States, but again Peking and Hanoi said no. They refused to allow a visit by U. Thant, who had offered his services to mediate. Patrick Gordon Walker offered to go to Asia to try to work out a settlement, but Peking and Hanoi refused. President Johnson, in his Mark Hopkins speech, offered to undertake unconditional discussion, but again Peking and Hanoi rejected this. The Seventeen Nation Disarmament Group suggested unconditional discussions, and again Peking and Hanoi refused. The president of India made a proposal for peace negotiations, and again they refused. The Canadian government contacted Hanoi and received no response.

The recent bombing pause of five and one-half days only led to an insulting response from Gromyko. The latest offer from U.K. Prime Minister Wilson, just a day ago, is that he and four other heads of state offered their services to negotiate a settlement, but it is not clear how this can be done. President Johnson welcomed this offer toward negotiation of a settlement. Many other contacts were undertaken, some of them covert, but none have led to positive results. The problem is not one of lack of contact. The problem is that we have made many, many contacts with no response. Peking has announced that Thailand is next on the list for conquest. There is a proclaimed appetite—a military doctrine of world revolution on the part of the Communists. This is just as clear as Hitler's Mein Kampf was previous to World War II. It

is not surprising that there are numerous governments in South Vietnam in view of the continuing crisis. During the Greek crisis, for example, which was finally weathered, there were eight different governments. It is not easy to maintain governments during such crises.

We have contacted governments all over the world, and 60–70 of them wish us well, 30 are giving us assistance in South Vietnam, 10 say that help is on the way, 25 are disinterested or neutral, and about 25 are opposed to our presence in South Vietnam. This is all part of an old problem—we have had differences with the Communist world since 1945, but there are limits to what we can do without cooperation from the other side and, of course, the free world has the problem of deciding what to do. There is, of course, some agreement between us and the Communist world—there is a common understanding that no nuclear weapons should be used. Also, there seems to be an understanding that in today's war we can't send mass troops across frontiers any more because of the danger of triggering a nuclear war. The United States is studying ways to meet these so-called wars of liberation. Moscow, however, is not converted to these views regarding the non-use of nuclear weapons and mass troops across frontiers. The Soviets are cautious because they know that NATO has spent a trillion dollars to be ready for defense. There is no doubt about the eventual outcome of this war; our philosophy will prevail, but in the meantime we are engaged in a dangerous conflict. We must keep all doors open to a peaceful settlement.

The president then called on McNamara to give a summary of the military issue. McNamara said we had just heard of the 12–14 steps in the diplomatic field that had been taken by our State Department in an effort to settle the Vietnam matter. He said all of these had been rebuffed because the Communists apparently are set on obtaining their military objective. He said that we could stop them by moving toward a stalemate. During the past year the Viet Cong have infiltrated large amounts of men and equipment in South Vietnam. This military equipment does not come from captured equipment in South Vietnam as is claimed, and it is a shame that Senator McGovern repeated this *falsehood again yesterday on the Senate floor. This is just sheer fantasy—the captured weapons are from China, and we have even identified the specific arsenals in China where the weapons are made.*

Out of the 150,000 captured weapons, about 135,000 have been identified as coming from China and 15,000 from North Vietnam. There are about 64,000 North Vietnam regular troops and about 80,000–100,000 irregulars, making a total of about 160,000 men facing the South Vietnam troops. Facing these is a South Vietnam force of about 550,000. Thus, the ratio is about 4 to 1, whereas a ratio of about 10 to 1 is required to offset guerrilla warfare. We are trying to augment the South Vietnam forces by about 100,000 men, but we need more. The Koreans are sending in men to help. The total U.S. military strength will be about 70,000–75,000 men, of which about 15,000–20,000 will be of combat strength. We ran a B-52 raid last night, striking a heavily canopied jungle area, demolishing stocks of food, and capturing 7500 lbs of rice and several hundred weapons. The alternatives are difficult. A forced win at any cost would expend too many lives, and withdrawal is impossible.

Wednesday, March 27, 1968

At 1:15 p.m. I attended a National Security Council meeting in the Cabinet Room of the White House. Present were President Johnson, Vice President Humphrey, Arthur J. Goldberg (U.S. representative to the UN), Clark Clifford (secretary of defense), Paul H. Nitze (secretary of the navy), Gen. Earle G. Wheeler (chairman, Joint Chiefs of Staff), and Gen. Creighton W. Abrams (vice chief of staff of the army), Hornig (OST), Katzenbach (undersecretary of state), Richard Helms (director, CIA), Price Daniel (director of the Office of Emergency Preparedness), Marks, Rostow, Bromley Smith (executive secretary of the National Security Council), and Spurgeon Keeny (technical assistant to the president's science adviser). The president opened the meeting by saying its purpose was to discuss the NPT.

The president said that first, however, he would ask Wheeler and Abrams to give a briefing on the situation in Vietnam. He pointed out that Wheeler had just visited Vietnam.

Wheeler said that when he was in Vietnam during the latter part of February he talked to Gen. William Westmoreland. [I had met Westmoreland, who was now so misleadingly optimistic about the outcome of the war in Vietnam, at Berkeley in fall 1960 when I was serving as chancellor at the University of California, at the time that he was serving as superintendent of the Military Academy at West Point.] He found what he characterized as a very fluid situation. He said that the Tet offensive had of course shaken their force there. He said he had suggested to President Nguyen Van Thieu of South Vietnam that the best way to protect the city was to get out of the city and drive the North Vietnamese away from it. He described a recent meeting on March 24 in the Philippines with Westmoreland. Westmoreland assured him that the South Vietnamese armed forces had performed well during the Tet offensive and, since that time, he is regaining the initiative. He said there is hard fighting yet to come, but he has absolutely no fear of a general defeat. He said that the enemy suffered more than 50,000 casualties in Tet and afterward.

The president then called on Abrams to tell us about the South Vietnamese, their plans to strengthen themselves, and what they are doing about drafting 18- and 19-year-olds. Abrams said that for the past 11 months, at the request of President Johnson and Westmoreland, he has devoted his attention to the South Vietnamese army and forces. He had come to know the senior, division, corps, and some regimental commanders. He said they were gaining confidence in themselves and continually performing better. There are, of course, some dark spots. Eight out of 149 battalions were not satisfactory; however, 30 have shown distinguished performance and the rest are satisfactory. He said in general the performance of the South Vietnamese army was good. President Thieu has said they will increase their force by 135,000 in the immediate future. They are drafting 19-year-olds now and will start drafting 18-year-olds when the supply of 19-year-olds is gone, perhaps by June.

President Johnson said that the average U.S. draft age last year was 20.7 years and this year is 20.4 years. Abrams said that, beginning last December, we had improved the weap-

ons available to the South Vietnamese. President Johnson then inquired about the performance of Koreans in Vietnam, and Abrams said that it was good. The president asked about the Australians, and Abrams described their performance as first-class. The president asked how much damage the Tet offensive did to the U.S. military and to the South Vietnamese military and what we could look forward to in April. Abrams said that the Tet offensive had given a quantum jump to the improvement of the South Vietnamese army morale: They now knew that they could meet and defeat the enemy. He said that the spirit of the U.S. troops has been good and is now great.

Abrams told the story of Hill 881, which is a hill 881 meters high that is occupied by U.S. troops who, every morning at 8, raise the American flag in a ceremonial manner while standing in their trenches. This draws fire from the North Vietnamese, and the story goes that this is as good a way as any to have them use their daily quota of ammunition. With respect to the question about April, Abrams just said it would be a month of more fighting. President Johnson asked Abrams whether he has adequate forces for the days ahead, and Abrams said that the forces that have been programmed can handle the situation.

Despite the continually optimistic observations of his military advisers, President Johnson had by this time begun to see the light. It was only four days later (March 31, 1968) that he made his dramatic televised announcement of his plans for de-escalation of the Vietnam war and the cessation of bombing of North Vietnam, and his decision to forego the Democratic nomination for president. McNamara had already become disaffected with the Vietnam war, and the president had eased him out of his job as secretary of defense, replacing him with Clark Clifford, who took office on Jan. 19, 1968. Clifford was instrumental in convincing the president to de-escalate the Vietnam war. My diary made reference to McNamara's

resignation following a meeting in the Cabinet Room on Tuesday, Nov. 28, 1967:

Last night the press carried reports of McNamara's resignation as secretary of defense and appointment as head of the World Bank. I asked him about this at the meeting in the Cabinet Room and he was noncommittal, saying he didn't have any new job yet.

Then, on Wednesday, Feb. 28, 1968:

At 1 p.m. Helen and I went to the ceremony in the East Room of the White House, where the president bestowed the Medal of Freedom on McNamara. Mrs. Johnson, Mr. and Mrs. Hubert Humphrey, congressmen, nearly all cabinet members and some of their wives, and others were there. It was a very moving ceremony, and McNamara was so choked up he couldn't give a speech. In the reception line he expressed appreciation for my coming and said, "Glenn, we have been here together since the beginning seven years ago."

I, too, was influenced by the constant favorable predictions by our military leaders for victory in the Vietnam war. My four sons and older daughter were strongly opposed to our continued involvement in the war. I came to this view by the time of President Johnson's televised announcement of his plans for de-escalation.

On Oct. 1, 1968, Francesco Castagliola, a JCAE staff member strongly recommended to President Johnson by Pastore, was sworn in as an AEC commissioner, replacing Nabrit; he remained only until June 30, 1969.

President Johnson did take pride in much that was accomplished by his administration in the domestic area. His commitment to The Great Society was very real and, obviously, he had never forgotten the poverty of his youth and his desire to improve the lot of poor or disenfranchised Americans. I sus-

AEC General Manager Robert E. Hollingsworth, AEC Director of Regulation Harold L. Price, and the five commissioners of the Atomic Energy Commission. Left to right: Hollingsworth, Francesco Costagliola (newly appointed), Wilfred E. Johnson, Seaborg, James T. Ramey, Gerald F. Tape, and Harold L. Price. Washington, D.C. Oct. 1, 1968. Courtesy of the U.S. Department of Energy, Germantown, MD. Photo by Ed Westcott.

pect that, in the long run, history will recognize these accomplishments and credit LBJ's dedication and determination. I know that I will long remember the genuine warmth of his friendship, which was based on the most solid foundation for friendship: mutual respect.

During the last month of the Johnson administration I was interviewed on audiotape by AEC historian Richard G. Hewlett, and the tape and a transcript are deposited in the Lyndon B. Johnson Library. Helen and I were present for the dedication of the Johnson Library on May 22, 1971:

We arrived at Bergstrom Air Force Base, Austin, at 10:40 a.m. and proceeded by bus to the University of Texas campus. Here we attended the dedication of the Lyndon Baines Johnson Library and the Lyndon Baines Johnson School of Public Affairs, conducted in the glade next to the circular water pool adjoining these buildings. The Rev. Billy Graham gave the invocation. Frank C. Erwin, Jr. (former chairman of the Board of Regents of the University of Texas) presided and

introduced those on the platform, including Mrs. Johnson, Mrs. Richard (Pat) M. Nixon, Vice President Spiro Agnew, House Speaker Carl Albert, Secretary William Rogers, Secretary John Connally, Preston Smith, Texas Legislature Speaker Gus Macher, Lieutenant Governor Ben Barnes, Robert L. Kunzig (General Services Administration administrator), Harry Middleton (director of the LBJ Library), John Gronouski (dean of the LBJ School of Public Affairs), Bryce Jordan (president of the University of Texas at Austin), and Harry Ransom (chancellor emeritus, the University of Texas System).

Erwin then called on Ransom, who spoke briefly about the background of the library.

Erwin then called on Lyndon Johnson, who described the contents of the library (which includes 31 million papers and 40 years' worth of his Washington public papers, gifts, historical documents, etc.) and its purpose. He expressed his philosophy in two words: "Man can."

Erwin then introduced President Nixon, who emphasized the value of the library and Johnson's great public service. President Nixon misspoke and said that President Johnson had just finished "throwing" him through the

Dedication of the Lyndon Baines Johnson Presidential Library and the Lyndon Baines Johnson School of Public Affairs, University of Texas, Austin. Lady Bird Johnson (left) and First Lady Pat Nixon stroll together ahead of President Richard M. Nixon and former President Lyndon B. Johnson (behind Mrs. Nixon). Once again, Seaborg captures the occasion with his Minox camera. Austin, TX. May 22, 1971.

library; he had meant to say "showing" him through the library, of course, and corrected himself immediately.

Erwin then introduced, as members of the audience, David and Julie Eisenhower and Johnson's daughters and their husbands, Chuck and Lynda Bird Robb and Patrick and Luci Nugent and their son Lyn.

At the conclusion of the dedication the Rev. George R. Davis (National City Christian Church of Washington, DC) gave the benediction.

Immediately after the ceremony I watched President and Mrs. Nixon and President and Mrs. Johnson as they walked up the path to the library and, as he passed, President Nixon said to me, "Hi, Glenn." I took some pictures with my Minox camera as they passed (as I had earlier, during the ceremony).

I met Lady Bird Johnson again when she spoke on the "Beautification of America" at a Commonwealth Club luncheon at the Palace Hotel in San Francisco on Feb. 19, 1988.

I met Hubert H. Humphrey a number of times, beginning when he was first elected as a U.S. senator from Minnesota and during his vice presidency under Lyndon Johnson,

when he served as chairman of the space council and of the marine resources council, upon which I served as a member. I recall that he had me assume the chairmanship of the marine resources council when he was unable to attend, and I introduced him for his talks to various organizations. Especially memorable was my introduction, as president of Science Service, for his talk at the Science Talent Search banquet at the Sheraton Park Hotel on March 6, 1967. Helen and I also met his wife, Muriel, on numerous occasions and became very well acquainted with her.

On Friday, Jan. 13, 1978, we received the sad news of Humphrey's death:

Helen, Dianne, Dave, and I had dinner in the playroom while watching TV news. We learned that Senator Humphrey is near death. We learned later that he died at 6:25 p.m.

On Monday, Jan. 22, 1973, during a meeting in my office, my administrative assistant, Sheila Saxby, came in to share with me the sad news she had just heard:

Sheila dropped in at about 5 p.m. to give me the bad news, which she had just learned from Elinor Potter (research assistant, Lawrence Berkeley Laboratory), that President Johnson had died a little earlier this afternoon. This was very sad news and made me despondent. I was very fond of him and feel that I have lost a very close friend. When I came home I found that Helen shared my deep feeling of grief.

Then on Thursday, Jan. 25, 1973, memorial services for President Johnson were held:

I watched on TV the memorial services for President Johnson, which took place at the National City Christian Church in Washington, DC. Marvin Watson spoke a moving eulogy, an outstanding performance.

Lady Bird Johnson and Seaborg at the Commonwealth Club luncheon held in the Palace Hotel for Johnson's talk on ''The Beautification of America.'' San Francisco. Feb. 19, 1988. Photo by Shirley Burton-Cohelan, San Francisco, CA.

Seaborg speaks with Vice President Hubert H. Humphrey at the Science Talent Search dinner at the Sheraton Park Hotel. Washington, D.C. March 6, 1967. Courtesy of Science Service, Inc., Washington, D.C. Copyright, Westinghouse Photographic Unit.

The NPT story, the major arms control achievement of the Johnson administration, can be instructive in a number of ways. It indicates that if major arms control agreements are to have any chance, the president of the United States must take an active, affirmative role. Real forward movement toward a compromise solution with the Soviet Union did not occur until he made it unmistakably clear to Rusk and others in the bureaucracy that that was what he wanted.

The NPT story also tells us something about how arms control agreements can best be negotiated. Most arms control proposals announced publicly by both sides have been insincere. They have included features known in advance to be unacceptable to the other side, thereby ensuring that they would be rejected. The cynical calculation has been that they would earn some credit in world opinion as peaceful initiatives while

the rejecting side would be condemned for being intransigent.

My book, *Stemming the Tide— Arms Control in the Johnson Years,* contains a number of examples of this kind of behavior on both sides and the resulting lack of progress. When there has been progress, it has come about through secret negotiations. These enable each side to make concessions without loss of face. They also enable each to recognize the sensibilities and irreducible needs of the other side. The device that was worked out was for us to issue an interpretation of the treaty and for them to be silent about it. Such a delicate deal could never have been worked out except in private negotiations between patriotic, yet sensitive, negotiators on both sides.

I think that President Johnson had the right approach to the problem of arms limitation. It was the approach that animated his efforts and those of his administration in arms control. It was the approach that led to two significant accomplishments—the NPT and the beginnings of SALT—and some lesser ones. That more was not accomplished in the Johnson years can be laid at the door of indifferent execution—some by the president himself—and inhospitable world conditions. But Johnson's address to the problem, as expressed in these words, was exactly right. He said, "While differing principles and differing values may always divide us, they should not, and must not, deter us from rational acts of common endeavor."

In summary, I would describe President Johnson as a master politician, the possessor of a very interesting and complex personality, a good friend or formidable adversary, and an energetic worker and effective performer. I am pleased to have been counted among his friends.

chapter 6 RICHARD MILHOUS NIXON

1969–1974

ADJUSTING TO TROUBLED TIMES

I continued to serve as chairman of the Atomic Energy

Commission (AEC) during the first two and a half years of the administration of Richard Milhous Nixon. I cannot give as complete an account of arms control developments as for the Kennedy and Johnson years because the Nixon administration chose to exclude AEC from much of its former participation in this area.

I shall describe my early contacts with Nixon, many years before his presidency; my first meeting with him as president, when he asked me to continue (as almost the only holdover from the Johnson administration) as AEC chairman; a confrontation with the Department of Justice in which AEC fought to defend the rights of an individual we believed was being falsely accused of disloyal behavior; my role as a member of numerous executive office councils; the efforts, probably unrealistic and largely frustrated, to move forward with a new generation of reactors (breeders) that would provide more fissionable material than they consumed; the Apollo 11 astronauts' feats (first landing on the moon); the college student "revolution"; a brief look at how the Nixon administration approached its arms control challenges; and the struggle (ultimately, partly successful) to maintain the integrity of AEC's basic structure against the consequences of drastic reorganization proposals that would have splintered it.

My acquaintance with Nixon began quite early in our respective careers, although not as a result of my coming to Washington to serve as AEC chairman, which had been the case for my first acquaintance with John Kennedy and with Lyndon Johnson.

I first met Nixon soon after he had been elected as a U.S. congressman from California. We met at the banquet for the U.S. Junior Chamber of Commerce's "Ten

Outstanding Young Men of 1947," held in Chattanooga, TN. Below is an excerpt from my diary for Jan. 21, 1948:

In Chattanooga. Some of the "Ten Outstanding Young Men," including me, were given a tour of sites of Civil War battles, such as that at Missionary Ridge, where an exceedingly bloody battle took place in November 1863.

The list of the "Ten Outstanding Young Men of 1947" includes Cord Meyer, Jr. (27, president of the United World Federalists, New York), Dr. Robert A. Hingson (34, among the first to use the new invention hypospray and a developer of caudal anesthesia to eliminate childbirth pain, Memphis), De Lesseps S. Morrison (35, mayor of New Orleans), Lavon P. Peterson (28, blind founder of an engineering school for the blind, Omaha), Glenn Robert Davis (33, congressman, Waukesha, WI), James Quigg Newton, Jr. (35, mayor of Denver), Richard M. Nixon (34, congressman, Whittier, CA), Adrian Sanford Fisher (33, AEC counselor, Washington, DC), Thomas R. Reid (33, human relations expert, Baltimore), and me (35).

All of us except Reid and Morrison, who were absent from the ceremonies, were honored, along with Harold Stassen (Republican candidate for president and the main speaker), by a reception in the Sun Room of the Read House at 5:40 p.m. At 7:30 p.m. we were introduced on ABC's "Vox Pop" radio program, which originated as a national broadcast from Chattanooga High School. Highlighting the program for me was Barbara Jo Walker of Memphis, "Miss America" for 1947. After the 8:30 p.m. banquet and Stassen's talk, we were presented at about 10:30 p.m. with ruby-studded distinguished service keys in a ceremony broadcast over the ABC radio network. John Shepperd, president of the U.S. Junior Chamber of Commerce, presented the keys individually at the center mike. We each expressed our appreciation, and then Cord Meyer gave the response from the honorees. Paramount News filmed the ceremony. It was a full evening, but it did have its lighter moments. A high school student asked me, "When you were doing that work, did you know you were going to make that discov-

U.S. Junior Chamber of Commerce's Ten Outstanding Young Men of 1947 trail Miss America, Barbara Walker. Left to right: Walker, Adrian Fisher, Cord Meyer, Jr., James Q. Newton, LaVon Peterson, Seaborg, Robert Hingson, Representative Richard M. Nixon, Representative Glenn Davis. Missing from photo: Thomas Reid and de Lesseps S. Morrison. Chattanooga, TN. Jan. 21, 1948.

A Chemist in the White House: From the Manhattan Project to the End of the Cold War

ery?" When I said, "No," he responded with, "Oh, you mean you were just piddling around?" I then replied, "I guess that's about the best definition of science yet."

Nixon suggested to me that we should remain in touch with each other and "stick together."

I next encountered Nixon when he was a candidate for the U.S. Senate and visited us in the Radiation Laboratory on Friday, May 26, 1950:

As scheduled, I gave Richard Nixon and his entourage a tour through our chemistry research labs. Even Al Ghiorso (my longtime colleague and co-worker) was polite (but cool). When one of Nixon's aides announced that he (Nixon) and I were going to the Claremont Hotel for a photo session, I realized that such photographs would constitute my endorsement of Nixon's candidacy. I promptly and ineptly refused—however, Nixon quickly and gracefully accepted my excuse.

I then encountered Nixon at the New Year's Day Rose Bowl game on Jan. 1, 1951:

Before taking our seats at the game, we visited the restrooms and, while waiting for Helen and Jeanette (my sister), I saw Nixon, whom I have known since 1948. For some reason, whenever I think of Richard Nixon, I am reminded of Dr. James Nickson, whom I knew during my years at the Metallurgical Labortory (Met Lab), so I introduced my friend, Bill Jenkins, to James Nixon. When Nixon moved on, Bill remarked that he thought Nixon's first name was Richard.

I had a number of contacts with Vice President Nixon while I was chancellor, including an exchange in March 1959, when he sent a message of congratulation to our University of California at Berkeley's NCAA championship basketball team. On April 30, 1959, I met with Nixon in his vice-presidential office in the Capitol Building.

We discussed a number of matters, including the student-exchange program with the Soviet Union, the widespread dissatisfaction in the academic community over the loyalty oath demanded by the National Defense Education Act (NDEA)—and the story on the naming of mendelevium, in case this might be useful on his forthcoming trip to Russia. As the discoverers of the chemical element with the atomic number 101, my colleagues and I gave it the name mendelevium in honor of the great Russian chemist, Dimitrii Mendeleev.

Vice President Nixon made good use of the information about the naming of mendelevium. In August I received a package containing a precious book autographed by Dmitrii Mendeleev from Emmanuel Tsipelzon with a translation from Russian to English by Edward L. Freers of the American embassy in Moscow of the following message:

During this visit to the USSR, Vice President, USA, R. Nixon informed us that before his departure for the Soviet Union he was visited by his friend, professor of chemistry, Mr. Seaborg, who named the 101st element, discovered by him, of the D.I. Mendeleev Periodic Table after this great Russian chemist.

In this friendly act of the American scientist each Soviet citizen discerns a great respect toward our people and its culture, as well as one of the steps toward the liquidation of the absurd, according to Nikita Sergeevich Khrushchev, tense state of "cold war" between two great nations.

May I present to you, in commemoration of your remarkable discovery and your noble act, the book by Dmitrii Ivanovich Mendeleev "Fundamentals of Chemistry" with his autograph.

[The autograph read] "To my deeply appreciated colleague, Dr. N.I. Bistrov, in commemoration of saving my son. D. Mendeleev 1889.''

This letter from Tsipelzon, who described himself as "an old Moscow second-hand book dealer," explains that Mendeleev presented this book to Bistrov when he arrived back in Moscow. Mendeleev left the meeting of the Royal London Society, where he was to read a paper, because he heard that his son was deathly ill. He discovered that, thanks to Bistrov, his son was already on the road to recovery.

I next met with Vice President Nixon in his office in the Capitol Building on Dec. 14, 1959, when I was in town for a President's Science Advisory Committee (PSAC) meeting, as I have described in Chapter 3. We discussed the problems in precollege education, the potential use of TV in education, the need to remove the disclaimer provision of the student loan section of the NDEA, and prospects for the 1960 presidential race. He believed that the Republican nomination would go to either himself or Nelson Rockefeller, and he thought the Democratic nomination would go to either Stuart Symington or Adlai Stevenson; surprisingly, he did not regard John F. Kennedy as a leading candidate for the Democratic nomination.

On May 16, 1960, again in connection with my visit to Washington for a PSAC meeting, I met with Nixon, who was then a clearly identified candidate for the presidency of the United States. We discussed the future of PSAC (he said he intended to keep the same members but would pick a new chairman), the need for federal support of private and state universities (he said that such support should extend beyond that for building programs to include faculty salary, scholarships, and so forth), and plans for support of the Lawrence Hall of Science (he said that he would see to it that this receives

federal support and that this would be his first campaign promise).

At the time of the June 1960 PSAC meeting in Washington, my two oldest children, Lynne and Peter, accompanied me on some sight-seeing in Washington and New York. We made a purely social visit to Nixon in his office on June 24, 1960.

As chancellor at Berkeley I invited him early in 1960 and again in the summer of 1960 to come and address our students, but in the first instance he had to decline because of vice presidential responsibilities and in the second he promised to keep it in mind but was unable to fit it into his schedule.

Another encounter was Dec. 12, 1963, at the annual Westinghouse Christmas party hosted by William Knox (vice president, Westinghouse Electric Corporation) at the Wall Street Club in New York.

On Sunday, July 23, 1967, I had an interesting breakfast at the Bohemian Grove with Nixon and Governor Ronald Reagan:

I attended a "Gin Fizz" breakfast at the Owl's Nest Camp as Ed Pauley's guest. I sat at the same table as Reagan, Dick Nixon, Preston Hotchkiss (Los Angeles business executive), Harvey Hancock, Jack Sparks Rannes, Pauley (chairman, Pauley Petroleum, Inc.), Art Linkletter (radio and TV broadcaster), and Herb Hoover, Jr. Harry Wellman (vice president, acting president, University of California), Franklin Murphy (chancellor, UCLA), Ed Strong (professor, University of California, Berkeley), Dan London (hotel executive, San Francisco), and Eddie Carlson (president, Western [name changed to Westin in the early 1970s] International Hotels, Seattle) were among those present. I discussed my recent South American trip with Nixon and invited Reagan to visit the Livermore Laboratory.

Pauley said privately to me that he was on the spot because he could not choose between Reagan and Nixon as the guest speaker. He asked if I would

Seaborg, daughter Lynne, son Peter, and Vice President (and presidential candidate) Richard M. Nixon in the vice president's office in the Capitol. Washington, D.C. June 24, 1960 Courtesy of the Office of the Vice President, Washington, D.C.

serve as the speaker, and I agreed. Pauley then introduced me, emphasizing my Watts connection. [I had attended David Starr Jordan High School in Watts.] I gave about a 30-minute talk on the peaceful uses of nuclear energy: electric power, desalting, energy complexes, maritime propulsion, nuclear rocket and auxiliary electric power, artificial heart, Plowshare, applications to dating, humanities, criminology, and so on. It was very well received, and when Reagan was called on he spoke briefly and humorously, saying he knew better than to follow an act like mine. In my question period Reagan asked about the use of neutron activation analysis for analysis of baby's disease through fingernail clippings. I admonished the group that California must solve its problems of nuclear power reactor siting and promised to send Hoover a copy of my Commonwealth Club speech. Nixon also spoke briefly and humorously, saying Reagan should become vice president so as to be elected to the Bohemian Club without a long wait.

My close connection with Nixon began in January 1969 after he became president following his victory over Hubert H. Humphrey

OFFICE OF THE VICE PRESIDENT

WASHINGTON

February 23, 1960

Dear Glenn:

This is the first opportunity I have had to thank you for your letter of February 1st inviting me to address an open meeting on the Berkeley campus sometime during the spring semester. I very much appreciate your cordial invitation and wish it were possible for me to accept.

Unfortunately, however, it is particularly difficult for me now to arrange my calendar very far in advance. With a busy Congressional session just getting under way and the President planning several foreign visits, during which time I will be expected to remain close to the Capital, I shall not be making very many trips away from Washington for the next few months.

For the above reasons I must, most regretfully, decline the invitation to participate in your program. I wish it were not necessary for me to make this decision but you can, I am sure, understand the many problems involved in planning my schedule.

With every good wish,

Sincerely,

Richard Nixon

Dr. Glenn T. Seaborg
The University of California
Berkeley 4, California

Vice President Nixon's letter declining with regrets University of California Chancellor Seaborg's invitation to speak before the University of California–Berkeley student body. Feb. 23, 1960.

fall of the breeder reactor; certain administrative matters, such as a confrontation with the Department of Justice over a security clearance case; and a successful struggle to maintain AEC's basic structure and my decreasing rapport with the president. [This is in contrast to our books, *Kennedy, Khrushchev, and the Test Ban* and *Stemming the Tide— Arms Control in the Johnson Years*, which focused on attainment of the Limited Test Ban Treaty (LTBT) and the Nonproliferation Treaty (NPT)].

Nixon's victory over Humphrey presented me with a personal problem. When President Johnson reappointed me AEC chairman in June 1968, I told him that I would leave the commission if a Republican became president and designated someone else to be chairman. As a result of this consideration and my desire to return to Berkeley, I agreed to swap a new five-year full term for Commissioner James Ramey's remaining two-year tenure ending 1970. Chairman Chet Holifield of the congressional Joint Committee on Atomic Energy (JCAE) had eagerly sought such an arrangement so that Ramey, JCAE's former staff director, could continue to hold the committee's line within the commission. With misgivings that such maneuvering tended to politicize the traditional nonpartisan AEC, Johnson and John Macy, head of the Civil Service Commission, acquiesced. Therefore, when President Nixon assumed office on Jan. 20, 1969, I had the potential of one year and six months in my term as a commissioner, but I had no idea whether I would continue as chairman.

After meeting with Nixon's assistant, Robert F. Ellsworth, on Jan. 22, I had my first meeting with the president on Jan. 28, 1969:

From 3:40 to 4:15 p.m. I had my first appointment with President Nixon with Lee A. DuBridge (his science

in the election held the previous November. In my book, *The Atomic Energy Commission Under Nixon: Adjusting to Troubled Times*, written with Benjamin Loeb, I covered my AEC chairmanship under President Nixon rather broadly, including the arms control topic Strategic Arms Limitation Treaty (SALT); the proposed civilian (Plowshare) and military (antiballistic missile [ABM]) use of nuclear explosives; standards for the emission of radiation from nuclear power plants; problems for AEC resulting from the rising environmental movement; the rise and

A Chemist in the White House: From the Manhattan Project to the End of the Cold War

OFFICE OF THE VICE PRESIDENT

WASHINGTON

August 30, 1960

Dear Glenn:

        I appreciated your writing as you did on August 17th, renewing your invitation to address the student body and faculty of the University of California at Berkeley this fall.

        Though my California schedule has not yet been finally determined, you may be sure that your generous invitation will be kept in mind for serious consideration, and we shall certainly be in touch with you at the appropriate time if it appears that this can be fitted into my schedule.

        With kind regards,

               Sincerely,

               Richard Nixon

Dr. Glenn T. Seaborg
The University of California
Berkeley 4, California

Nixon's response to Seaborg's invitation to speak before the Berkeley student body during his California presidential campaign swing. Aug. 30, 1960.

adviser), Henry A. Kissinger (his national security adviser), Ellsworth, and H.R. Haldeman (his personal adviser).

The president asked us to enter his office before his preceding appointment—with Washington Senator Scoop Jackson, Kissinger, Gen. Andrew Goodpaster (supreme allied commander, Europe), and Bryce Harlow (presidential assistant)—had ended. The president introduced us to these people and, as Senator Jackson was leaving, asked him in a semi-joking way what his recommendation on the Supersonic Transport (SST) would be. Jackson replied lightly that he had sort of a conflict of interest there and didn't reply directly, but the president reiterated his interest in the SST.

We sat down on the two couches facing each other, in front of the fireplace, which had a crackling fire. I sat on the couch next to the president; DuBridge, Ellsworth, and Kissinger sat on the opposite couch; and Haldeman sat on a chair in between the couches.

The president began by handing me a letter, with an accompanying envelope, saying it was self-explanatory. (This was the letter asking me to stay on as AEC chairman.) He jokingly said it was a close decision, but that he had the final say. . . .

The president then raised the question of Plowshare in the peaceful uses of nuclear explosives and said he is very interested in this subject. He wants this to have a high priority in his administration; in fact, he has a special prejudice for this program—the way all people have special quirks and prejudices—and hopes it can go forward expeditiously. He asked me to describe the Australian project.

The president had a two-page memorandum before him, which he glanced at occasionally as we spoke. (I wouldn't be surprised if this might not have been a memorandum written by Ellsworth covering the conversation I had with Ellsworth and his assistants, Thomas Whitehead and Daniel W. Hofgren, last Wednesday, Jan. 22). . . .

The president reiterated his interest in the peaceful uses of atomic energy and said he thought this was something that should be accelerated. As he was speaking in broad terms beyond the Plowshare aspect, I asked whether he had in mind power reactor development, and especially breeder development, and he said he did. He thus expanded his request for information on the peaceful uses of atomic energy on a scale rather broader than just the Plowshare program.

The president said he would like to set up a briefing session in which he and others might be informed about the peaceful uses of atomic energy. He recalled in this connection the talk I had given at the Bohemian Grove (the talk at the Owl's Nest in July 1967) and how impressed he was by my description of the great potential in this field. I immediately suggested that he come to our headquarters in Germantown and

Richard M. Nixon, William Martin, and Seaborg at William Knox's Christmas luncheon at the Wall Street Club. New York. Dec. 12, 1963. Courtesy of William Knox (deceased).

meet some of the key staff and be briefed on our program. He said he would like to do this and suggested that we cover weapons and national security aspects as well as peaceful uses of atomic energy. At first he said he might come out in about two months but later suggested it might be March. When I hinted it might be useful to come even earlier, he said he could possibly do it in February. He asked Haldeman to

include AEC among the departments and agencies that he was scheduling for visits and to include Secretary of State William Rogers in the briefing at Germantown. . . .

Somebody raised the question of whether I should see the press; although I indicated it would probably be best if I left it low key without doing this, the president immediately said he wanted me to meet with the press—because he

Breakfast at the Owl's Nest Camp, Bohemian Grove, CA. July 23, 1967. Around table, clockwise: Preston Hotchkis, California Governor Ronald W. Reagan, Harvey Hancock (standing), Richard M. Nixon, Seaborg, Jack Sparks, (unidentified individual), (unidentified individual), and Edwin Pauley (mostly out of the picture). Courtesy of William Carter, Palo Alto, CA.

was making such meetings a part of his mode of operation. He suggested I might indicate that he had asked me to stay on as chairman and that we had discussed a number of items, especially the peaceful uses of atomic energy. He suggested I indicate his interest in this field, mentioning the Plowshare program, the Australian harbor experiment, and the breeder reactors. He also suggested I mention that he was going to visit AEC at Germantown. [This visit never took place.]

As the meeting broke up, the president suggested that a photographer come in to take some pictures for the historical record. The president and I recalled our meeting at Chattanooga, when we were both in the group of the "Ten Outstanding Young Men of 1947." I said I had a photograph of us in that context and asked whether he would autograph it, and he said he would be glad to do so. I also said I had a picture of him with my kids, taken in June 1960, and he indicated very cordially that he would be glad to autograph this also.

After my appointment with the president, Ellsworth took me to the Press Room, where we met Presidential Press Secretary Ron Ziegler. We waited for a few minutes outside while Interior Secretary Walter Hickel was finishing his press conference; when he came out, I met him. Ziegler then took me to the packed Press Room to a microphone.

I began by saying that I had just finished meeting with the president and that the president has asked me to tell the press about our conversation. I said we had talked about some national security matters and then had gone on to discuss AEC's program for the peaceful uses of atomic energy. I said the president had expressed a great interest in this program, notably Plowshare and the power reactor breeder program. I said he has asked that these programs receive adequate attention and be accelerated by AEC. I said he had indicated a special interest in the use of nuclear explosives for building a harbor in Australia. I also mentioned the president's intention to visit AEC headquarters sometime in February to receive a more thorough briefing on our program. . . .

In answer to a question about whether the president had asked me to continue as AEC chairman, I indicated that he had. I was asked how long my term is, and I indicated that I voluntarily accepted a short term, which expires in June 1970. I indicated that my term as commissioner was determined by law, and that my designation as chairman was a separate matter, done by the president.

I continued the practice of sending a biweekly status report to DuBridge and Ellsworth on significant developments in the atomic energy program. This was done under a different system from that used by Presidents Kennedy and Johnson, to whom I sent my status reports directly:

AEC Biweekly Status Report for March 11, 1969

1. A delegation of U.S. physicists visited the USSR from February 28 to March 5 to discuss possible future experiments involving collaboration of U.S. and Soviet scientists using the Soviet's new 76 BeV proton accelerator at Serpukhov. The Soviets took the position that a major equipment contribution by the U.S., preferably in the form of a large computer, would be required before large-scale cooperative experiments could take place.

2. AEC endorsed the proposed Water Quality Improvement Act of 1969 (H.R. 5511, S.7) as a significant step toward solving the complex problems of water quality in testimony before the Senate Committee on Public Works, Subcommittee on Air and Water Pollution (on March 3) and the House Committee on Public Works (on March 6). The Act would require that before a construction permit or operating license for a nuclear power plant can be issued AEC must receive reasonable assurance that the plant activity will be conducted in a manner that will insure compliance with water quality standards established by the states and approved by the Secretary of Interior in accordance with the Water Quality Act of 1965.

3. Among my activities during the past two weeks were my appear-

Henry Kissinger (National Security Adviser), Robert F. Ellsworth, Patrick Haggerty, President Richard M. Nixon, Seaborg, and Lee A. DuBridge (Presidential Science Adviser) pose for a photo opportunity in the Oval Office at the White House. Washington, D.C. Jan. 28, 1969. Courtesy of the Nixon Project, The National Archives.

ance before the Senate Armed Services Committee on February 28 to support the Nonproliferation Treaty and a visit to the Lawrence Radiation Laboratory on March 6 to participate in the dedication of a new Bio-Medical Laboratory.

4. Secretary of State Rogers, in his testimony before the Senate Foreign Relations Committee on February 18, stated in response to a question by Senator Mansfield that the Department of State, the Atomic Energy Commission and the Arms Control and Disarmament Agency would be involved in the review of the deployment of the antiballistic missile system.

5. Arrangements are in progress for E.I. DuPont de Nemours, Inc., a major AEC operating contractor, to provide limited technical assistance to Deuterium of Canada, a company engaged in constructing a heavy water plant, which has been experiencing difficulties in the startup of the plant.

6. Congressional hearings on uranium mining radiation standards will be held by the Joint Committee on Atomic Energy on March 17–18. AEC testimony will be given on the first day, together with that

from the Federal Radiation Council and the Departments of Labor and Interior. Representatives from labor and the uranium mining industry will be heard on the second day. The hearings will be conducted in open sessions by the Subcommittee on Research, Development, and Radiation.

7. AEC presented testimony on March 10 in an Executive Session of the Joint Committee on Atomic Energy concerning renewal of our agreement with the United Kingdom to provide enriched uranium for use in military propulsion reactors.

8. Senate consideration of AEC's appropriations will now begin in the full Subcommittee on Public Works Appropriations instead of in the AEC-TVA component of the Subcommittee.

9. The new chairman of AEC's General Advisory Committee (which advises on scientific and technical matters relating to materials, production, and research and development) is Howard Vesper, a retired director and vice president of Standard Oil of California.

10. The Western Interstate Nuclear Compact became effective last

week. California, Colorado, Idaho, Washington, and Wyoming have ratified the compact, which establishes an interstate board for promoting the development of nuclear technology and commerce within the region. The compact could eventually include 13 western states. It will be necessary to enact federal legislation if the compact is to have official federal recognition, as has been done for the Southern Interstate Nuclear Board.

11. The new inspector general of the International Atomic Energy Agency is Rudolph Rometsch (a Swiss), who thus becomes the primary official for the IAEA's safeguards program for assuring that nuclear materials supplied for peaceful uses are not diverted for military purposes. The IAEA now has 40 safeguards agreements involving 69 reactors and other facilities in 30 countries. The agency's safeguards function would be greatly expanded under the Nonproliferation Treaty.

12. Three representatives of the French government will meet with the Commission on March 17 for a general exchange of views on cooperation in the civil applications of atomic energy and a briefing on AEC programs. They will then visit the Linear Accelerator Center at Stanford University and a number of atomic industrial groups.

13. The West German government has decided to accelerate development of the advanced fast breeder reactor with the goal of improving the international competitive position of that country's nuclear power reactor industry.

14. Often claims submitted for alleged minor damage from the BENHAM weapons test conducted in Nevada last December, two were withdrawn, and five were found not attributable to the event. Three others involving cracks in walls and a shifted roof line were settled for a total of $575 after investigators decided it was credible that the damage could have resulted from the approximately one-megaton underground detonation.

President Nixon also instigated a new reporting system, which I learned about on Feb. 28, 1969:

I received a call from John Whitaker, secretary to the cabinet, who said that the president has instituted a system in which he will receive, on a daily basis, staff notes, as well as agency and department notes, consisting of one to one and a half pages and containing 5–10 items. The ground rules are that each should be a problem that is really a nagging one but has not yet found its way into the newspapers, so that the

Richard M. Nixon's letter affirming Seaborg's continuance as chairman of the Atomic Energy Commission soon after Nixon's inauguration as president. Jan. 28, 1969.

THE WHITE HOUSE
WASHINGTON

January 28, 1969

Dear Dr. Seaborg:

This is to assure you that I wish you to continue as Chairman of the Atomic Energy Commission. I believe that, according to the law, no formal appointment is required, but I want you to know of my desire to have you remain and continue your important work.

Sincerely yours,

Richard Nixon

The Honorable Glenn T. Seaborg
Chairman
U. S. Atomic Energy Commission
Washington, D. C.

president isn't surprised. Al Toner (Whitaker's assistant) will be responsible for this project. I designated Julius Rubin (my special assistant) as the key contact for AEC, and Whitaker asked that Julie call Toner next week.

I attended the new administration's first meeting of the National Council on Marine Resources and Engineering Development on Wednesday, Feb. 26, 1969:

We were all sitting in our designated places in the Cabinet Room when Vice President Spiro Agnew entered at 9:30 a.m. We rose to greet him. Present at the table were Agnew, sitting in the chair usually occupied by the president at cabinet meetings, Edward Wenk (executive secretary to the council), DuBridge, Robert Mayo (director, Bureau of Budget [BOB]), Lee Haworth (director, National Science Foundation [NSF]), William H. Stewart (Public Health Service, Department of Health, Education, and Welfare [HEW]), John A. Volpe (secretary of transportation), Russell E. Train (undersecretary, Department of the Interior), John H. Chaffee (secretary of the navy), Paul W. McCracken (chairman, Council of Economic Advisors), Alexis Johnson (undersecretary for political affairs, State Department), Thomas O. Paine (administrator, National Aeronautics and Space Administration [NASA]), Rutherford Poats (deputy administrator, Agency for International Development [AID]), Dillon Ripley (administrator, Smithsonian), Rocco C. Siciliano (undersecretary, Commerce Department), and I. J.A. Stratton (chairman, Ford Foundation), Jerry Tape, Herman Pollack (director, International Scientific and Technical Affairs, State), Glenn Schweitzer (marine council staff), David Adams (marine council staff), and others were sitting in chairs around the wall. This was the first meeting of the marine council in the Nixon administration, and it is interesting that the only continuing members (heads of agencies) from the Johnson administration were Paine, Haworth, and myself, all scientists.

Vice President Agnew opened the meeting by saying that this was his first meeting of the marine council. He pointed out that, although the council

has a statutory life of only four more months, President Nixon gives high priority to its work and wants to move forward deliberately with its program. He said the work of the council is very important because no single agency has overall responsibility in this field, and he intends to give it the attention it deserves.

He then called on Wenk, who gave a very well organized slide talk on the responsibilities, scope, and cognizance of the council. He pointed out that the total budget in related marine council matters spread throughout the federal government is $528 million for FY 1970. He then discussed the present and future big business of offshore oil drilling; the importance of the continental shelf (which extends out to where the depth of the ocean is about 600 feet and is equivalent in area to about one-fourth of the land area of the Earth); the status of the U.S. fishing industry, which is losing its relative position with respect to foreign fishing activities; the ocean-borne trade and the diminishing role of the U.S. Merchant Fleet; the navy's role in national security; the potential oceanographic observation by spacecraft; and the role and magnitude of the worldwide oceanographic research fleet.

Vice President Agnew then called on Robert Packard of the State Department, who spoke on the international legal regime for the seabed and the role of the United Nations (UN). He pointed out that there is a UN agreement whereby countries are entitled to the ocean's resources extending from their shores to a point where the ocean has a depth of about 200 meters or, in some cases, where resources are recoverable extending somewhat farther. A number of nations, such as Canada, Italy, Germany, Japan, and Peru, do not adhere to this agreement. To make the cognizant group more manageable, there has been created a UN Seabeds Committee consisting of 42 nations. The United States is undecided about the best definition of the marine and seabed boundaries and the legal regime. Agnew said these questions regarding the extent of the legal regime were among the most important areas requiring decisions by President Nixon—because of their relevance to national security and defense and because of the need to prevent escalation of fixed weapons on the

---

"He pointed out that, although the council has a statutory life of only four more months, President Nixon gives high priority to its work and wants to move forward deliberately with its program."

ocean floor. These questions will require much of the council's attention.

The vice president then called on Schweitzer, who spoke on the international decade of ocean exploration. He said that the United States has one-third and the USSR one-third of the world's capability for ocean exploration and that 100 organizations worldwide, government and nongovernment alike, are concerned with this activity. The vice president said that the council will need to make some recommendations to the president regarding the international decade and that he will create an interagency task force to bring recommendations to the council.

The vice president then called on Train, who spoke about oil pollution control. He said that a plan is evolving for federal regional control that would provide cooperative forces of federal, state, and local governments along with private organizations to help clean up any oil spills. Federal participants will be the Departments of Interior, Transportation, Defense, and HEW, and the Office of Emergency Planning. The vice president asked that DuBridge and Wenk be invited as consultants in these efforts.

The vice president then called on Adams, who spoke on the management of coastal zone activities.

Then the vice president introduced Dr. Julius Stratton, chairman of the commission on Marine Science, Engineering, and Resources. Stratton said that his commission has been hard at work for two years and has just issued a report whose contents he described. One proposal of his commission is to form an interagency group to handle marine matters, and this might have the name National Oceanic and Atmospheric Agency. The vice president said that he would like to have recommendations on the commission's report from the agencies by March 10, which would then be combined into an overall commentary by March 15. This would be used as a basis for recommendations to the president. In commenting on the commission's recommendation for the creation of the new agency (NOAA), Volpe and Train expressed doubts about whether it should be created (NOAA would include the Coast Guard, to be taken from the Department of Transportation, and a number of other functions, to be taken from the Department of the Interior). Agnew recognized the problems in creating such an agency and said that often the hoped-for advantages could be obtained by assembling ad hoc groups from the different agencies to solve the individual problems. He said that he hadn't yet decided his position on the creation of this new agency.

The vice president then closed the meeting by commenting on how the council can carry on its statutory responsibilities. He said that in the near future it will operate much as it has in the past, but he won't activate the council committees for the time being; he will rely on task forces. He asked Wenk to check with members and to come up with recommendations by the next meeting concerning the council's methods of operation.

I attended my first meeting of Nixon's National Security Council on Saturday, March 15, 1969:

I attended a meeting of the National Security Council in the Cabinet Room at the White House, from approximately 10 to 11:15 a.m. Present were President Nixon, Vice President Agnew, Gerard Smith (director, Arms Control and Disarmament Agency [ACDA]), DuBridge, Kissinger, Gen. Earle G. Wheeler, Secretary of Defense Melvin Laird, Secretary of State William Rogers, Richard M. Helms (director, Central Intelligence Agency [CIA]), Elliot Richardson (undersecretary, State), George A. Lincoln (director, Office of Emergency Planning [OEP]), and I, plus Spurgeon Keeny (senior staff member, National Security Council) and Col. Alexander Haig (military assistant to assistant to the president for national security affairs) on the side line.

When President Nixon entered the room and we all stood up, he noticed me and said, "Hi, Glenn." After we had all been seated, he called on Kissinger to set the basis for the meeting. Kissinger outlined the three items up for discussion: (1) the Comprehensive Test Ban (CTB), (2) the Cutoff of Production of Fissionable Material for Weapons, and (3) the Proposal for Seabed Arms Control. He said that on each of these items the United States has a present position at the Eighteen Nation Disarmament Conference (ENDC), and the question will be whether the proposal is in our net security interest and whether we should maintain our present position or take a modified position. As an argument for the CTB, he pointed out that such a test ban would inhibit the Soviet progress on the MIRV and the ABM, that the previous problem on this has been a disagreement between the United States and the Soviet Union about on-site inspections, and that our own testing requirements are such that we could accept a CTB some three years from now. Arguments against the CTB include our immediate need for testing, our continuing need to validate our weapons stockpile by testing, some inhibition to the peaceful uses of nuclear explosives, and our inability to test new weapons concepts.

President Nixon then called on Smith to explain the nature of ENDC and inquired whether the NPT originated from ENDC. He also indicated that he would want Smith to take positive positions at ENDC and not merely indulge in gamesmanship. Smith explained the history of ENDC and pointed out the role it had played in developing the NPT. President Nixon inquired specifically about Soviet interest in participating in ENDC and asked if they sent high-ranking people to the deliberations. Smith said the Soviets take it very seriously and do send high-ranking representatives. He said that ENDC is a very useful forum and results in much publicity for arms control negotiations. He also pointed out that there is a commitment in Article VI of the NPT to carry on negotiations in the arms limitation field. President Nixon asked what the Soviets are proposing in ENDC, and Smith cited such aims as banning nuclear weapons, cutting back on strategic nuclear weapons, prohibiting flights of bombers carrying nuclear weapons, banning underground tests (which raises the problem of on-site inspections), prohibiting chemical and biological warfare agents, dismantling military bases, creating nuclear-free zones, and developing peaceful uses of the seabed.

The president then asked Smith for a summary of his general views of what the U.S. philosophy should be, and he said that he tended toward a program with the following four objectives: (1) a ban on the transfer of nuclear weapons from one country to another (for which the NPT represents great progress), (2) an eventual ban on all nuclear weapons testing, (3) the cutoff in production of fissionable material for use in nuclear weapons, and (4) a ban on the deployment of nuclear weapons (for which the Antarctic Treaty, the Outer Space Treaty, and now the Seabed Arms Control proposal are examples of progress). Smith said that we should probably emphasize the seabed proposal, seek to define what is fixed, and focus on the feasibility of verification. Rogers urged that we reaffirm our support for the Seabed Arms Control proposal and indicted that, in any case, it would probably take four or five years of negotiation to accomplish such a treaty. Smith said it would take as long as he was in office, and the president indicated that it might even take longer than that.

The president then asked for the position of the Department of Defense (DOD). Secretary Laird said that some elements of DOD have traditionally taken a rather negative position. The department doesn't have any definite program for fixing weapons systems in the sea, but the Joint Chiefs of Staff have urged that we not make any commitment that would forbid this. Laird then called on Wheeler, who said that they have some interest and equipment in the seas, including fixed equipment, but this is non-nuclear. He doesn't actually oppose the ultimate attainment of a Seabed Arms Control Treaty, but there has never been a serious study of its implications, and we don't know too much about the ocean and the ocean beds. He feels that, in connection with negotiating such a treaty, we could talk about boundaries and verification procedures, and he feels that Smith could be occupied with this aspect for some months. Laird said that he doesn't oppose the concept of such a treaty but feels there should be a go-ahead for general discussion, although we shouldn't agree to all the details of the suggested proposals at the present time.

The president reiterated that our general posture is against proliferation and the three areas under discussion should be pursued. He said we must not present a picture of dragging our feet. Vice President Agnew said that

"The president reiterated that our general posture is against proliferation and the three areas under discussion should be pursued."

there is strong support for such a seabed treaty in the marine council and in the Senate by such senators as Claiborne Pell (Rhode Island). DuBridge supported Agnew, saying that such a treaty is almost a prerequisite for the large international program for scientific cooperation concerning matters of the ocean, which is getting started. He said we would certainly need to agree not to use the ocean for military bases if these programs are going to be successful.

The vice president reiterated that we should go ahead with the Seabed Treaty because this would be the safest area in which to begin arms talks. The president asked whether Senator Pell is actually so interested, and the vice president indicated that Pell already has a treaty drawn up for consideration. The president stated that he agrees in principle with going ahead with such a treaty, and that he feels strongly that we should "take a positive position on this." We should be very specific and not indulge in general malarkey in our negotiations; we should identify and pick out hard items, make our position clear, and negotiate on these.

DuBridge indicated again that this would advance the general concept of an International Decade of Ocean Exploration, which has been put forth by the United States, and the president said that perhaps this thought should be used in Smith's statement at ENDC. Agnew again emphasized that the International Decade of Ocean Exploration is our initiative. This led to a general discussion of the international controversy over the 12-mile limit.

I said that I am in complete agreement on the desirability of a Seabed Arms Control Treaty. We then turned to the other items. . . .

Kissinger said that it certainly wouldn't be feasible to walk away from our previous positions. The president said he agreed, and we should maintain the three proposals we are discussing and not withdraw them from ENDC. He said that, with respect to the CTB and the Seabed Arms Control proposal, we should lay out our former position and "let it rest there." This, then, might be accomplished by a full review within the U.S. government of the consequences of these proposals with a view toward determining or modifying our future position. He indicated that if our national security depended in a very serious way on making weapons tests, we would have to do so. . . .

Thus, in summary, it was agreed that the U.S. position will remain with all three proposals at ENDC, recognizing that long negotiations will be required and that there will be opportunities to re-evaluate our position as negotiations proceed.

Although a discussion of the Vietnam situation was scheduled, the president said that he was running out of time and would have to postpone this item for the next meeting. He was scheduled to talk within the next half-hour at a luncheon meeting of the National Alliance of Businessmen.

My notes on a meeting of the National Security Council on Wednesday, June 25, 1969, contain some rather revealing statements about President Nixon's attitudes toward national security, negotiations for arms control, relations with our allies, interactions between the executive and legislative branches, and influences on public opinion:

From 11 a.m. to 1 p.m. I attended a meeting of the National Security Council in the Cabinet Room of the White House, which was held to continue consideration of the U.S. position for the SALT with regard to the Soviet Union. The president began by stating, quite forcibly, "There is only one person responsible for the security of our nation, and that's me. My actions, in addition to their immediate impact, will greatly affect the options available to our next president at a period when some of these armament matters may be even more critical. I shall listen carefully to all the viewpoints expressed, but in the end, when I lay it down, I expect it to be followed. . . ."

Next came a discussion of how we would handle consultation with the allies. President Nixon wondered whether consultation wasn't a matter of "therapy" for the allies and therefore if we needed to do more than indicate to them the options we were considering.

If we needed their support in order to go ahead, we would have to sell them on a specific course of action. As it was, we should make it clear that it was our decision to make after we consulted with them. Kissinger interposed a caution that the North Atlantic Treaty Organization (NATO) countries were legitimately concerned about the future of the nuclear umbrella that protected them.

Laird pointed out that, within 15–20 hours after we consulted with our allies, all the information would be in the hands of the Soviets. The president agreed, saying, "My God, with the Norwegians, Danes, and Swedes sitting there, what else would you expect?" (There was a certain imprecision here; the Swedes, not members of NATO, would not be "sitting there.") He said that everything leaks also at the Senate Foreign Relations Committee and that it would not be safe to give them anything but a sanitized version of our position. He said we could give more to the Senate Armed Services Committee. It was then pointed out that Senator Stuart Symington sat on both committees, and the president admitted this was a problem. He said it was more important to succeed in the negotiations than to brief Congress. The president concluded that discussions with NATO should be kept rather loose and probably should include two or three ridiculous things in order to throw off the Soviets. He said this might also be the way to handle Congress.

The president asked whether we needed to open the negotiations by making a definite proposal. Smith said he would certainly be more confident if he knew what our position was at the start, even if it was not revealed. Nixon pressed Smith on why it was desirable to make a full proposal at the start if all we could expect from the Soviets was a propaganda proposal. Smith replied that there were two reasons: We had proposed SALT in the first instance, and the Soviets historically start such discussions with broad, propaganda-oriented proposals. The president alluded to the thesis of Ambassador Llewellyn Thompson—that the Soviets would lose interest if we did not start out with a definite proposal—and said he disagreed with Thompson. He thought a better reason for starting with a substantive proposal

was the favorable effect this would have on American public opinion. Secretary Rogers suggested that perhaps we should start by tabling a very comprehensive agreement, including verification requirements that the Soviets wouldn't accept. We could then fall back from these requirements later. Kissinger objected that, historically, when we fell back from our opening positions we tended to fall back too far. Richardson suggested that we beat the Soviets to the punch by doing what they do: start with our best propaganda position of broad, general principles.

Vice President Agnew said that, whatever position we adopted, some people would find it unreasonable. The president said that whether this was important depended on who thought the position was unreasonable. He said he would be horrified if the New York Times endorsed our position and that we mustn't try to be fashionable. Rogers said we weren't talking about the New York Times but about the American people. Agnew replied that the situation was so complex that perhaps it was not reasonable to expect the American people to understand it. . . .

My first meeting with PSAC under the Nixon administration took place on Tuesday, March 18, 1969:

I attended a PSAC meeting in Room 208 of the Executive Office Building, from 8:30 a.m. to 1:30 p.m. DuBridge presided.

The first item was a briefing by AEC people on Plowshare for which John Kelly (director, Division of Peaceful Nuclear Explosives), Dr. Michael May (director, Lawrence Radiation Laboratory [LRL], Livermore), Glenn Werth (associate director, LRL, Livermore), Gary Higgins (director, Division of Peaceful Nuclear Explosives, LRL, Livermore), Ted Cherry (group leader, geophysical computations, LRL), Dave Dorn (staff assistant to associate director, nuclear design, LRL, Livermore), and Carl Gerber (AEC staff) were present. Dr. Gerald Tape was present part of the time. After a general introduction of the Plowshare topic by me, Kelly and May described the program broadly. Dorn described developments in the special explosives required, and Cherry described the dynamics of cratering. Higgins described the fallout pattern and, in connection with this, PSAC members requested that we release the data on fallout in the western states to the authorities in those states. Higgins, May, and Werth then described the Cape Keraudren Australian project and the Panama Canal project. I concluded by describing the problems concerning interpreting what constitutes radioactive debris in connection with possible violations of the LTBT and said our proposed interpretation is that radioactive debris is not present when concentrations fall below the levels of not present as defined by the International Radiological Commission.

Following this briefing we were rejoined by Tape and joined by Paul McDaniel, Herb Kinney, William Wallenmeyer, and other members of the AEC Division of Research. We discussed the status of funding in the fields of high-, medium-, and low-energy nuclear physics. Victor Weisskopf gave an overview of the field, and Pief Panofsky outlined the status of current methods of experimentation. There was a broad discussion of the inadequacy of funding, with a great deal of talk about whether more flexibility for exchanging funds between operations and equipment would help. However, it was clear that the main problem is lack of funds. I emphasized the large number of low-energy accelerators that have been requested and for which we have not been able to obtain funding. DuBridge suggested the creation of a PSAC–GAC (General Advisory Committee) panel to study the situation and make recommendations in all of these areas of nuclear physics.

My first meeting with the new Federal Council on Science and Technology (FCST) came on Tuesday, April 1, 1969:

From 2:30 to 4:30 p.m. I attended the FCST meeting in Room 208 of the Executive Office Building. This meeting, presided over by DuBridge (whose multihat capacity includes the chairmanship of FCST), was the first such meeting under the Nixon administration.

Present were the following statutory members or their representatives: DuBridge (chairman), Secor Brown (Transportation), Donald Dunlop (Interior), Lee Haworth (NSF), Thomas Paine (NASA), Herman Pollack (State), Thomas Rogers (Department of Housing and Urban Development [HUD]), Myron Tribus (Commerce), and I. Others present were Richard Ottman (OEP); Saul Nelson (Council of Economic Advisors); Hugh Loweth (BOB); Dr. Allen Astin (National Bureau of Standards [NBS]); Lionel Bernstein (Veterans Administration); S. W. Betts (Commerce); Dr. Harvey Brooks (Harvard); David Challinor (Smithsonian); Dr. Philip Corfman (National Institutes of Health [NIH]); Dr. Charles Falk (NSF); Benson Gammon (NASA); Gerald Garmatz (Defense); Robert Green (NAS); Albert Heyward (Defense); Dr. Norman Hilnar (NIH); Dr. Leon Jacobs (HEW); O. A. Neumann (Commerce); Rodney Nichols (Defense); Sidney Passman (ACDA); Carl Shultz and Ernest Pierkil (HEW); Rolf Verstieg (NIH); Russell Hale (space council); Patrick Moynihan (White House); P.N. Whitaker (NASA); Clifford Berg, Peter Rumsey, and Robert Howard (BOB); Dr. Charles Kidd; Dr. Donald King; Col. Andrew Aines; Dr. John Buckley; and Paul Anderson, David Beckler, and William Hooper (Office of Science and Technology [OST]).

DuBridge made an opening summary statement to put into perspective the role of FCST in the Nixon administration. He emphasized that the council would play an important role; President Nixon is very interested in the status of science and technology, and how FCST members can help to obtain the objectives of his administration. DuBridge mentioned the importance of such problems as those connected with urban development (Moynihan's area) and foreign aid (John Hannah's area, AID). He also mentioned the importance of arms control activities (Smith's area); he mentioned Secretary Rogers' interest in the role of science and reminded us that the very first reception in which Rogers participated was one for the scientific attachés from the various embassies in Washington. DuBridge then mentioned the report of the marine commission, prepared under Stratton, and emphasized that FCST

> "Such undertakings must encompass the whole area of social science and social policy, an area that began to have substantial practical effects in the 1960s. The real question is how social science theory can be transformed into public policy."

would have a role to play in its implementation.

DuBridge then pointed out that this administrative interest in science and technology unfortunately comes at a time when budget cuts are necessary. He said that President Nixon has directed that there be minimal cuts in basic science in the course of this exercise.

DuBridge then talked about the future roles of various FCST committees and said many of these will have different members from what was indicated in the 1967 report of the council (the last report issued). He indicated that he would continue the Committee on Academic Science and Engineering (CASE) under the chairmanship of Lee Haworth. He proposed that the Committee on Atmospheric Sciences be expanded into a Committee on Atmospheric and Marine Sciences to accommodate action on the Stratton committee report. He emphasized the importance of the Committee on Environmental Quality and indicated that Vice President Agnew will head a new committee on this. He said there should be coordination in this area. He also mentioned the Committee on Federal Laboratories (Allen Astin, chairman) and indicated, as a goal, further university participation in federal laboratories. He asked me whether the technical Committee on High Energy Physics might be expanded to include medium- and low-energy physics; I agreed that it should be. He also indicated that the Committee on International Programs should continue.

DuBridge then called on Moynihan (assistant to the president for urban affairs) to give his report, "DOD Programs as a Base for Innovation and R&D on Civilian Problems." Moynihan indicated that an Urban Affairs Council, established by President Nixon's first executive order, consists of seven secretaries (cabinet officials) representing seven departments. The DOD's effort in the urban area is perhaps the most substantial of any. Secretary Laird uses this as a focus for wider federal action in the urban development area. Moynihan said this is a good start, but we need to do more. He suggested that FCST might establish an ad hoc committee to help determine the joint projects that might be undertaken by DOD and various

other departments during the coming fiscal year. Such undertakings must encompass the whole area of social science and social policy, an area that began to have substantial practical effects in the 1960s. The real question is how social science theory can be transformed into public policy. Moynihan emphasized that getting from a technical solution to a problem to public adoption of that solution is difficult.

As a result of a query from Moynihan, Garmatz briefly described the role of DOD in urban development. He said the department has to deal with many relevant problems, such as where to build defense bases and how to dispose of them. DOD also builds housing on a large scale and has a great deal of experience in this endeavor.

Nichols expanded on Garmatz's statement by saying that joint research and development funding by DOD and another department might be good, and he endorses the idea of an ad hoc committee, as suggested by Moynihan. Tribus said the ad hoc committee would need to be composed of people with technical backgrounds, social scientists, and colored people in order to be effective.

Passman (ACDA) called attention to the fact that a number of agencies, such as NASA and AEC, have joint technology utilization programs, and advantage should be taken of these in the composition and operation of the ad hoc committee. Paine indicated that NASA has such technology utilization programs, and I said that AEC does, too. I emphasized that DOD is the only agency that appears to have money to carry on activities of the type suggested by Moynihan and that the AEC line item budgeting process wouldn't allow the diversion of much money to such a program. I indicated, however, that AEC would be glad to cooperate in connection with their technology utilization programs. DuBridge said he will set up a small working group to recommend ways of setting up the ad hoc committee.

DuBridge then called on Brooks, chairman of the Committee on Science and Public Policy (COSPUP) of the National Academy of Sciences (NAS), who was scheduled to present the next item on the agenda, "Current and Prospective Activities of COSPUP and

Their Relationships with FCST." He pointed out that COSPUP had been formed in 1962, largely as the result of efforts by Kistiakowsky, and he gave a history of its activities in the intervening years. He described the many reports that have been issued under the aegis of COSPUP and NAS.

DuBridge summarized this area by saying he hopes to be able to continue to count on the efforts of COSPUP and NAS.

DuBridge then spoke a little more about the philosophy of the operation of FCST. He said the administration is interested in maintaining an adequate academic science budget. He wants FCST to address policy questions, and he wants these to be sufficiently important that the principals will attend the meetings and provide their views and suggestions. He emphasized that we need to tell Congress something about what science does instead of always telling them only about the funding that science needs.

DuBridge then called on Corfman, who spoke about the population problem. He called for comments on the paper concerned with the definition of the population problem, which had been handed out earlier. He also commented briefly on the Stevens report (Nixon's panel on science and technology) on the world population problem, an excerpt of which had also been distributed.

DuBridge then called on Loweth, who said BOB is studying the terms and conditions under which research grants and contracts might be made more uniform. They are finally getting around to this as a result of long, continuing pressure from the academic community. He said that letters requesting comments on this have gone out to the agencies, and the first step is getting replies from the agencies designating representatives. This representative group would discuss whether it is desirable to unify these grants and contracts. He said BOB has no preconceived notion that there should be a standardized contract but merely wants to make an unbiased study of this as a possibility. He estimates that the study will take about four months. When asked whether the academic community is familiar with this undertaking, he said that it is familiar with it through their business officers

group. Various FCST members indicated that this might not be a sufficient channel of communication to reach the academic officials, and other members wanted better university communication. Loweth then said that he would inform the American Council on Education and use it as a channel to universities. He also said he would use other avenues such as visits to the universities.

DuBridge then called on Kidd for some final remarks. Kidd handed out a memorandum on strengthening biomaterials research and indicated that FCST members who wish to comment on it should get in touch with Leon Jacobs before the next meeting.

DuBridge brought the meeting to a close by saying that FCST will meet on an average of every two or three months and that the subgroups or committees will hold meetings between FCST meetings.

My first meeting, during the Nixon administration, with AEC's GAC took place on Wednesday, April 23, 1969:

At 11:15 a.m. I attended the GAC meeting (all members present except Lombard Squires) with the commissioners (except Ramey). I gave my usual report covering important items since the last meeting: (1) the commission's intention to ask President Nixon to give a new atomic pioneer award to Vannevar Bush, James B. Conant, and Gen. Leslie R. Groves; (2) the consternation created in the Rocky Mountain states by our need to shut down the MTR (Materials Testing Reactor) for financial reasons; (3) the issue of transferring the gaseous diffusion plants to private ownership and the involvement of the White House committee in this and its relations to AEC and JCAE; (4) changes in the fiscal year 1970 budget proposed by President Nixon; (5) the mounting public criticism of nuclear power plants and AEC's plans to counter this; (6) the mounting criticism of high-yield underground testing in Nevada and the commission's program to counter this; (7) Commissioner Tape's visit to Vienna for technical discussions on Plowshare with the Soviets; (8) the deferral of the Plowshare Rulison Agreement in order *to complete a safety study; (9) a number of International Atomic Energy Agency (IAEA) organizational matters, including the decision of the United States to support the reappointment of Sigvard Eklund as director general and President Nixon's reappointment of Harry Smyth as the U.S. representative; and (10) the recent definite identification of isotopes of element 104 by Ghiorso and his co-workers in Berkeley.*

On March 10, 1969, Vice President Agnew issued a statement establishing a Space Task Group.

Office of the Vice President
Washington
March 10, 1969

Memorandum for
Secretary of State

Director, Bureau of the Budget

Chairman, Atomic Energy Commission

Subject: Space Task Group

A Space Task Group (STG) has been formed to prepare for the president, by September 1, 1969, a coordinated program and budget proposal for the Post-Apollo period. This program will recommend the scope, direction and goals of the Space Program for the next decade. The president has requested that I chair this group, with the Secretary of Defense, the Administrator of NASA, and the Science Adviser to the President as principals. In addition, I hereby request the Secretary of State, the Director of the Bureau of the Budget and chairman of the Atomic Energy Commission to participate on the Task Group as observers. . . .

This task force held numerous meetings and submitted a report to President Nixon in September 1969, which was issued, after incorporating numerous changes by President Nixon, on March 7, 1970.

Statement About the Future of the United States Space Program.
March 7, 1970

Over the last decade, the principal goal of our Nation's space pro-

gram has been the moon. By the end of that decade men from our planet had traveled to the moon on four occasions and twice they had walked on its surface. With these unforgettable experiences, we have gained a new perspective of ourselves and our world. . . .

When this administration came into office, there were no clear, comprehensive plans for our space program after the first Apollo landing. To help remedy this situation, I established in February of 1969 a Space Task Group, headed by the vice president, to study possibilities for the future of that program. Their report was presented to me in September. After reviewing that report and considering our national priorities, I have reached a number of conclusions concerning the future pace and direction of the Nation's space efforts. . . .

In my judgment, three general purposes should guide our space program.

One purpose is exploration. From time immemorial, man has insisted on venturing into the unknown despite his inability to predict precisely the value of any given exploration. He has been willing to take risks, willing to be surprised, willing to adapt to new experiences. Man has come to feel that such quests are worthwhile in and of themselves—for they represent one way in which he expands his vision and expresses the human spirit. A great nation must always be an exploring nation if it wishes to remain great.

A second purpose of our space program is scientific knowledge—a greater systematic understanding about ourselves and our universe. . . .

A third purpose of the United States space effort is that of practical applications—turning the lessons we learn in space to the early benefit of life on earth. . . .

With these general considerations in mind, I have concluded that our space program should work toward the following specific objectives:

1. We should continue to explore the moon. Future Apollo manned lunar landings will be spaced so as to maximize our scientific return from each mission, always providing, of course, for the safety of those who undertake these ventures. Our decisions about manned and unmanned lunar voyages beyond the Apollo program will be based on the results of these missions.

2. We should move ahead with bold exploration of the planets and the universe. In the next few years, scientific satellites of many types will be launched into earth orbit to bring us new information about the universe, the solar system, and even our own planet. During the next decade, we will also launch unmanned spacecraft to all the planets of our solar system, including an unmanned vehicle which will be sent to land on Mars and to investigate its surface. In the late 1970s, the "Grand Tour" missions will study the mysterious outer planets of the solar system—Jupiter, Saturn, Uranus, Neptune, and Pluto. . . .

There is one major but longer-range goal we should keep in mind as we proceed with our exploration of the planets. As a part of this program we will eventually send men to explore the planet Mars.

3. We should work to reduce substantially the cost of space operations. Our present rocket technology will provide a reliable launch capability for some time. But as we build for the longer-range future, we must devise less costly and less complicated ways of transporting payloads into space. . . . We are currently examining in greater detail the feasibility of reusable space shuttles as one way of achieving this objective.

4. We should seek to extend man's capability to live and work in space. The Experimental Space Station (XSSS)—a large orbiting workshop—will be an important part of this effort. We are now building such a station—using systems originally developed for the Apollo program—and plan to begin using it for operational missions in the next few years. We expect that men will be working in space for months at a time during the coming decade. . . .

5. We should hasten and expand the practical applications of space technology. The development of earth resources satellites—platforms which can help in such varied tasks as surveying crops, locating mineral deposits, and measuring water resources—will enable us to assess our environment and use our resources more effectively. We should continue to pursue other applications of space-related technology in a wide variety of fields, including meteorology, communications, navigation, air traffic control, education, and national defense. The very act of reaching into space can help man improve the quality of life on earth.

6. We should encourage greater international cooperation in space. In my address to the United Nations last September, I indicated that the United States will take positive, concrete steps "toward internationalizing man's epic venture into space—an adventure that belongs not to one nation but to all mankind. . . ."

As we enter a new decade, we are conscious of the fact that man is also entering a new historic era. For the first time, he has reached beyond his planet; for the rest of time, we will think of ourselves as men from the planet earth. It is my hope that as we go forward with our space program, we can plan and work in a way which makes us proud both of the planet from which we come and of our ability to travel beyond it.

Many of the goals highlighted in President Nixon's "Statement about the Future of the United States Space Program" were achieved. Five Apollo manned lunar landings and the Viking unmanned Mars lander were successfully launched. The "Grand Tour" missions to the outer

"It is my hope that as we go forward with our space program, we can plan and work in a way which makes us proud both of the planet from which we come and of our ability to travel beyond it."

plants—Jupiter, Saturn, Uranus, Neptune, Pluto, and beyond—were also carried out. The success of these ventures depended on the use of the Space Nuclear Auxiliary Power (SNAP) units for communication purposes powered by radioactive plutonium-238. Many "earth resources satellites" have been launched, a program that has recently included "international cooperation in space" with Russia, and plans are still active for a "large orbiting workshop," (space station).

An interesting meeting of the Space Task Group was held on July 7, 1969:

From 10 a.m. to 1:15 p.m. I attended a meeting of the Space Task Group and "Invited Contributors" in the Indian Treaty Room of the Executive Office Building. Distinguished citizens had been invited to contribute their ideas to the Space Task Group concerning the general benefits, both direct and indirect, that the space program may produce; the hope is to obtain an assessment of the more general and possibly politically significant benefits.

Present were Dr. Jaime Benitez (University of Puerto Rico), Mrs. Shirley Temple Black, Dr. Albert A. Campbell (University of Michigan), Dr. C. Stark Draper (MIT), William C. Foster, Dr. T. Keith Glennan, Dr. Peter C. Goldmark (CBS Laboratories), Najeeb E. Halaby (Pan Am), Frederick R. Kappel (retired from AT&T), Foy D. Kohler (University of Miami), Dr. Willard D. Lewis (president, Lehigh University), Governor John A. Love (State of Colorado), Dr. Richard Meyer (Harvard School of Business), Governor Richard V. Ogilvie (State of Illinois), Henry S. Rowen (Rand Corporation), Leon Schachter (Amalgamated Meat Cutters and Butchers Workmen), Dan Seymour (J. Walter Thompson Company), and Frank Stanton (CBS). Also present were Vice President Agnew, Jerome B. Wolff (Office of the Vice President), DuBridge, Dr. Russell C. Drew (OST), Dr. Robert C. Seamans, Brig. Gen. Walter R. Hedrick (Defense), I. Nevin Palley (Defense), Paine, Dr. Homer Newell (NASA), Robert F. Packard (State), Donald Derman (BOB), Dr. Lewis M. Branscomb (PSAC), Dr. Thomas Gold (PSAC), and Milton Klein and I (AEC).

Vice President Agnew opened the meeting by emphasizing its importance and indicated that an opinion in writing is desired from those present and should be sent in sometime after they have returned home. He thought there should be no debate over whether we should have a space program; the technological gains and the contributions to the spirit of our country make it clear that we should have such a program, and the only question is the level. The dynamics of discovery cannot be quantified. We have just begun to apply space technology to our domestic needs, and the potential is great. He raised such possibilities as our potential for suborbital space travel and mentioned the potential for learning more about our weather. He said that we must structure our space program so that the next generation will respect our efforts. When the vice president concluded, he went around the table and shook hands with a number of people.

Wolff took over when the vice president left for another engagement. He called on Paine and Seamans, who each gave very short greetings.

Wolff then called on DuBridge who, in turn, introduced the visitors and many others in the room. He described the Space Task Group and its assignment. There will be thousands of years of space exploration ahead, and the Apollo program is of historic importance. Other projects are also important, and he mentioned the two space probes now on the way to Mars. Human travel outside our solar system might be out of the question because travel to the nearest stars takes thousands of years. He mentioned the many uses of satellites, including synchronous, for navigation, communications, geologic measurements, survey of agricultural status, meteorology, and weather. But, above all, he said, exploration lifts people's spirits and minds. He also mentioned the defense and security aspects of the program. In his view, space activities do not compete with welfare programs; in the past 10 years only about $34 billion has been spent on space versus about $524 billion on welfare programs. He said space is a subject that requires education of *the people and a propaganda effort. The Space Task Group has not yet set a dollar value for the space program but rather is concerned with setting its goals. . . .*

Illinois Governor Ogilvie emphasized the tremendous amount of goodwill shown toward the U.S. astronauts in their travels. He also mentioned the significant economic impact of the NASA program on industry. Paine agreed but said this economic impact alone isn't sufficient; the program must be worthwhile. He did mention, however, the great impact on some communities in the South when NASA programs are cut out.

As the morning part of the meeting came to a close, Paine and DuBridge indicated that transcripts or summaries of statements will be sent to participants. The meeting continued in the afternoon, but I had other commitments and left after lunch. After the morning meeting, sandwiches and cake were brought in for the group, and we had a standup lunch during which we talked in small groups.

In my conversation with Mrs. Shirley Temple Black, I recalled our meeting in June 1934 at Lake Arrowhead when she was filming her movie "Now and Forever" with Gary Cooper. We also recalled the University of California Alumni Association annual dinner that we attended at the Palace Hotel in San Francisco about 10 years ago, while I was chancellor at Berkeley.

I mentioned to Frank Stanton my interview with Steve Rowan of CBS in which I reminisced about discussions held with President Kennedy and others during March, April, and May 1961, while the man-on-the-moon program was being developed. He told me that Walter Cronkite had a very successful interview with President Johnson (at his ranch in Texas) about these discussions, which led to the adoption of our national objective of putting a man on the moon before the end of the decade. This interview was part of a series of some 12 interviews—a sort of verbal diary—for which they have contracted with President Johnson. . . .

From my diary, I recount my attendance at the first meeting of the National Aeronautics and Space Council (NASC) held under the chairmanship of Vice President Agnew on Oct. 28, 1969:

From 2:35 to 3:25 p.m. I attended the NASC meeting in Room 170 of the Executive Office Building. Present were Agnew, William A. Anders (executive secretary), U. Alexis Johnson (State), DuBridge (OST), Seamans (secretary, U.S. Air Force), Paine (NASA), James R. Schlesinger (BOB), and I. Others present were Herman Pollack (State); I. Nevin Palley (Defense); Milton W. Rosen (NASA); Milton Klein (AEC); C. Stanley Blair and Jerome B. Wolff (Vice President's Office); and Russell W. Hale, Winfred E. Berg, William E. Thurman, and Alfred C. Barbee (NASC staff).

Vice President Agnew opened the meeting with a short statement, recognizing that this was the first meeting of the reorganized NASC and suggesting that it has an important role to play. He said it will be necessary to find the proper balance between attention to astronautics and aeronautics. He also mentioned that the fate of the Space

National Aeronautics and Space Council meeting chaired by Vice President Spiro T. Agnew, Executive Office Building, Washington, D.C. Oct. 28, 1969. Left to right: Lee A. DuBridge (Presidential Science Adviser), Seaborg, Alexis Johnson (State Department), Agnew, William A. Anders (Executive Secretary, Space Council), Robert Seamans (Defense Department), Thomas Paine (Administrator, NASA), and James Schlesinger (Bureau of the Budget). Courtesy of the National Archives and Records Administration, Washington, D.C.

Task Group report and its recommendations have not yet been determined. He then called on Anders.

Anders described the changes in NASC that have taken place (mentioning that I am the only original member still present), the objectives of the NASC Act of 1958, and the objectives outlined in the Space Task Group report. He also discussed his plans for staffing NASC. He is asking BOB for six more people in the 1971 budget and has received the loan of a number of people from DOD, NASA, AEC, and other agencies.

Agnew complimented Anders for his fine report and for the progress that is being made under his leadership. He then called on him to describe items for future NASC meeting agenda. Anders mentioned the following items: (1) develop activities of the council and staff in relation to Space Task Group recommendations; (2) survey aeronautical policy from a national viewpoint; (3) define policy implications of navigational satellite systems for civil, military, and international use; (4) develop recommendations for national and international policy for the operational use of earth resource satellite programs; (5) assist in defining government's responsibility in the development of commercial domestic satellite systems; (6) provide a recommendation with respect to national policy on international cooperation in both aeronautics and space; and (7) develop a mechanism to provide recognition of noncouncil member agencies responsible for or affected by aeronautical and space activities.

Agnew then called on Johnson, who emphasized the importance of items 6 and 7, especially the international aspects. In response to a request for my comments, I made the point that item 1 is a broad assignment. DuBridge commented on the important role of international cooperation, both bilaterally and multilaterally. He said that there is a sort of crisis in the science component of the program and that some scientists have even criticized the STG report. Clearly there is a public relations job to do, and he emphasized the need to get academic scientists to help bridge the gap by working in the space program. Schlesinger emphasized the budgetary aspects and said that the president has suggested budget sharing in the international program. Paine emphasized the international flavor of all the agenda items.

I asked Paine whether the Soviets actually have made a lunar landing program a very low priority, as recently stated in the newspapers. Paine replied that intelligence reports show that the Soviets definitely do have a moon program and that recent statements by Mstislav Keldysh (president, Soviet Academy of Sciences), who often shoots from the hip, were probably meant as a cover for some of their failures.

Agnew finally called on Seamans for his comments, and he emphasized the importance of the space shuttle and the satellites.

Agnew then drew the meeting to a close. He said that today's meeting only served as a testing ground for future meetings, which will be devoted to more specific topics.

President Nixon responded to our invitation to meet with NAS and NSF representatives to discuss a touchy subject on Monday, April 28, 1969:

Members of the NAS council and the NSF board met with President Nixon at 3:30 p.m. to discuss the Franklin A. Long affair. The president thinks he made a mistake in not appointing Long (Cornell University, former assistant director, ACDA) as NSF director (because of Long's opposition to the ABM). He said that, therefore, he offered the position to Long but Long has decided not to accept because he finds the political situation distasteful and prefers not to have it reopened. President Nixon referred to his interest in the international exchange of scientists and mentioned that the 100th anniversary of the Mendeleev Periodic Table was about to be observed. (This had been called to his attention in one of our reports to his office.) The president mentioned my role in this and made reference to the name of Element 101 after Mendeleev.

President Nixon had rejected the suggested nomination of Long as NSF director because he objected to Long's vocal opposition to his plans to initiate an ABM program. In retrospect, Nixon's insistence that his appointees agree with him on all issues should have been a warning for me of things to come. I also opposed the ABM program and was not willing to make speeches in support of it, as Nixon wanted me to do. This, combined with his basic lack of respect for the potential contributions of prominent scientists like Long, made my job considerably more difficult.

I soon found that President Nixon's modus operandi differed substantially from those of Presidents Kennedy and Johnson. Rather than report directly to the president, as I had with Kennedy and especially with Johnson, I reported to Nixon through a sequential series of intermediaries. Also, I attended cabinet meetings and National Security Council meetings only on rare occasions. Nixon also eliminated the Committee of Principals, which had dealt with arms control matters in the Eisenhower, Kennedy, and Johnson administrations. DuBridge had less and less authority and was soon bypassed on matters that had been within the province of his predecessors. Although not cut off as completely as DuBridge, I soon found that I was often bypassed in discussions of arms control matters. There appeared to be an attempt to change, as much as possible, all modes of operation used by previous administrations—as illustrated by the fact that the National Security Action Memoranda (NSAM) issued by Presidents Kennedy and Johnson were replaced by National Security Study Memoranda (NSSM) and National Security Decision Memoranda (NSDM). As another example of such change, BOB was given the new name Office of Management and Budget (OMB). President Nixon's desire to tinker with systems that were working well, simply for the sake of change itself, would lead to a proposal to reorga-nize AEC in a potentially disastrous way. (I will describe this in greater detail later in this chapter.)

Illustrative of my mounting problem was the admonition I received from President Nixon at an early meeting on arms control (April 30, 1969). He asked me to confine my advice to scientific matters and not include matters with political implications:

. . . President Nixon recalled that, in the case of the LTBT, he thought it was a close question but decided, on balance, to come out in favor of it. He wasn't sure whether this had been the correct decision and asked my opinion. I said that I thought the LTBT had been clearly to the advantage of the United States, to the Soviet Union, and to the world. I said that its main benefit had been its effect in slowing the arms race. Another beneficial effect was in stopping radioactive fallout from atmospheric testing; had atmospheric testing continued unabated, fallout could have reached undesirable levels.

President Nixon then asked my view on the Seabed Treaty, and I indicated that, on balance, I was in favor of it. I said I thought it would be a mistake to reverse our position on this at ENDC now, after we had come out in favor of it. The president indicated that he was most interested in my technical judgment and not my political judgment. (I believe he was speaking here largely for the benefit of Secretary Laird and Vice President Agnew, who had spoken so definitely against the Seabed Treaty.) I said that even in confining my judgment to the technical aspects, which is difficult to do in this case, I thought the Seabed Treaty is to our national advantage. . . .

The president's comment, that he was more interested in my technical judgment than my political judgment, gave me much to ponder. Apparently, my possession of scientific expertise disqualified me as a source of political judgment. As AEC chairman during the Kennedy and Johnson years, I had been an active participant in all

"I soon found that President Nixon's modus operandi differed substantially from that of Presidents Kennedy and Johnson. Rather than report directly to the president, as I had done with Kennedy and especially with Johnson, I reported to Nixon through a sequential series of intermediaries."

> "I learned a few days later that President Nixon had ruled against me on practically every item. I was allowed no more appeal sessions with the president."

aspects of formulating arms control policy. President Nixon was telling me that this probably would not continue, and I later found that I was kept informed but no longer participated in policy deliberations. DuBridge later found himself cut off completely even from information about such deliberations.

In contrast to my successes with President Kennedy, and especially with President Johnson, in appealing budgetary decisions, I had essentially no success with President Nixon. My first and last appeal session with President Nixon was held on Tuesday, Dec. 23, 1969:

I then went to the White House to meet with President Nixon concerning our appeal on the FY 1971 budget. The meeting, held in the Oval Office, lasted from 4 to 4:30 p.m. Others present were Robert P. Mayo (director, OMB), James R. Schlesinger (deputy director, OMB), John D. Ehrlichman (assistant to the president for domestic affairs), and Henry A. Kissinger (assistant to the president for national security affairs). The president began by saying that this is a tight budget year, particularly because of the need to make up a shortfall of $2.5 billion as a result of recent congressional action. He then called on me to present my case. He referred to some briefing papers as I talked. Actually, it was pretty much a monologue; I presented arguments with very little comment from the others. Mayo and Schlesinger made essentially no comments, and Ehrlichman and Kissinger made a couple of comments. The president pretty much limited his remarks to calling for the next item as I proceeded and indicated at the end of each item that he understood the issue, making notes on his legal pad as I talked. . . .

To conclude the meeting, the president repeated that this is a tough budget year and that things might be better next year.

He referred again to the $2.5 billion shortfall. He indicated, somewhat enigmatically, that the AEC budget had been well prepared and [inferred] well treated. The president thanked me for my presentation. This budget appeal session was in sharp contrast to my

meetings with Presidents Kennedy and Johnson, with their give-and-take discussions. I have the impression that President Nixon had his mind made up before I came and will rule against me on practically all of my appeal items.

A grim-faced Paine [administrator of NASA] was in the waiting room and was asked to enter just as I was leaving.

I learned a few days later that President Nixon had ruled against me on practically every item. I was allowed no more appeal sessions with the president.

Another area in which I did not endear myself to President Nixon was my attitude toward his planned ABM safeguard system. I received word that the White House wanted me to make speeches promoting this program. I believed that such a program would be ineffective, excessively expensive, and dangerously provocative. As AEC chairman, I felt it my duty to provide for the testing of any weapons adopted as part of our national program; however, I did not consider it my duty to make speeches in favor of a policy being debated by Congress.

To his credit, Nixon later changed his mind, which led to an ABM treaty that limited the testing and development of ABMs and prevented the deployment of nationwide ballistic missile defense systems. He signed the Instrument of Ratification of the NPT and negotiated the SALT I Treaty which, although more limited in scope than some of us wanted, was a move in the right direction.

The intensity of the nuclear arms race at this time is dramatized by this entry from my diary for Wednesday, Oct. 14, 1970:

We learned that the Chinese exploded a 3-megaton nuclear weapon in the atmosphere at Lop Nor this morning and the Soviets a 6-megaton underground nuclear explosion at Novaya Zemlya. TIJERAS, the second test of the EMERY series, was detonated at 10:30 a.m. at the Nevada Test Site today.

Illustrative of our problems is the phone conversation I had with Don Rice (OMB) on May 3, 1971:

I called Don Rice about our budget amendment, on which we received word last Friday. I told him we are in trouble and really need something for water reactor safety and the salt mine nuclear waste disposal problem, which were cut out entirely. I asked if it would be possible to allow us something for these items. He asked if the water reactor safety is research work, and I said that it is and must be done on a crash basis because we have run into some problems. On the salt mine disposal problem, I said we are being pressured by the Kansas people, and he asked if there are other waste disposal areas to use. I said that in the long run the salt mine is the only solution.

The salt mine we were investigating was situated near Lyons, KS, and Kansas Senator Bob Dole gave us a lot of trouble in this connection. It was particularly unfortunate that we weren't allowed funds for research on water reactor safety in order to further improve their safe operation. It is interesting to speculate that if we had been given more support for our effort to solve the nuclear waste disposal problem at this time, our country might have achieved an acceptable solution during the intervening 20-some years. Now it appears that a solution will not be available for at least another 20 years. Another episode that illustrates the occasional rough treatment administered by the Nixon hierarchy involved the attempted removal of the security clearance of a respected scientist without benefit of the customary hearing process.

My diary for Feb. 18, 1969, contained the following entry:

I discussed with members of my staff the growing concern that an officer of a certain industrial nuclear facility may have diverted appreciable amounts of enriched uranium-235 to Israel over the last several years. This possibility has apparently been brought to the attention of the president.

A letter from FBI Director J. Edgar Hoover to AEC Director of Security William T. Riley, strongly suggesting that AEC might wish to revoke the individual's security clearance and cancel the facility's classified AEC contracts, attracted our attention. Hoover made it clear that the suspected offenses included not only the diversion of material but also the divulging of classified information to representatives of Israel.

As was true of virtually all communications about the matter at the time, neither my diary entry nor Hoover's letter identified the facility or the individual by name. The case seemed so sensitive that we sought to avoid all risk of public disclosure. But the story has now been the subject of extensive public comment, and the need for caution no longer exists. The company involved was the Nuclear Materials and Equipment Corporation (NUMEC) of Apollo, PA, located 30 miles northeast of Pittsburgh. The individual in question was NUMEC's founder and president, Dr. Zalman Mordecai Shapiro.

On the basis of all available information, AEC commissioners unanimously believed that Shapiro and NUMEC had not diverted any nuclear material to Israel or to any other country for transshipment to Israel. We found no evidence to support a charge of diversion. It was hard to believe that such an effort could have occurred without some trace of evidence for AEC, joint committee, and FBI investigators. The idea that many of the employees in the plant would have been in the dark about the diversion seemed impossible. For Shapiro to have committed an act of disloyalty to the United States was inconsistent with his character as we came to know it. We had a simpler explanation for the losses. This was that NUMEC had subordinated other considerations to the pursuit of profit and, encouraged by AEC's lax enforcement, had adopted shortcuts in its processing that led to excessive and irretrievable losses of material. Though we solved the case to our satisfaction, we still had to confront the fact that powerful forces in the government had reached an opposite conclusion.

Beginning about the time when J. Edgar Hoover wrote his letter to AEC on Feb. 18, 1969, news about the NUMEC case had spread widely in the upper echelons of the Nixon administration.

On Feb. 25, 1969, I went to the Justice Department to discuss Hoover's letter with Attorney General John M. Mitchell. Mitchell immediately noted that President Nixon was personally interested in the case. I opined that it would be a mistake to prosecute Shapiro on the assumption that an adequate case could be made against him.

I wrote the attorney general a long letter on April 3. I noted that it would be necessary under AEC regulations to prepare a letter of notification to Shapiro before we could revoke his security clearance as the FBI director had suggested. The letter would set forth any derogatory information and offer him a hearing before a personnel security board.

I wrote to Mitchell on Aug. 27, 1969, stating that, on the basis of the information developed during an interview with Shapiro, AEC "does not contemplate further action in this matter at this time."

When Hoover replied a week later, he strongly implied that AEC was making a mistake in not revoking Shapiro's security clearance. He stated that the FBI's "thorough and extended investigation of Shapiro for more than a year" had "developed information clearly [pointing] to Shapiro's pronounced pro-Israeli sympathies and close contacts with Israeli officials, including several Israeli intelli-

gence officers. . . . The basis of the security risk posed by the subject lies in his continued access to sensitive information and material. . . and the only effective way to counter this risk would be to preclude Shapiro from such access." But, because AEC planned no further action, the FBI would also discontinue its active investigation.

In October 1970, Shapiro accepted a position with Kawecki Berylco Industries, Inc. (KBI), a Pennsylvania-based metals-processing company that required him to have access to weapons information. Shapiro applied to AEC for the upgraded ("Sigma") security clearance that would permit such access.

AEC security chief William Riley and AEC attorney Sidney Kingsley met on Dec. 8 with Assistant Attorney General Robert Mardian. After returning from the meeting, Riley and Kingsley briefed the members of the commission. As I noted in my diary:

Mardian had told them that it was the considered opinion of the Department of Justice, including Mitchell, and of the White House, perhaps including Kissinger and even the president, that the Sigma clearance should be denied to Shapiro without a hearing.

This caused consternation among the commissioners because it would be the first instance of such a peremptory action in AEC's history. It was agreed that I would talk to Mitchell, apprise him of our views, and seek to determine whether some way might be found to handle the situation through informal contacts with the subject or his company.

Soon afterward, we learned that Shapiro had hired one of the most skillful attorneys in the country, Edward Bennett Williams, and his associate, Harold Ungar. It was evident that Shapiro intended to fight to clear his name.

Accompanied by Riley and AEC general counsel Joseph Hennessey,

I met with Mitchell on Jan. 21, 1971. Mardian was present. It was clear that a gulf existed between AEC and the Justice Department.

I said I thought the charges were essentially without substance and that I strongly opposed denying a clearance without going through the hearing process. I said this had never been done by any government agency. Mitchell said he felt the charges were serious enough that the man should not have access to sensitive weapons information. He thought the case should be settled by the courts. I emphasized that this could lead to a sensational public relations problem because the man was being defended by very prominent counsel and that they intended to put up a public fight to defend his honor. I suggested that other executive departments be consulted for advice, and he suggested that I get in touch with Kissinger and Secretary Rogers.

I saw Rogers, Kissinger, and Science Adviser Edward E. David within the next few days. All three seemed to agree that it would be inappropriate to deny clearance without a hearing. David mentioned the danger that scientists, still smarting from the 1954 proceedings in which J. Robert Oppenheimer's security clearance had been revoked, might protest yet another mistreatment of one of their number. He said he would be ready to talk to Mitchell about it.

Williams called me on Feb. 3, urging us to reach our clearance decision before his client faced the possibility of losing his job with KBI.

The next day, David called. He and Peter Flanigan (presidential assistant) had urged Mitchell to work in accordance with tradition and accepted procedures in determining clearance status. We also learned that Kissinger had called Mitchell to the same effect. The result of all these interventions was communicated by Mardian on Feb. 5: Mitchell, he said, still believed that Shapiro should be denied a clearance without a hearing.

The Justice Department's position seemed to harden despite overtures for a compromise. The commissioners then decided that the only way out of the dilemma was to find a job for Shapiro that did not require a Sigma clearance. On April 1 we learned that John Simpson, president of power systems of Westinghouse, and another Westinghouse executive had talked to Shapiro and offered him a position in a senior technical-advisory capacity. For the first time, Shapiro learned of the difficulties involved in upgrading his security clearance. He accepted the Westinghouse offer and, about two weeks later, formally withdrew his request for the Sigma clearance.

Six months later, after I had left the AEC, I chanced to talk to Williams on another matter.

He expressed satisfaction with the way I had handled the case involving his client. I told Williams there was more to the case than he knew and that I would reveal more details to him someday. Knowing of my interest in the Washington Redskins, Williams, the team's owner, then expressed his opinion that quarterback Billy Kilmer was a better team leader than Sonny Jurgensen had been, even though he was not as good a passer.

After 1971 Shapiro occupied positions of increasing responsibility with Westinghouse until his retirement in 1983, after which he continued to work as a consultant. Though he has had a distinguished career, one wonders how even more illustrious it might have been without these unjust charges leveled against him.

Even the White House dinners were different, as illustrated by the one that Helen and I attended on May 6, 1969:

Helen and I attended a dinner (white tie) at the White House given by President and Mrs. Nixon in honor of the Australian Prime Minister John Gorton and Mrs. Gorton. The arrangements were quite different from those for corresponding events during the Johnson administration. Upon arrival, the guests mingled in the main foyer and the Green Room. Then President and Mrs. Nixon and Prime Minister and Mrs. Gorton formed a reception line in the Blue Room, and the guests went through the reception line but not in a designated order, as had been the custom in the past. When the president introduced me to the prime minister, we briefly discussed my visit to Australia and to the nuclear laboratory at Lucas Heights and mentioned the Australian interest in Plowshare projects. The president and party then emerged through the Red Room and went directly to the State Dining Room for dinner.

A new table arrangement was used—a horseshoe with a third long table bisecting the horseshoe (sort of an "E"-shaped table). I sat next to Mrs. Marshall Green (he is assistant secretary of state for far eastern and Asian affairs) and Mrs. Thomas Moorer (he is chief of Naval Operations) and near Shelby C. Davis (ambassador-designate to Switzerland). Helen sat across from me and next to David Packard (deputy secretary, DOD, and a longtime California friend), and Donald Clausen (congressman from California). Toward the end of the meal we were serenaded by the violins of an air force musical group. At the end of the dinner, President Nixon spoke extemporaneously in praise of the historic and fine relationship between the United States and Australia and gave a toast to that relationship, to Australia, and to Prime Minister Gorton. Gorton responded in a very eloquent, in fact, impressive manner, referring to the huge picture of Lincoln on the wall to his right and drawing an analogy to President Lincoln's and President Nixon's problems with a strong description of the extraordinary relationship between the United States and Australia.

After dinner, the guests assembled in the Red and Blue Rooms. Helen and I had a rather long talk with Mrs. Gorton and Mrs. Nixon in the Blue Room. Mrs. Gorton told us that she had been raised in Maine, finished her schooling there, and then went to Paris, where she met Mr. Gorton. Since that time she has lived in Australia. I told Mrs. Gorton about the fine nuclear laboratory at Lucas Heights that I had visited and suggested that she might like to visit it sometime. We discussed with Mrs. Nixon the new arrangements under which the dinner was carried out, and we also discussed President Nixon's heavy schedule.

The guests then assembled in the East Room for the entertainment: a concert by Grant Johannesen, pianist, and his wife Zara Nelsova, violoncellist. Dancing in the main foyer followed the entertainment, and Helen and I participated in this for awhile before leaving.

Another new experience for Helen and me was our attendance at the annual Republican party dinner on Thursday, Oct. 30, 1969:

Helen and I attended the 1969 annual Republican dinner, held in the Presidential Ballroom of the Statler Hilton Hotel. Preceding the dinner we attended a reception in the New York Room, which was given by Mr. and Mrs. Edmund Pendleton, Jr. (chairman, DC Republican National Committee), and the general reception in the Congressional Room.

We sat with Gen. and Mrs. Elwood R. Quesada (he is a retired lieutenant general, U.S. Air Force; chairman, L'Enfant Plaza Corporation; and longtime friend), Mr. and Mrs. James Lemon (former owner of the Washington Senators), Mr. and Mrs. Colburn (a banker), Mr. Whitney Gilliland (member of the Civil Aeronautics Board. I sat next to Mrs. Quesada, which gave me an opportunity to get better acquainted with her.

Others present included Treasury Secretary and Mrs. David M. Kennedy, HUD Secretary and Mrs. George W. Romney, The Honorable Rogers C.B. Morton (chairman, Republican National Committee), Washington Mayor Walter E. Washington (who left early after making a few welcoming remarks), Congressman and Mrs. Joel Broyhill (Virginia), Congressman and Mrs. Fletcher Thompson (Georgia), Congressman and Mrs. Clark MacGregor (Minnesota),

Congressman and Mrs. Donald Brotzman (Colorado), Mr. and Mrs. Russell Train (undersecretary of the interior), Mr. and Mrs. Rocco C. Siciliano (undersecretary of commerce), Mr. Robert Mayo (director, OMB), Mr. Paul McCracken (chairman, Council of Economic Advisors), Mr. and Mrs. Bryce Harlow (congressional relations adviser to the president), Mr. and Mrs. Charles A. Peacock (National Mediation Board), Mr. and Mrs. Lewis Strauss, C. Thomas Clagett, Jr. (chairman of the dinner committee), Mr. and Mrs. J. Willard Marriott (restaurant and hotel executive), and Mr. and Mrs. Carl L. Shipley.

Clagett, serving as master of ceremonies, introduced Pendleton, who made a short welcoming speech and introduced Morton, who gave the main talk. The thrust of his talk was somewhat defensive; he indicated that the Republican party should be more positive in its outlook than it had been in the past. The whole evening was more staid and less exciting than dinners sponsored by the Democratic National Committee.

On occasion President Nixon could be very gracious, as illustrated by the ceremony that took place in his office on Feb. 27, 1970, when, at my suggestion, he conferred Special Atomic Pioneer Awards upon Dr. Vannevar Bush, Dr. James B. Conant, and General Leslie R. Groves:

I then went to the president's Oval Office in the White House to attend a ceremony for the presentation, by the president, of the Atomic Pioneer Awards to Bush, Conant, and Groves. Others present were AEC Commissioners Ramey and Larson, Drs. John H. Bush and Richard Bush (sons of Vannevar Bush), Mrs. James B. Conant, Mr. and Mrs. James R. Conant (son and daughter-in-law), Mrs. Leslie R. Groves and Brig. Gen. and Mrs. Richard H. Groves (son and daughter-in-law), Senator John O. Pastore, and Congressman Craig Hosmer. Among those helping with the arrangements were Will Kriegsman (White House assistant to John Whitaker), Steve Bull (White House staff assistant), Edward Brennan (director, Office of Multilateral Policy and Pro-

grams, State), Bruce Whelihan (White House press attaché), and John Nidecker (White House congressional liaison). A number of movie and still photographers were present during the ceremony.

After we entered the Oval Office I greeted the president and then introduced him to Bush, Conant, and Groves. I also introduced him to Ramey and Larson, who followed the awardees into the room. The president also greeted Pastore and Hosmer.

The award ceremony took place in front of the president's desk. I stood at the president's right, and the awardees were at his left. I made a few introductory remarks about the achievements that had led to these awards and emphasized that they were the first and only Atomic Pioneer Awards, which could not be duplicated because of the unique nature of the recipients. The president took up this theme in his initial remarks, emphasizing that the awards were unique and especially distinguished. I read the citation for Bush, and the president presented him with a medal. The same sequence was followed for the other two presentations.

At the conclusion of the presentations, I remarked to the president that these three distinguished gentlemen had been in the Oval Office many times before to confer with his predecessors, notably Presidents Roosevelt and Truman. This led the president to inquire of the recipients whether they at any time had felt any doubts about the successful outcome of their endeavors to produce the nuclear weapons. The general tenor of their responses was that they had entertained such doubts, and each of them reminisced briefly. Bush recalled the story of an official who, having been queried about the possibility of failure of the initial atomic bomb test, responded that he had worried lest the calculations were wrong, leading to excess energy release that would have caused trouble; but he had been just as worried that the test might be a complete failure, which would have caused even more trouble.

President Nixon encouraged Groves to reminisce in a similar fashion, whereupon Groves recalled that Gen. Jack Madigan (assistant to Groves during World War II) had rented a double house near the Capitol near the end of the war and had invited Groves to

> "Bush recalled the story of an official who, having been queried about the possibility of failure of the initial atomic bomb test, responded that he had worried lest the calculations were wrong, leading to excess energy release that would have caused trouble; but he had been just as worried that the test might be a complete failure, which would have caused even more trouble."

move into the other half because it was evident that he would be testifying continuously and indefinitely before Congress in defense of the project.

The president then turned to Conant, who recalled that Undersecretary of War Robert P. Patterson had asked Gen. Brehon Somervell to make an investigation of where all the money was going in the Manhattan Project and to try to determine its chances for success. Somervell was given only a few minutes to make his report to the busy Patterson, but he said this would be sufficient time because his report would be simple—should the project be a success, there would be no need for an investigation as the war would end; and, should it be a failure, there would be no need to investigate anything else.

The president then asked Pastore to say a few words. Pastore spoke of the important contribution to free-world security that the three recipients had made and the debt that we all owe them.

The president then called on Hosmer, who expressed the regrets of Holifield (who couldn't be present) and spoke in laudatory terms about the contributions of the three recipients.

The president then had pictures taken, individually with each of the

Atomic Pioneer Awards ceremony at the White House. Washington, DC Feb. 27, 1970. Left to right: Seaborg, President Nixon, General Leslie R. Groves, Vannevar Bush, and James B. Conant. Courtesy of the U.S. Department of Energy, Germantown, MD. Photo by Ed Westcott.

three recipients and their families and with Ramey, Larson, and me and the three recipients.

The president then showed us some of the items in his office, singling out the tonged device the Apollo 12 astronauts had used to pick up rocks on the moon and that they had mistakenly taken with them back to Earth. The president said that he had kept them out of trouble by taking the device from them and displaying it in his office.

The president very graciously said individual goodbyes to those present. He was very relaxed and friendly throughout the ceremony.

The timing of this award was very fortunate for Groves; he died just a few months later, on July 13, 1970.

Another time when I benefited from President Nixon's hospitality was when I was privileged to watch the launch of Apollo 11 (for the first manned moon landing) at Cape Kennedy:

Wednesday, July 16, 1969

A number of NASA buses then took us to the grandstand area, where we were to view the launch of Apollo 11 (at a distance of about three and one half miles). I sat in a section of the grandstand with Mr. and Mrs. Lyndon B. Johnson, Jim and Patsy Webb, Vice President and Mrs. Agnew, Mrs. Thomas Paine, Mrs. George Mueller, Ramey, Flanigan, Tom Whitehead, and William Anders (the nuclear engineer astronaut who will be executive secretary of the space council). Hornig and DuBridge sat nearby in an adjoining section. Among the 5000 people in the general area were about 600 members of Congress (including one family member in most cases), several cabinet officers, about half the governors of the states, a number of science ministers from other nations, and the passengers of the Super-DC 8 that I flew down on yesterday and a number those from the Super-DC 8s that flew down this morning. . . .

Along with most of our nation and the world, I viewed on TV the astronauts landing on the moon and their return to Earth:

Former President Lyndon B. and Lady Bird Johnson at the Apollo 11 launch, Cape Kennedy, FL. July 16, 1969.

Sunday, July 20, 1969

I viewed the first steps on the moon by Neil Armstrong at 10:56 p.m., followed by Edwin (Buzz) Aldrin, on a TV set at the airport. I then took a taxi home and watched with my family on TV the various activities of Armstrong and Aldrin—gathering moon samples at the Sea of Tranquillity, and so on—until they returned to the Lunar Module about 1 a.m.

Monday, July 21, 1969

I continued to watch the Apollo 11 astronauts on TV, in particular, their lift-off from the moon at 1:54 p.m. to *rejoin the command module circulating around the moon.*

Tuesday, July 22, 1969

Between 9:10 and 9:30 p.m. I saw the TV broadcast in which the astronauts, in the returning Apollo 11, showed us containers holding samples of rock from the moon, a number of actions illustrating zero gravity, and a magnificent picture of the Earth.

Wednesday, July 23, 1969

At 7:30 p.m. I watched the TV broadcast from Apollo 11 in which each of the returning astronauts—

Apollo 11 launch site viewing area. Lady Bird and Lyndon Johnson are in the front row; James Webb (second from left) and Seaborg (extreme right) are in the second row behind the Johnsons. Cape Kennedy, FL. July 16, 1969. Photo by Herm Pollack.

Michael Collins, Aldrin, and Armstrong—made short statements describing in broad terms the support that had gone into their flight and the nature of the difficulties that had been surmounted. They thanked their colleagues and the American public for their help.

In recent years I have become well acquainted with Buzz Aldrin in connection with a business venture in which we both became involved. Helen and I were pleased to attend the formal state dinner in Los Angeles in honor of the Apollo 11 astronauts:

Wednesday, August 13, 1969

In the evening Helen and I attended the formal state dinner that President and Mrs. Nixon hosted at the Century Plaza Hotel to honor the Apollo 11 astronauts (Neil A. Armstrong, Lt. Col. Michael Collins, and U.S. Air Force Col. Edwin E. Aldrin, Jr.), their wives, and the historic achievement of the first manned landing on the moon. Some 1400 guests were present. The guests assembled in the foyer of the California Room for a reception before the dinner, which was held in the Los Angeles Ballroom. Here we met a large number of people, including Mr. and Mrs. Ed Nixon (he is president Nixon's brother), Mr. and Mrs. Tom Ryan (he is

Pat Ryan Nixon's brother), Buzz Aldrin's father and his sister, Walter Cronkite, and Mr. and Mrs. Bob Hope. Helen and I were interviewed by ABC-TV (on the nuclear rocket and the role of astronauts wives), which a number of our friends saw broadcast. I also was interviewed by CBS-TV on the nuclear rocket; this was shown on the 11 p.m. CBS news.

At about 8:15 p.m. we entered the ballroom for dinner. The tables were placed on three levels. The lower level was in the center of the room, with two semicircular tiers at higher levels. We sat at table number 98 (middle tier) with Ward Quaal (an Ishpeming-born friend, now chief executive officer of WGN, Chicago), Neil and Bess Reagan (brother of Governor Reagan), Mr. and Mrs. Gerhard Neumann (group vice president for Aeronautics, General Electric), Mr. and Mrs. Harold McClelland (dean faculty, Claremont Men's College), and Chow Shu-Kai, the Chinese ambassador. Mr. and Mrs. Bob Hope, Red Skelton, and David and Julie Eisenhower were seated near us on the upper tier.

Others present (usually with their wives) were the governors of 44 states, 14 members of the president's cabinet (all except Mitchell), eight senators (Senate Appropriations Committee for Space), 18 representatives (House Appropriations Committee for Space), joint chiefs of staff, many U.S. govern-

ment officials, White House staff, essentially all the astronauts (about 65), and foreign ambassadors or their representatives from about 85 countries.

The head table guests—President and Mrs. Nixon, Vice President and Mrs. Agnew, Mr. and Mrs. Neil Armstrong, Mr. and Mrs. Buzz Aldrin, and Mr. and Mrs. Michael Collins—entered at about 8:30 p.m. to "Ruffles and Flourishes" and "Hail to the Chief" by the Marine Band. White-uniformed Army Herald Trumpets signaled their entry to the ballroom.

The first toast—to the president—was proposed by Governor Reagan, who called the occasion "the nicest birthday present California has received in 200 years." The governor welcomed everyone to California and introduced President Nixon. The president began by introducing some of the more distinguished guests. During his remarks the president paid honor to Astronauts Virgil I. (Gus) Grissom, Edward H. White, II, and Roger B. Chaffee, who died in making this achievement possible. He called on Paine to read citations to these three men, a posthumous award of NASA's highest decoration, the Distinguished Service Medal. Paine also read the citation for a Group Achievement Award representing all those on the ground who made the venture to the moon possible; this award was accepted by Steve Bales, the flight control engineer on this project.

The president then called on Vice President Agnew, who presented the Apollo 11 astronauts with the Medal of Freedom, the highest civilian honor that can be presented to an American citizen, and read the citations. All three of the astronauts responded; first Collins, then Aldrin, and then Armstrong. I thought that Aldrin spoke best. After the astronauts finished speaking, the president asked everyone to remain standing and said it was his high honor and privilege at this point to propose a toast. He wanted to say very simply to our three astronauts, "We thank you for your courage. We thank you for raising our sights, the sights of men and women throughout the world to a new dimension—the sky is no longer the limit."

The program ended at 11 p.m.

My assistant, Justin Bloom, and I were given the opportunity to watch the second manned landing on the moon at NASA's Manned Spacecraft Flight Center (MSFC) near Houston. This was the occasion when AEC's SNAP-27 radioactive isotope (plutonium-238) was used as an electric power source:

Wednesday, Nov. 19, 1969

We were met by Robert Carpenter of the AEC Space Nuclear Systems Division and the man he has stationed at Houston, William Remini. We were given access to the "viewing room" of the Control Center, which was handling the Apollo 12 mission, and observed the color TV broadcast of Pete Conrad's and Alan Bean's first steps on the moon. We heard the voice communications concerned with the difficulty in removing the SNAP-27 heat source from its transport container and other events associated with the deployment of ALSEP (Apollo Lunar Surface Experiments Package). During this period I met Dr. Robert R. Gilruth (director, Mannei Spacecraft Flight Center) and Neil Armstrong, the command pilot for Apollo-11.

On Wednesday, Dec. 9, 1970, we celebrated the first anniversary of this use of radioisotopic power on the moon:

Helen and I attended a reception in the Arts and Industries Building of the Smithsonian Institution, given by the secretary of the Smithsonian Institution with AEC and General Electric Company in honor of the successful operation for one year of the first nuclear electric power device on the moon. The program consisted of a general introduction by Dillon Ripley, secretary of the Smithsonian. He then introduced me and I made a few extemporaneous remarks. I expressed my pleasure on this occasion and recalled that I had been at the NASA Manned Spacecraft Flight Center in Houston when Apollo 12 astronauts Bean and Conrad placed the SNAP-27 generator on the moon. I also mentioned my moments of concern when Bean had trouble removing the Pu-238 capsule from its transport container and my relief when this problem was solved through the use of a hammer by Conrad. I said that this

SNAP-27 project is in addition to the use of the SNAP devices in NASA weather satellites and will be followed by the use of SNAP devices in the Pioneer spacecraft of the Jupiter probe and on the generators that will land on Mars.

President Nixon fell heir to the U.S. involvement in the war in Vietnam. Although President Johnson had decided in spring 1968 to de-escalate the war, the hostilities persisted in a frustrating manner. I had an interesting talk with Averell Harriman about the Vietnam situation at the annual banquet of the International Platform Association (IPA) in Washington, DC, on July 24, 1969:

I talked with Averell Harriman. He told me that he is quite frustrated by the Nixon administration's attitude at the Vietnam peace talks in Paris. He believes the U.S. program of taking the offensive in the action in Vietnam has stopped progress in the Paris talks and is jeopardizing their success. He feels that, if he had stayed on the job as negotiator under a president sympathetic to his philosophy, he might have made a great deal of progress toward bringing the war to an end. He believes the administration has been following an erroneous and dangerous military and diplomatic strategy in Saigon. He deplores the policy of applying continuing pressure on the enemy to try to expand control of Vietnam territory. He believes that the U.S. should accept the present military and political status quo and not try to improve the military and political power of the Saigon government. He said that until we change our tactics there will be no serious negotiations in Paris. He would reduce the violence and accept the present military–political power balance and quit boasting about how many enemy we kill.

In spring 1970, after two years of indecisive action, Nixon decided to resume the bombing of North Vietnam and to invade neighboring Cambodia in order to neutralize action supporting Vietnamese in that region. This strategy, designed

to smash his way to victory in Indochina, provoked a storm in U.S. universities, a constitutional crisis in the Congress, and outcries across the world. This adverse reaction reached crisis proportions on Monday, May 4, 1970:

We received the shocking news this afternoon that national guardsmen had shot and killed four students at Kent State University in Ohio, during a student strike and riot directed against the U.S. sending troops into Cambodia.

Tuesday, May 5, 1970

My son Eric and his friend Scott Luria told me that they hadn't been in school today because Wilson High School is on strike as an aftermath of national guardsmen killing four students at Kent State University yesterday. There is widespread striking and rioting by college and even high school students as a result of this tragic incident.

Wednesday, May 6, 1970

While I was on my way home, my assistant, Julie Rubin, called me on the car phone to inform me that Ellison Shute (area manager of AEC's San Francisco office) has advised headquarters that Governor Reagan has just announced that the University of California and the California State College system will be closed for the balance of the week.

The situation required a briefing of federal officials by leaders in the Nixon administration, which took place on Wednesday, May 20, 1970:

From 2:40 to 4:20 p.m. I attended a briefing on the Cambodian situation in the West Auditorium of the State Department.
The briefing began with the entry of Vice President Agnew, Ehrlichman, Secretary Rogers, and Secretary Laird.
The vice president opened the meeting with general remarks on the campaign commitments of President Nixon and the policy of the administration. He said he became convinced yesterday that we could win a land war in Southeast Asia—the president asked him to go over there with his four wood

and his tennis racket. He said that, unfortunately, we can't rely on the news media to present an unbiased picture and that we can't have our policy made on the editorial pages or even entirely in Congress. He said this is a most important meeting and he has full confidence in the fairness of the people conducting the briefing. Then he had to leave and turned the meeting over to Ehrlichman.

Ehrlichman said the governors have received part of the information that we will receive this afternoon; also, we will hear from those who briefed the cabinet recently, so we will get the best of both briefings. He then introduced Rogers, who said he would have to leave soon, but Assistant Secretary of State Marshall Green would continue on behalf of the State Department. He said we had inherited the war in Vietnam and the foreign policy and that people who ask why we are there and why we do not leave show ignorance. He said we have security treaties with Thailand, the Philippines, South Korea, Japan, and other countries, and we have a commitment to live up to the treaty obligations. Since we couldn't just walk out of the Vietnam situation and we couldn't bombard North Vietnam, the president decided on the Vietnamization procedures as the only sensible choice. After making that decision, we can't just get out. He said when the end of June comes and we get out of Cambodia, the suspicion will disappear. The results will show the young people that we were right. The young people don't seem to want to concede that the president may be right, but they will know this by the end of June. He said all nations in the area support the president. The foreign minister of Germany made a supporting speech after his return to Germany from a visit to Cambodia. He said we should support the president. . . .

Ehrlichman then introduced Laird, who made the following remarks. In January 1969 the new administration found a plan to turn over responsibility to South Vietnam for meeting the Viet Cong threat to South Vietnam. But this was premature and didn't sufficiently take into account the North Vietnamese. So the administration developed the Vietnamization program in which the United States will help South Viet-

nam to become self-sufficient by the summer of 1971. But a great proportion of North Vietnam's supplies came through Cambodia. The North Vietnamese had plans to attack U.S. forces from these sanctuaries in order to cause huge casualties. To protect U.S. forces in South Vietnam, to help the Vietnamization program, and to speed the date when the U.S. forces could withdraw from South Vietnam, it was decided that it would be in the best interests of the United States to attack the sanctuaries in Cambodia. The success should not be judged on the basis of tactical success, but rather on the strategic, more long-range success of the reduced casualties in the last two quarters of this year and on our increased ability to withdraw U.S. forces. There will be some reverses, but these difficulties will be small compared with our strategic successes. . . .

These presentations were followed by a question-and-answer period. Laird said we will definitely be out by June 30; however, there will be continued incursions, solely by the South Vietnamese, after June 30. Green said we were not asked by the Cambodians to come in as this would have impaired their neutrality. They are pleased but dare not say so because of the effect on their neutrality. We had absolutely nothing to do with Norodom Sihanouk's (prince of Cambodia) ouster. . . .

Within three months the U.S. campaign in Cambodia ended, but the bombing of North Vietnam was renewed. South Vietnamese forces played an increasingly larger role in the war, as evidenced by casualty photos that were more numerous in 1971 for Vietnamese troops than for U.S. soldiers. By fall 1972, Kissinger and Le Duc Tho (secretary general of the Communist party of Vietnam) were negotiating for peace in Paris. Talks were broken off in December 1972, when the parties were unable to separate military and political issues (which would prove to be necessary for a peace agreement); and, by Dec. 16, 1972, President Nixon ordered the most extensive bombing of North Vietnam ever to take place. In spite

> "Talks were broken off in December 1972, when the parties were unable to separate military and political issues . . . and by December 16, 1972, President Nixon ordered the most extensive bombing of North Vietnam ever to take place."

of this, talks were resumed, a peace was forged, and all U.S. troops were finally out of Vietnam by March 1973.

I had an interesting discussion with former Secretary of Defense Robert S. McNamara at about this time (Thursday, April 16, 1970):

McNamara said that he is interested in the role of nuclear weapons, arms limitation, and SALT talks in the new administration. I told him that these matters are handled somewhat differently and that the Committee of Principals has now been disbanded. I indicated that the incentive to move

forward in these matters is not as great as in the previous administration, but that it is changing, partly as a result of the changing attitude of Congress and the American people. He expressed great satisfaction at the achievement of the LTBT in the Kennedy administration. We agreed that at that time it was a matter of the administration leading the Congress and the American people, whereas now it is largely the other way around. I said I feel that the attitude of the present administration has undergone quite a change just within the past year as a result of the general movement toward arms limitation in the Congress and by the American people and, as a result, perhaps within another year

President Richard M. Nixon's memorandum reappointing Seaborg as chairman of the Atomic Energy Commission. July 24, 1970.

ORDER

Pursuant to the provisions of the Atomic Energy Act of 1954, I hereby designate Glenn T. Seaborg as Chairman of the Atomic Energy Commission.

[signature: Richard Nixon]

THE WHITE HOUSE,
July 24, 1970.

XBL 923-497

the position of the administration will be quite satisfactory on these issues.

Because I had requested from President Johnson, at the expiration of my term as AEC chairman in 1968, a reappointment for a two-year term, this term came to an end on Aug. 1, 1970. Encouraged by strong endorsements by industrial and other leaders in the atomic energy community, President Nixon reappointed me as AEC chairman with the understanding that I intended to leave the position after one year, in summer 1971.

Theos J. Thompson (MIT nuclear engineer) was sworn in as commissioner on June 12, 1969 (he had a tragic death in a plane crash in the line of duty on Nov. 25, 1970), and Clarence E. Larson (president, Union Carbide nuclear division) was sworn in on Sept. 2, 1969, replacing Commissioners Tape and Francesco Castagliola.

DuBridge's position as science adviser to President Nixon and director of OST became untenable because he was increasingly ignored, and he resigned in 1970. President Nixon preferred a scientist with an industrial background as DuBridge's replacement and therefore appointed David, who had been serving as the executive director of research for the Bell Telephone Laboratories in Murray Hill, NJ. I found that I was able to establish a reasonably satisfactorily working arrangement with David.

President Nixon initiated plans for the reorganization of the executive branch of the federal government that threatened the very existence of AEC. For this, he appointed Roy Ash (president of Litton Industries) to serve as chairman of the President's Advisory Council on Executive Organization. The plans, as they developed, would have had a disastrous effect on AEC. The AEC program of civilian nuclear power would be placed in a new Department of Natural Resources. The regulatory function of AEC would be placed into something like an Energy Regulatory Agency. Still further steps would involve taking the weapons function and putting it into DOD, the research function into NSF, and so forth.

This reorganization could have been implemented only as a result of congressional action. I participated in a series of meetings with the Ash council and its subcommittees and was able, step by step, to convince council members and the decision-making members of the Nixon administration that dismemberment of AEC was both unwise and probably not feasible because it would not get congressional approval. I persisted in recommending that a more sensible reorganization would be to place the governmental energy functions in a single agency that might be built around AEC. In this defensive

Clarence E. Larson is sworn in as a new AEC commissioner. Germantown, MD. Sept. 2, 1969. Left to right: Commissioners Theos J. Thompson, Wilfred E. Johnson, Seaborg, Larson, and James T. Ramey. Courtesy of the U.S. Department of Energy, Germantown, MD. Photo by Ed Westcott.

"A few years later, in January 1975, a reorganization was effected, which, incidentally, was along the lines that I had advocated."

stand, I had the support of many of my friends in Congress, especially JCAE members. The plans of the Ash council never came to fruition. A few years later, in January 1975, a reorganization was effected, which, incidentally, was along the lines that I had advocated. The centralized energy agency was created in the form of the Energy Research and Development Agency (ERDA), and AEC's regulatory function was placed in a new Nuclear Regulatory Commission (NRC). A couple of years later, effective Oct. 1, 1977, ERDA was expanded with a wider range of functions into the Department of Energy (DOE).

It is interesting that ERDA and DOE were, indeed, built around AEC, much along the lines of an editorial, "For a U.S. Energy Agency," that I published in the journal *Science* on June 16, 1972, soon after I left the AEC chairmanship.

Of the many social and technological challenges facing the American people, none is more central to our short- and long-term welfare than that of energy: We must have sufficient energy to meet our legitimate needs, and it must be clean energy. It is essential that our view of, and attack on, the energy problem be commensurate with its magnitude and character. Besides the social aspects of the problem—which include the necessity to change the energy-wasteful habits of our people—there remain diverse technical problems related to resource assessment and to the development of efficient and environmentally sound energy technologies and energy-storage and transmission systems.

A number of energy sources are available, and each one must be explored and developed. Each presents its own advantages and problems. Oil and natural gas are relatively clean, but the supply is dwindling and will be required as chemical raw materials. Coal is more abundant, but it is difficult to mine and burn without degrading the environment. We must im-

prove our fossil fuel technology—for example, coal gasification and liquefaction, fluidized-bed combustion, and oil-shale processing.

Nuclear energy is available because the nation has committed substantial resources to its development. The supply of nuclear fuel will last for hundreds of years if it is efficiently used in breeder systems. Solutions to the two major problems associated with nuclear power—isolation of waste products from the environment and adequate safeguards against a major accident—are being pursued vigorously.

Other clean and abundant sources of energy await development: solar radiation, the earth's heat, and the fusion of light nuclei. Only in the last of these, nuclear fusion, is the United States engaged in a serious, although not yet adequate, development effort. The heat of the earth's crust is sufficient to satisfy much of our energy requirements for hundreds of years if it can be extracted efficiently. For some of the near-surface geothermal sources, the present state of technology may suffice; for deeper-lying sources of hot rock, new technologies will be required. Much research and development will be required to determine whether it is feasible to collect and convert the enormous, but dilute, flux of solar radiation. However, if society chooses to invest sufficiently in solar and geothermal energy, it is possible that these technologies might be in widespread use by the end of the century.

Some problems of development are common to all of these diverse energy sources: namely, resource assessment; plant siting; the technologies of cooling, energy storage, and conversion; power transmission; and waste disposal. To develop and utilize these technologies in the most economic and expeditious manner, the coordination of U.S. energy programs must be the responsibility of a single government agency.

The Atomic Energy Commission has developed, over several

decades, a superb research base, with excellent laboratories and a tradition of successfully managing large projects in the public interest. It is already developing two of the principal sources of energy—fission and fusion—and in these programs has maintained close liaison and cooperation with the industrial sector. The Commission has the scientific expertise, technical capability, and organizational strength to develop the other energy sources as well.

No other agency of the federal government is in a more favorable position to launch a unified program for meeting the energy needs of the American people than is the Atomic Energy Commission. It should be transformed into the U.S. Energy Agency.
—Glenn T. Seaborg

On April 13, 1971, I was asked to brief the president's cabinet on breeder reactors:

I went to a meeting of the president's cabinet in the Cabinet Room of the White House, which began just after 8 a.m. and ran until 9:20 a.m. Present were the president, Laird, Postmaster General Winton Blount, Agriculture Secretary Clifford Hardin, Romney, Commerce Secretary Maurice Stans, Volpe, State Undersecretary John Irwin, OMB Director George Shultz, Counselor Bob Finch, Holifield, Anderson, Pastore, Baker, and I, around the main table. Also present were Cap Weinberger (deputy director, OMB), Flanigan, Ambassador David Kennedy (secretary, Treasury), McCracken, Ehrlichman, Herb Klein (director communications, Executive Branch), Ziegler, David, Dave Freeman, William Ruckelshaus (administrator, Environmental Protection Agency), Abe Lincoln, Clark MacGregor (presidential counselor for legislative and congressional affairs), John Whitaker, Hollis Dole (assistant secretary, mineral resources, Interior), Will Kriegsman, Richard Kleindienst (deputy attorney general), Milt Shaw, and others.

The purpose was to discuss the future of the fast breeder reactor, and it may well be recorded as a very historic meeting. The idea originated as a result of the discussion between Holifield and the president on an airplane trip to California. Although the primary reason, based on that discussion, is a consideration of the fast breeder [this is Holifield's point of view], the president also has in mind his reorganization plan and the possibility of enlisting Holifield's support.

When the president came into the room and after we took our seats, he immediately opened the meeting. (I had a designated seat at the end of the table to the left of the president, and the congressional members sat at the opposite end of the table to the right of the president.) The president referred to his discussion with Holifield and described Holifield's feeling that the breeder reactor should have top priority during the next decade. He also referred to the "early morning" meeting that he attended at the Bohemian Grove at which I had held the group in the camp spellbound with my talk on the peaceful uses of atomic energy. He recalled that the Atoms for Peace program had been launched in 1953 by President Eisenhower. He then called on me to make my presentation.

I opened with a reference to the Bohemian Grove gathering and added that since President Nixon and Governor Reagan were both present, the host had solved the problem of who should make some remarks by calling on me to speak on the peaceful uses of atomic energy. I indicated that the Atoms for Peace program had been very successful in giving the United States an entree into many countries of the world by helping them, and had also contributed to diminishing the threat of nuclear proliferation. I said that 25 countries are involved now in the development of nuclear power for the production of electricity. . . .

I then went on to my discussion of the fast breeder reactor. I explained briefly the role of plutonium in serving as a catalyst to make it possible to burn nonfissionable, fertile U-238 and emphasized that in this manner there would be enough uranium to serve as a fuel for thousands of years. I described the sodium metal-cooling aspects and indicated that the liquid metal-cooled fast breeder reactor, which I emphasized is becoming widely known as the LMFBR, is our main program. This is also the case in the Soviet Union, England, France, Japan, and other countries. I described the advantages of the breeder: It utilizes the uranium some 30-fold more efficiently, it will be more economic than even water-cooled reactors, it will have fewer radioactive effluents because it is hermetically sealed, and it will discharge less waste heat. . . .

I indicated that we are creating advisory committees to help us generally and to help raise funds. Members include Jack Horton (chairman, Southern California Edison), Shermer Sibley (president, Pacific Gas & Electric), Red Wagner (chairman, TVA), B.B. Parker (executive vice president, Duke Power Company), and others. I emphasized that both private and publicly owned utilities are represented. I said that the problem is to come up with the funding required to build this first demonstration reactor. I then mentioned the backup reactors: the molten salt reactor at Senator Baker's laboratory, the light water breeder reactor being developed by Adm. Rickover (both of which use slow neutrons and thorium as a fuel), and the helium gas-cooled alternate to the liquid metal-cooled fast breeder reactor. I then indicated that Milt Shaw was present to help answer other questions and said that I would rest my presentation at that point.

The president then called on Holifield, who emphasized the importance of electrical energy in cleaning up our environment and keeping our industrial economy going. He said the fast breeder should be the number one priority in the 1970s and that it is much more important to spend $2 billion to $3 billion this way than it had been to spend $50 billion to reach the moon.

The president then called on Pastore, who also indicated strong support for the fast breeder reactor. He said he thought that all 18 members of JCAE support this reactor and that we couldn't overestimate the importance to our future of having an adequate electrical energy source.

The president asked me how other countries are coming along on the breeder reactor, and I said that they are ahead of us. England is building a demonstration reactor (at Dounreay, Scotland, which I visited last September) of

250,000-kilowatt capacity, to be put into operation next year, and France is building a 250,000-kilowatt demonstration plant to come into operation in 1973. The Soviet Union is well along on a 350,000-kilowatt demonstration reactor and is also building a 600,000-kilowatt demonstration reactor. All of these will be of the liquid metal-cooled fast breeder type. The president asked me how it happened that these countries are ahead of us, and I indicated that they are only ahead with the construction of their first demonstration reactors (because they started construction before we did), but that if we speed up our program with the 1980 objective, we would get there before these others. In answer to another question by the president, I described the objective as a 1-million-kilowatt, reasonably economical, and reasonably low-doubling-time reactor. I explained that a fast breeder reactor forms about 1.4 kilograms of plutonium in the blanket for each kilogram of plutonium burned in the core, thus leading to a doubling time of about 10 years.

The president then called on Baker, who strongly supported the need for electric power via the fast breeder reactor and mentioned the possibilities of fusion. The president called on Anderson, who also strongly supported the fast breeder reactor. Both emphasized the need for a speedup in the program.

At one stage the president expressed the opinion that an increased effort in the breeder reactor might result in increased employment for some of the scientists and engineers who have become unemployed recently. . . .

The question of fusion came up, and I indicated that although research support for this should be increased, it is a problem of such complexity that we wouldn't have a practical plant operating before the year 2000. I said, however, that many people who talk with scientists (not so much scientists themselves) feel that we should skip the fast breeder and go directly to fusion power. I indicated that this is totally unrealistic.

Irwin also indicated strong support for nuclear power and the breeder approach to the generation of electric power. . . .

In summing up the meeting, the president indicated that he would discuss this further within the administration and would try to come up with a decision soon.

Beginning late in 1970, I served as a member of the domestic council subcommittee on the national energy situation. After numerous meetings, our output served as the basis for President Nixon's energy message to Congress. I participated in the press briefing on Friday, June 4, 1971:

At 11 a.m. we went across the street to the White House, and after Whitaker, David, Morton and I had waited awhile in the Cabinet Room, we went into the press briefing room with Zeigler and Neil Ball (Zeigler's assistant), where we were joined by others who had been present in Whitaker's office. The president entered about 11:05 a.m. and asked Morton, David, and me to join him on the platform. He made some opening remarks, saying that he was today sending Congress an historic energy message, the first to be sent by a president covering a program of providing enough energy for our future needs and providing it in a way that will protect our environment. He said that in the past these two goals have clashed, but in the future he feels it will be possible to provide sufficient energy while protecting the environment.

As he left the platform, the president called on Morton, who briefly described the energy message, following which questions were put to Morton, David, and me. I was asked about the projected electrical requirements in the future (including the nuclear component), the role of the breeder, and the planned program for the development of nuclear fusion. David gave an excellent statement on the importance of developing the breeder. I pointed out the particular role of the fast breeder demonstration plant and the need for the breeder. I emphasized that if we are going to meet our future electrical energy requirements, the breeder is a clean, safe, and economical energy source. I pointed out that AEC is developing fusion on a vigorous basis, but this energy source will not play an

THE WHITE HOUSE
WASHINGTON

June 11, 1971

Dear Glenn:

This is just a note to tell you how much I appreciated your participation at the recent press briefing last Friday. Your incisive responses to highly technical and difficult questions added greatly to the effectiveness and success of the briefing, and I thought your remarks about the breeder reactor were especially telling. As you know, I regard the matter of our future energy requirements as being particularly worthy of prompt consideration by the Congress, and I wanted to take this opportunity to express my personal thanks for your contributions on this important issue.

With my best wishes,

Sincerely,

Honorable Glenn T. Seaborg
Chairman
Atomic Energy Commission
Washington, D.C. 20545

President Richard M. Nixon's letter of appreciation to Seaborg for the AEC chairman's contributions to a press briefing on energy. June 11, 1971.

important role until the turn of the century. Following the briefing, I answered some questions from reporters who crowded around the platform. The questions concerned the emergency core cooling problem, the role of CIP [cascade improvement program], the tying-in of water-cooled reactors with breeders through their plutonium production, and other topics.

The president sent me a nice letter expressing his appreciation.

This optimistic outlook for developing the breeder reactor in the United States did not materialize, and the program was later terminated. (I should say, I was overly optimistic.) Although conventional nuclear power has grown to a point where it produces about 20% of the electricity in the United States, further growth in the total has virtually stopped because of a combination of low load growth, environmentalist pressure, regulatory uncertainty, and increased costs.

AEC and its staff enjoyed a good relationship with the U.S. nuclear industry throughout the 1960s. Illustrative of our communications is the meeting we held with the membership of the Atomic Industrial Forum (AIF) at the "Western White House" in San Clemente, CA, on Friday, May 8, 1970:

I had breakfast in the coffee shop of the inn with the AIF–AEC group, in a booth with Howard Nason (president, Monsanto Research Corporation), AEC Commissioner Clarence Larson, and Julie Rubin.

We then rode over to the Western White House in a number of cars. I rode in a car (driven by Leonard Lanni of the AEC Canoga Park Office) with Sherman Knapp (chairman, Northeast Utilities), Charlie Able (vice president, Douglas Aircraft), Dave Shaw (chairman, United Nuclear Corporation), and Julie Rubin.

We then met in the conference room. Ernest Gabarino [representing General Services Administration (GSA), which operates the premises], Knapp, and I gave opening remarks.

Knapp opened the meeting by calling on Merril Eisenbud (New York University Medical Center), who spoke on environmental law technology. He encouraged us to hold a conference between pro- and antinuclear people.

Philip N. Powers [president, Argonne Universities Association (AUA)] spoke on related research and development activities. Hollingsworth (general manager) and Joe DiNunno (special assistant to general manager for environmental affairs) commented on this and described our program and problems. DiNunno emphasized the importance of solving long-range prob-

lems such as the accumulation of nuclear power plants on Lake Michigan.

William R. Gould (senior vice president, Southern California Edison) discussed AEC responsibility under the National Environmental Policy Act.

Gabriel O. Wessenauer (consultant, TVA) spoke on various environmental bills.

We then took a coffee break, during which some snapshots were taken.

We resumed the meeting with a discussion of the plans for and future of uranium enrichment. Howard Winterson (vice president, Combustion Engineering Company) was the first speaker and he gave forecasts of future enriching requirements.

Manson Benedict (head, Nuclear Engineering Department, MIT), described the future enriched uranium production plans, including the use of CIP and the Cascade Upgrade Program. John F. Bonner (executive vice president, PG&E) and W. Kenneth Davis (vice president, Bechtel Corporation) discussed the Uranium Enrichment Directorate. Davis suggested that the directorate be replaced by a convertible corporation. Wilfrid Johnson replied, saying AEC tried to get authorization for a corporation but got no industrial or White House support. He said we will try to proceed with the directorate.

Johnson mentioned the gas centrifuge approach to enrichment, and then Kenneth R. Osborn (general manager, Nuclear Fuels Department, Allied Chemical Corporation) spoke about such alternate enrichment approaches. Osborn questioned the wisdom of the AEC policy banning industrial participation in gas centrifuge development. He said industry needs to know about this before it can invest in the gaseous diffusion process. Larson, Johnson, and I complimented Osborn on his statement, while Ramey said, sarcastically, that it was "a fine ideological statement." I asked that a copy of this statement be sent to the commission for its study. Osborn then spoke on "Nuclear Materials Safeguards."

John Landis (group vice president, Engineering and Manufacturing, Gulf General Atomic Inc., and a guest of AIF) made a statement on the Geneva IV Conference, and George Gleason (legal projects manager, AIF) spoke on "Prac-

During a break in a joint Atomic Energy Commission–Atomic Industrial Forum meeting, Seaborg and AIF Chairman Sherman Knapp step outside for some coffee. May 8, 1970. Western White House, San Clemente, CA. Courtesy of the U.S. Department of Energy, Germantown, MD.

tical Value and Regulatory Matters." He said AIF will comment on our recent change in Part 20 to keep radioactive effluents "as low as practicable."

At the end of the meeting, Knapp and I made a few comments expressing satisfaction with its results.

I rode back to the San Clemente Inn with Knapp and Rubin and told Knapp something about the debate among the commissioners on changing our policy of barring industry from gas centrifuge technology.

I had many contacts with Vice President Agnew during meetings of the space council (and the Space Task Group) and the marine council, on which I served as a member and he served as chairman.

Agnew resigned from his vice presidency in October 1973 and was immediately replaced by Gerald Ford.

The last biweekly report during my tenure as chairman was sent to the White House on July 27, 1971:

AEC Biweekly Report for July 27, 1971

1. Chairman Seaborg's resignation, to become effective at the conve-

nience of President Nixon, and his forthcoming return to the University of California, Berkeley, were announced on July 21, when the president also announced the nomination of Dr. Schlesinger and Mr. Daub as members of the Atomic Energy Commission.

2. A U.S. scientific delegation, headed by Chairman Seaborg, will visit peaceful nuclear energy facilities in the Soviet Union during August 4–20 at the invitation of the chairman of the USSR State Committee on Atomic Energy, who led a Soviet scientific group on a similar tour of the United States last April. The U.S. group will visit research centers, nuclear power stations, and universities in various parts of the USSR.

3. In connection with the 25th anniversary of the signing of the Atomic Energy Act of 1946, Chairman Seaborg will appear on "Meet the Press" on August 1. (The time of the broadcast may be delayed because of Apollo 15 coverage.) Later the same day, he and the other Commissioners will host a reception at the Department of State to which government offi-

cials, former Commissioners, and other dignitaries have been invited. Also a brief film clip on the anniversary will be distributed by AEC to television networks and local stations around the country.

4. The willingness of the United States government to undertake exploratory multilateral discussions with other nations interested in constructing uranium enrichment facilities based on U.S. gaseous diffusion technology has been made known to the European Communities office in Washington and the Washington embassies of appropriate foreign governments. Studies on additional enrichment facilities to be located abroad are now being made by various countries. The purpose of the talks would be to determine if possibilities exist for multi-national projects using gaseous diffusion technology under arrangements that would include appropriate security value of the technology. These talks would not involve a commitment at this time on the part of the U.S. to make such technology available for use abroad.

5. The third nuclear powered SNAP 27 thermoelectric generator will be deployed on the moon during the Apollo 15 mission. The first two such generators are still operating perfectly and transmitting signals from experimental packages left on the moon from previous Apollo missions.

6. The Commission testified on June 20 at hearings on the nation's energy and fuel resources before the Subcommittee on Special Small Business Problems of the House Select Committee on Small Business. These hearings have dealt in part with the growing interest of oil companies in uranium.

7. Current magazine articles in the July *Nation's Business* and the August *Better Homes and Gardens* reflect favorable viewpoints toward the important role of nuclear power in helping to meet the increasing demand for electricity. Also, an article on the program to develop peaceful uses of nuclear

explosions by Walter Sullivan of the *New York Times* is expected to be published any day.

8. Chairman Seaborg is speaking at the annual meeting of the International Platform Association in Washington, D.C., on July 27. This is an organization of professional speakers, lecturers, program chairmen, and others interested in this kind of activity. Chairman Seaborg is a former president of the association.

Adm. Hyman Rickover briefed the commissioners and other AEC staff on Friday, July 30, 1971, on the accomplishments in the nuclear navy since I had taken over as chairman in March 1961:

He pointed out that at that time there were 15 operational nuclear-powered submarines, whereas there are 95 now. Today we have four operational nuclear-powered surface ships. Our nuclear submarine fleet had steamed 800,000 miles then, and during my stewardship our submarines and surface ships have steamed 17.7 million additional miles. In March 1961 we were operating cores that had an expected life of about five years. Today we have some cores with an expected life of up to 13 years. The total investment in a nuclear navy is $17 billion, including vessels authorized and R&D, and this is to be compared with the $24 billion for the development of the Apollo capsule.

My last days in Washington as AEC chairman contained these highlights:

Saturday, July 31, 1971

I sent a letter to President Nixon advising him that I am sending him a special commemorative medal of the 25th anniversary observance of AEC:

United States
Atomic Energy Commission
Washington, D.C. 20545
Office of the Chairman
The President
The White House
Dear Mr. President:

We are sorry to learn that you and Mrs. Nixon will be unable to join us in our observance of the 25th Anniversary of the signing of the Atomic Energy Act at the Department of State, Sunday afternoon, August 1, 1971. Under your administration, the Commission has become increasingly aware of its responsibilities in promoting and supporting research in the nuclear sciences, physical and biological sciences, in medicine and in industrial and space applications while making the atom ecologically safe for the world. It is therefore with the deepest respect and pleasure that we are sending to you a Special Commemorative Medal as a memento of this occasion.

My colleagues on the Commission join me in extending to you and Mrs. Nixon our very best wishes.

Respectfully,
Glenn T. Seaborg

Sunday, August 1, 1971

At 1 p.m. I appeared on "Meet the Press," which was taped for later showing because we were preempted by the TV broadcasts of the astronauts on the moon. Participating with me, besides Lawrence Spivak, were Irving R. Levine (NBC), John Finney (New York Times), Howard Simons (Washington Post), and Burt Schorr (Wall Street Journal). I met a number of other people, including Mr. and Mrs. Bill Leeds (NBC, New York), Mrs. Lawrence Spivak, and Mr. and Mrs. Joseph Dukert (she is coproducer of "Meet the Press," is a freelance writer, and was ranking aide to Agnew

before Agnew became vice president). The show seemed to go quite well. There were questions on the continued need for nuclear weapons testing, the need for more and more weapons, and particularly the need to modernize our tactical weapons; these were somewhat difficult to handle in view of the conflict between how I feel personally and my role as AEC chairman. Other questions involved the need for, and the safety of, the CANNIKIN test and the recent decision by the DC U.S. Court of Appeals concerning the need for more AEC involvement in the environmental aspects in the reactor licensing process.

Helen, Dave, Steve, Dianne, and I arrived at the State Department a little after 3:30 p.m. for the 25th anniversary observance of the signing of the Atomic Energy Act. This took place in the diplomatic functions area. Helen and I met and talked to nearly all of the 500 people present. It was particularly gratifying to have former Chairman David E. Lilienthal present as he had not been very much in evidence in AEC functions for many years. It was also gratifying to have former Chairman Lewis L. Strauss present, as well as Mrs. Gordon E. Dean, representing former Chairman Gordon E. Dean. The program went rather well, although it was a little long. Because the crowd had to stand throughout the ceremony, the people in the back of the room, who could not see the participants, got a little restless and their conversations could be heard during the speeches.

Seaborg reappears on "Meet the Press." Washington, D.C. Aug. 1, 1971. Interviewers, from left to right: John Finney (*New York Times*), Burt Schorr (*Wall Street Journal*), Howard Simons (*Washington Post*), and Irving R. Levine (NBC). "Meet the Press" host Lawrence E. Spivak and Seaborg are to the right. Courtesy of the NBC, New York.

Seaborg at the lectern during 25th anniversary observance of the signing of the Atomic Energy Act held at the Department of State. Washington, DC Aug. 1, 1971. Seated behind Seaborg, left to right (partial list): Wilfrid E. Johnson, Melvin Price, M. Sterling Cole, Lewis L. Strauss, Mrs. Gordon Dean, David E. Lilienthal, Edward E. David, John O. Pastore, Bourke Hickenlooper, Clinton P. Anderson, James T. Ramey, and Chet Holifield. Courtesy of the United States Department of Energy, Germantown, MD. Photo by Ed Westcott.

I opened the ceremony with some prepared remarks in which I reviewed the original objectives of the Atomic Energy Act of 1946 and assessed the degree to which these objectives have been fulfilled. I then called on Dr. Edward E. David, who gave some excellent words of greeting, including a message from President Nixon. I then introduced JCAE Chairman Pastore with an account of his contributions to the field of atomic energy and, just before he spoke, presented him with a commemorative plaque. Senator Pastore responded and included in his remarks some kind statements about my service as AEC chairman, with special emphasis on my role in international activities. I then called on Commissioner Ramey, who made some descriptive remarks concerning Chet Holifield and Mel Price preparatory to presenting them, as charter members of JCAE, with commemorative plaques. Holifield responded at some length, and Price made a very short response. I then introduced Commissioner Wilfrid Johnson, who made presentations of commemorative plaques to Mrs. Richard Lane (daughter of the late Senator Brian McMahon) and to former Senator Hickenlooper in recognition of their roles as former chairmen of JCAE. I introduced Commissioner Larson, who presented a commemorative plaque to Sterling Cole, a former JCAE chairman, and to Senator Clinton Anderson.

Finally I presented, with suitable comments concerning their contributions, commemorative plaques to former Chairman David E. Lilienthal, Mrs. Gordon E. Dean (on behalf of her husband, the late Chairman Gordon E. Dean), and former Chairman Lewis L. Strauss. (Former Chairman John McCone was unable to be present.) In addition to these prepared remarks, I made a number of extemporaneous comments.

The Nixon years were difficult ones for AEC. In part, this may have been due to the special foibles of this president and his administration. More significantly, however, our difficulties can be attributed to the spirit of the times, particularly the opposition to the Vietnam War and a rising environmental consciousness. Such factors produced an atmosphere that was not friendly to large-scale science

25th anniversary observance of the signing of the Atomic Energy Act held at the Department of State. Washington, D.C. Aug. 1, 1971. Left to right: David E. Lilienthal, Seaborg, and Lewis L. Strauss. Courtesy of the U.S. Department of Energy, Germantown, MD. Photo by Ed Westcott.

and technology initiatives, particularly those that involved some government participation. In this uncongenial atmosphere, AEC sustained some frustrations and defeats. Not all of these were due to circumstance. AEC made its share of mistakes, some of which I freely acknowledge herein. On the other hand, we did some things right,

25th anniversary observance of the signing of the Atomic Energy Act held at the Department of State. Washington, D.C. Aug. 1, 1971. Left to right: Joint Committee on Atomic Energy members Chet Holifield, Melvin Price, and John O. Pastore and former AEC Chairman Lewis Strauss. Courtesy of the U.S. Department of Energy, Germantown, MD.

RICHARD NIXON

October 15, 1992

577 CHESTNUT RIDGE ROAD
WOODCLIFF LAKE, NEW JERSEY

Dear Glenn,

I am delighted to join with your colleagues and friends all over the world in congratulating you as you celebrate your 80th birthday.

I vividly recall our first meeting when we were honored among the ten outstanding young men of the United States by the Junior Chamber of Commerce over forty years ago. I have followed your career since that time with enormous appreciation for your achievements in the scientific and educational community.

I doubt if either of us believed when we met so long ago that one day we would serve together at the highest level in Washington when I was in the White House and you were chairman of the Atomic Energy Commission.

The nation is in your debt for your outstanding leadership in education and in science. As one who will catch up with you at 80 years of age on January 9, 1993, I congratulate you on reaching that milestone and wish you the best in the years ahead.

With warm regards,

Sincerely,

Richard Nixon

Dr. Glenn T. Seaborg

Richard M. Nixon's letter congratulating Seaborg on his 80th birthday. October 15, 1992.

and for this the agency deserves some credit.

I am pleased that President Nixon served as co-chairman of the belated banquet in honor of my 80th birthday held in the Sheraton Palace Hotel in San Francisco on Nov. 9, 1992, and wrote me a gracious letter of congratulations on that occasion.

Richard Nixon, after his forced resignation from his presidency on Aug. 9, 1974, spent the next 20 years until his death on April 22, 1994, with notable accomplishments as an elder statesman, political analyst, author, commentator, and wise counselor on foreign policy.

I knew Richard Nixon, as a U.S. congressman, U.S. senator, vice president, president, and ex-president, for more than 45 years. Clearly, he was a man with a complex personality and many contradictions—alternately friendly and harsh. Overall, I would, nevertheless, regard him as a friend.

chapter 7   GERALD RUDOLPH FORD, JR.

1974–1977

A LONGTIME FRIEND

My contacts with Gerald Rudolph Ford, Jr., started

during the 1960s when he served as U.S. Congressman from Michigan, including his service as minority leader of the House of Representatives, and continued during his vice presidency, presidency, and after his presidency. I urged upon him the development by our country of a realistic energy policy, which I described as an "energy budget" with a program to produce the energy required to meet that budget.

Helen and I met Gerald and Betty Ford at numerous dinners and other social functions during our stay in Washington. I recall that on one occasion, Gerald, Betty, Helen, and I had a very enjoyable dinner together at the Madison Hotel, enlivened by pleasant conversation sparked by our mutual interest in athletics.

On one memorable occasion at the famous annual Gridiron Club dinner held on March 20, 1965, at the Statler Hilton Hotel, Jerry and I were seated at the head table, and we both were very restless. Upon comparing notes, we learned that we were anxious about the NCAA basketball final that was being played between UCLA, my alma mater, and the University of Michigan, Jerry's alma mater. We decided to steal away from the banquet and go to an upstairs room that had a TV set, where we were joined by Michigan Governor George Romney. Romney, and especially Ford, agonized as it became increasingly apparent that the Michigan team was no match for the UCLA team, which won by a score of 91–80. I did my best to console them; but, of course, I was obviously happy that my alma mater had won. We returned to the banquet, which was still in progress and where no one showed any sign of having missed us.

President Nixon announced on a TV broadcast on the evening of Friday, Oct. 12, 1973, that he had chosen Gerald Ford to be vice president of the United States.

On April 25, 1974, I had the privilege of introducing Vice President Ford for his talk at a conference sponsored by the World Future Society in the International Ballroom Center of the Washington Hilton Hotel. He spoke on the energy problem, a central issue at that time. In fact, we were in the middle of what came to be called

Seaborg introduces Vice President Gerald R. Ford for talk at a World Future Society conference at the Washington Hilton Hotel, Washington, D.C. April 25, 1974. Courtesy of the World Future Society, Bethesda, MD.

the Energy Crisis. Throughout 1973 and into 1974, the shortage of energy and fuel in the United States became increasingly acute. Lines of cars snaked around gasoline stations as operating hours shortened and the amount of fuel sold to consumers became limited. Sales of domestic, gas-guzzling cars declined, forcing layoffs of thousands of automobile industry workers. Airlines canceled flights because of lack of fuel. Congress passed legislation that lowered the maximum speed limit to 55 mph in all 50 states. From Jan. 6, 1974, the United States went on daylight savings time for two years to save electricity.

This energy shortage affected not only the United States but also Japan and Western Europe. By 1973 the industrialized nations were consuming and demanding greater amounts of petroleum products such as oil and gasoline than in the past. The rate of demand was increasing faster than the rate of oil refining production. Spurred by the shortage, the Middle Eastern countries, which supplied the bulk of the petroleum products to the world, increased the cost of crude oil by 470%. In October 1973, the Organization of Arab Petroleum Exporting Countries cut production by 5% in retaliation for U.S. support of Israel in the Arab–Israeli war, which further exacerbated the energy shortage. Although the embargo was lifted in March 1974, fuel prices remained high and energy issues continued to be a top priority in the new Ford administration, which began a few months later (August) upon the resignation of Richard Nixon.

During 1973 and 1974 alone I gave 30 lectures on the energy crisis and energy resources for the future (and I have continued to speak on this subject since then). In these talks I emphasized the importance of conservation, of becoming a "recycle society," and of develop-ing new technologies for more efficient use of fuel. I also continued to advocate the development of nuclear power and to discuss emerging technologies that would allow us to use alternative sources such as solar and geothermal energy. I consider the attention I devoted to this emerging world problem, and my efforts to participate in the definition of solutions, to be an example of my continuing national service after I left Washington.

On Oct. 11, 1974, President Ford signed the Energy Reorganization Act of 1974. This was the end of the Atomic Energy Commission (AEC), which I had served so long. The reorganization plan made some real sense, though. It enabled energy policies and programs to come under a unified umbrella through the creation of the Energy Research and Development Administration (ERDA) and the Nuclear Regulatory Commission (which took over licensing and regulatory functions formerly performed by AEC).

I had an interesting conversation in Washington, DC, with ex-Secretary of Defense Melvin Laird on Aug. 12, 1974, very soon after Nixon's resignation as president:

Over lunch, I talked with Laird about his plans. He will be an intimate adviser to President Ford but will not accept a position in the White House; he would rather remain in the independent and more comfortable position of being a private citizen after so many years in Congress and the federal government. He told me that he convinced Nixon to nominate Ford as vice president; Nixon preferred John Connally. He also convinced Henry Kissinger to suggest to Nixon that resignation was the best route available to him, and this Nixon–Kissinger conversation was decisive. Laird said that, when he suggested to Nixon a year ago that he come clean on the Watergate affair, Nixon reacted very emotionally and nearly threw him out of his office.

Laird said that, when he became secretary of defense in January 1969, he inquired of the military whether the tape-recording system they had installed (activated by manual control, used by President Johnson to record cabinet meetings above-board) was still in the White House. He had trouble finding out, but when he learned it was still operable, he suggested that it be removed, and it was. About a year later, H. R. Haldeman suggested the voice-activated system, which was then installed. They didn't want to go back to the military to ask them to do it, so the purchase and installation job was given to the Secret Service; they went to stores in downtown Washington and bought commercial equipment.

Although Laird doesn't believe Nixon should be prosecuted further in criminal action, he feels he should be required to pay some additional price and make some more explicit acknowledgment of wrongdoing. He will soon be faced by about 75 civil suits for damages. (Laird is in favor of amnesty for Vietnam War defectors and draft dodgers, especially if it is accompanied by a small amount of civil service of some kind.) Laird is obviously a trusted, personal adviser to President Ford but is not a candidate for appointment as vice president because his background is too much like that of Ford.

On July 7, 1975, as president-elect of the American Chemical Society (ACS), I wrote a letter with ACS President William J. Bailey to invite President Ford to be the principal speaker at the Society's Centennial Banquet in New York City on April 6, 1976:

American Chemical Society
1155 Sixteenth Street, N.W.
Washington, D.C. 20036

July 7, 1975
The President
The White House
Washington, D.C. 20500
Dear Mr. President:

Last year the American Chemical Society invited you to present the principal address at the Society's Centennial Banquet in New York City on April 6, 1976. Replying on your behalf, your appointments secretary explained that you could not commit yourself to a date so far in the future; he suggested that we renew the invitation ten to twelve weeks in advance of the banquet.

As explained in our previous letter, the Centennial Banquet, which will be held exactly 100 years from the day the Society was founded, will be the most important of the many special events planned for the Society's anniversary observance. It will be attended by more than 3,000 people, including representatives of many scientific and engineering societies in the United States and abroad.

The Society is understandably anxious to make the banquet a truly memorable event. For this reason, we have not considered other banquet speakers. We are aware, of course, of the many problems you face in developing your schedule. However, the Centennial Banquet is so central a part of our anniversary observance, that we feel we must formalize the program considerably earlier than January 1976. We are therefore hopeful that you will find it possible to give us a favorable decision, even a tentative acceptance, by October 1975.

The American Chemical Society operates under a Congressional Charter which was signed into law by President Roosevelt in 1937. Since then, the Society has become the world's largest membership organization devoted to a single science. Our 110,000 members have built up unusual enthusiasm for our Centennial celebration and want to make it as impressive and meaningful as possible. Your participation would be a most fitting climax.

We hope you will accept our invitation to address the Centennial Banquet of the Society.

Very truly yours,
William J. Bailey
President
Glenn T. Seaborg
President-Elect

Although President Ford gave our invitation serious consideration, he finally decided that his schedule for his presidential campaign would make it impossible to accept. I was able to persuade Senator Edward M. (Ted) Kennedy to accept this role instead.

It was during my conversation on April 25, 1974, with Jerry Ford, that I suggested that the United States should develop as national policy an energy budget and then proceed with a program to produce the needed energy to satisfy that budget. Despite the lessons of the Energy Crisis of 1973–74, between 1975 and 1977 U.S. dependence on imported oil increased while domestic oil production decreased. President Ford cited a need for a "national energy plan." He proposed a 13-part Energy Independence Act, which eventually became the Energy Policy and Conservation Act. The effect of the act was to continue price controls on domestic oil, mandate federal fuel economy standards for new automobiles, and authorize the creation of a petroleum reserve. He proposed the creation of the Energy Independence Authority, which would assist in the construction of nuclear power plants, coal-fired power plants, oil refineries, synthetic fuel plants, and other energy production facilities. These proposals were fiercely debated in Congress; deliberation lasted almost a year. In late spring 1976, I wrote to the president in my role as ACS president to spell out the need to develop a coherent national energy policy:

May 28, 1976
The President
The White House
Washington, D.C. 20500
Dear Mr. President:

The American Chemical Society strongly urges the development of a coherent and realistic national energy policy and its implementation as a matter of high national priority. The foundation for such an energy policy should be (a) the definition of our national energy requirements, and (b) the establishment of definite programs, including research and development, with appropriate timetables for meeting those requirements. Inherent in such a policy should be an equitable balance of energy conservation with the development of adequate energy sources through a judicious application of science and technology.

The American Chemical Society pledges its scientific and technical resources to assist in the development and in the implementation of such a national energy policy.

Sincerely yours,
Glenn T. Seaborg

Although I didn't receive a specific response to this letter, I have the impression that it was seriously considered. At the close of his term in January 1977, Ford submitted his final energy-related proposal to Congress: the creation of a Department of Energy. We did receive an excellent response to a statement by the Committee of Scientific Society Presidents to our suggestions regarding the role of science and technology.

When Ford became president, he appointed Nelson Rockefeller as vice president (announcement on Aug. 20, 1974, and swearing-in on Dec. 19, 1974).

During my tenure as AEC chairman, I met Rockefeller on numerous occasions when he was governor of New York. Earlier, during my chancellorship of the University of California, Berkeley, I invited him to speak on campus on Sept. 14, 1960, in his role as a supporter of the Republican presidential ticket (Richard M. Nixon and Henry Cabot Lodge):

Rockefeller gave a speech in Dwinelle Plaza about liberty, freedom, and protection of the individual's rights, and how the Republican party and Richard Nixon would uphold these policies. He emphasized the danger of the Communist ideology. He disagreed with their concept of the individual "as a tool of the state with no human dignity, and the loss of the capacity for independent thought and spiritual realization." He advocated the joining of countries in larger entities to protect the freedom of the individual to develop intellectually, spiritually, and materially.

"It was during my conversation on April 25, 1974, with Jerry Ford, that I suggested that the United States should develop as national policy an energy budget and then proceed with a program to produce the needed energy to satisfy that budget."

THE WHITE HOUSE
WASHINGTON

December 17, 1976

Dear Dr. Seaborg:

I deeply appreciate the statement by the Committee
of Scientific Society Presidents commenting on my
efforts to increase the contributions of science and
technology.

Together, we have made considerable progress during
the past two years in creating an awareness of the
important contributions that science and technology
can make in achieving our important national objectives. I am particularly pleased that we have been
able to establish an Office of Science and Technology
Policy in the Executive Office of the President. I
am also pleased by the increase in Federal support
for research and development, particularly in the
basic sciences; and by the increased recognition of
contributions made by individual scientists and
engineers.

This progress is due in large part to the excellent
support provided by the Committee of Scientific
Society Presidents and the leaders and members of the
professional societies that make up the CSSP.

My continuing interest in science and technology will
be reflected in my 1978 Budget. I urge you to join
in supporting, before the Congress, these proposals
so that we can make further progress in implementing
the policies and programs we have developed together.

Sincerely,

Gerald R. Ford

Dr. Glenn T. Seaborg
President
American Chemical Society
1155 16th Street, NW.
Washington, D.C. 20036

President Gerald R. Ford's response to
the Committee of Scientific Society
Presidents' suggestions on the role of
science and technology. Dec. 17, 1976.

*He said that the biggest problem facing
the United States is the "need for excellent leadership" and that the major
problem lies in the international challenge "for freedom to be everywhere in
the world."*

*He answered questions from the
audience following his speech. He outlined the differences between Democrats and Republicans, saying that they
are different because of their quest for
civil rights. When asked about the
activities of the House Committee on
Un-American Activities, he answered,
"Americans must face the fact there is a
place for this kind of committee." Seg-
ments of the audience booed at this
remark, and he commented that there*
*were times that the kind of investigation
of the committee was not good.*

*On Red China, he said that some
kind of recognition must be made of
Red China activities in Tibet, Laos, and
Vietnam.*

On a more personal level, I
remember meeting Rockefeller on
Nov. 25, 1963, at President
Kennedy's funeral:

*At noon I attended President
Kennedy's funeral at St. Matthew's
Cathedral. After the funeral I was joined
by Pete, Dave, Steve, and Eric in front
of the cathedral, and we were driven in*

the funeral procession by Jim Haddow (my AEC security escort and driver) to Arlington National Cemetery, where we witnessed the burial service.

The kids met Governor Rockefeller, Governor Romney, John Glenn, Adlai Stevenson, Soviet Ambassador Anatoli Dobrynin, Minnesota Senator Eugene McCarthy, Speaker of the House John McCormack, U. Thant, Ralph Bunche, California Congressman Jeffrey Cohelan, and others.

Television and radio continued all-day programs about President Kennedy. People at the cathedral, the cemetery, and lining the way en route all displayed great emotion. The television and radio programs broadcast in our country and all over the world (except China) reflect the tremendous impact of President Kennedy's death on people. Our entire family shares this feeling of tremendous loss—even Dianne [my youngest daughter] seems to feel it.

What made this encounter with Rockefeller so memorable was his friendly act of carrying my nine-year-old son Eric on his shoulders to enable Eric to have a good view of the proceedings.

I had the pleasure of many encounters with Ford following the days of his presidency. On one par-ticularly noteworthy occasion, Helen and I attended the dinner tribute to Adm. Hyman G. Rickover, held on Feb. 28, 1983, which was also attended by Presidents Nixon and Carter.

Helen and I changed into our formal clothes and took a taxi to the Sheraton Washington Hotel to attend the dinner of tribute to Adm. Hyman G. Rickover. We attended the reception, where we talked to Chet Holifield and others.

At about 8 p.m. we went to the ballroom for dinner and sat at Table 5 with Dr. and Mrs. James Schlesinger, Mr. and Mrs. Yoshio Okawara (Japanese ambassador to the United States), Mr. and Mrs. Bob Wilson (former U.S. representative from California), and two others.

I visited some of the other tables and talked to Al Haig (about the need to upgrade the White House on arms limitation measures), Ed Teller, and others. Seated at the head table to the right of the speaker's lectern, left to right (as we saw them), were Adm. Rickover; ex-presidents Nixon, Ford, and Carter; and Joann DiGennaro (assistant to Rickover and later director of the Rickover Foundation). To the left of the lectern were Senators Scoop Jackson, John Warner, and Strom Thurmond, and Mrs. Rickover.

New York Governor Nelson Rockefeller speaking on Dwinelle Plaza, University of California–Berkeley. Peter Seaborg is sitting in center foreground. Glenn T. Seaborg is sitting right, next to Rockefeller, and to the right of Seaborg are former University of California president Robert Gordon Sproul, Vice Chancellor for Student Affairs Alex C. Sherriffs, and Mayor of the City of Berkeley Claude B. Hutchison. Student body president George H. Link is seated immediately to the left of Rockefeller. Berkeley, CA. Sept. 14, 1960.

Former Presidents Gerald R. Ford, Jimmy Carter, and Richard M. Nixon with Admiral Hyman G. Rickover at a dinner in honor of Rickover. The dinner was hosted by the Rickover Foundation (Center for Excellence in Education) at the Sheraton Washington Hotel, Washington, DC. Feb. 28, 1983. Courtesy of the Center for Excellence in Education, McLean, VA.

After dinner DiGennaro opened the program by introducing members of the Board of Directors of the Rickover Foundation, both those present (who stood) and those who were absent. Senator Jackson made some remarks about Rickover and then introduced President Carter, who spoke in a humorous way about how he had worked for Rickover both before and after he became president. (Both used prepared texts, with numerous interpolations.) Senator Warner then spoke about Rickover and introduced President Ford, who also spoke humorously about Rickover's having outranked in the navy all of the three ex-presidents present that evening. (Again, both used prepared texts, with interpolations.) Senator Thurmond spoke about Rickover and introduced President Nixon, who spoke about three personal encounters with Rickover, including two during his visit to the Soviet Union in 1959 (at the time of his famous kitchen debate with Khrushchev). Thurmond spoke largely extemporaneously but very clumsily, with each successive thought even worse than the preceding one; he indicated that Nixon's presidency was not all bad and that he did some good things in foreign affairs (opening the People's Republic of China), economic policy, and other areas. Nixon spoke extemporaneously without notes.

Adm. Rickover then responded with a prepared talk (which he didn't read very well) that recounted a number of incidents in his life as a Polish emigrant to the United States. (He and his mother and older sister left Poland under duress when he was four years old, and they had much trouble entering the United States through Ellis Island.) Congressman Charles E. Bennett (Florida) then made some remarks, announcing on behalf of the secretary of the navy that a nuclear submarine is going to be named after Hyman G. Rickover. A surprise, huge birthday cake was then wheeled onto the stage. Nixon played "Happy Birthday to You" and then "God Bless America" on the piano on stage. The three ex-presidents then joined Rickover at the birthday cake, waving to the audience of about 500 diners, after which the head table guests left the stage through a back curtain.

As Helen and I were leaving the ballroom, the popular TV commentator Martin Agronsky remarked to us that he had thought he would never be in a position to feel sorry for Richard Nixon (as he was that night with respect to Thurmond's unflattering introduction of Nixon).

Former First Lady Betty Ford receives a Golden Plate award from Ernest Hahn at the American Academy of Achievement's "Salute to Excellence." Denver, CO. June 29, 1985. Courtesy of the American Academy of Achievement, Malibu, CA.

I met First Lady Betty Ford on a number of occasions. One of the last times I saw her again was when she received a distinguished volunteer service award at the American Academy of Achievement's annual banquet of the Golden Plate dinner in Denver, CO, on June 29, 1985.

As I indicated earlier, I knew Ford during his service as congressman, vice president, and president and during post-presidential years. I believe he served his country well, covering some difficult times.

President Ford's main contribution to arms control was a communiqué with a treaty formula to follow the Strategic Arms Limitation Talks (SALT) I that he and President Brezhnev signed at Vladivostok in November 1974. This served as an important step toward SALT II, which President Carter tried but failed to achieve (as I describe in Chapter 8).

I feel that I have had an especially good rapport with Gerald Ford and am pleased that he served as co-chairman of the banquet in honor of my 80th birthday, which was held in the Palace Hotel in San Francisco on Nov. 9, 1992.

chapter 8 JAMES EARL (JIMMY) CARTER, JR.

1977–1981

A FELLOW NUCLEAR SCIENTIST

My first contacts with James Earl Carter, Jr., were indirect through such mutual friends as Adm. Hyman Rickover, and related to Jimmy Carter's early career in the nuclear navy and

my involvement in the development of nuclear energy. Later, when Carter became president, I often kept in touch with him through our mutual friend Sol Linowitz (public-spirited lawyer, adviser to President Carter, and ambassador to the Organization of American States) as a way to pass on suggestions regarding the development of nuclear power; steps to be taken in arms limitation, including the promotion of a comprehensive nuclear test ban; and so forth. Although President Carter's early stint as a nuclear engineer should have given us a common bond (and I believe it did), he was not an advocate for civilian nuclear power during or after his presidency.

I first met President Carter when he spoke to the finalists of the annual Westinghouse Science Talent Search (STS) on Friday, March 6, 1978:

President Carter entered at precisely 1:40 p.m., as scheduled, in the middle of William Perry's (undersecretary for research and development, Department of Defense [DOD]) talk, which was interrupted at this point. The president had left the meetings on the coal strike problem in which he was engaged.

The president began by congratulating the STS contestants. He said 40 had been selected from an original total of 13,000 and that 10 of the 40 will be chosen and announced at the dinner tonight at which Frank Press will speak. He emphasized the importance of science and young scientists in our society as well as the decline in the support of them in recent years, a trend that is being reversed in his administration. He commended the government representatives (NASA [National Aeronautics and Space Administration], DOD, NSF [National Science Foundation], NIH [National Institutes of Health]) who had taken time to speak to the STS winners (he failed to mention DOE [Department of Energy]). He said 99% of the previous STS finalists had finished their undergraduate education, 70% had gone on to graduate school, and one person had won the Nobel Prize (he first said Nobel Peace Prize but then corrected himself). As he concluded, speaking at a lower and lower voice level, he again emphasized that the future of our country depends on people like the STS winners, one of whom might become president someday. He said he worked on the first nuclear submarine reactor and sometimes looks back with nostalgia to those happier days as a scientist. He talked for about five minutes.

At the conclusion of his remarks, the president came by to shake hands with government representatives and with the STS people in the first row (and some in the second row). When he reached me, I introduced myself, and he responded, "Yes, I know; you're a very famous man."

One might assume that because Carter had worked as a nuclear engineer, he would consider nuclear power a viable option; but he did not (or at least he did not support its continuing development). President Carter inherited in 1977 the task of creating a national energy policy to deal with the ongoing energy problem. He formulated a national energy plan that consisted of proposals to reduce energy consumption and develop alternative technologies. The major outcome of this plan was the founding of the Department of Energy in fall 1977. The new department took over the functions of the Federal Energy Administration, the Energy Research and Development Administration, and the Federal Power Commission.

By 1977 some of the pressure was off. The first deliveries of crude oil were transported to California via the trans-Alaska pipeline. The national Strategic Petroleum Reserve, an effort to store quantities of crude oil as an emergency stockpile, began its buildup with the Alaskan oil shipments. However, the long-range problem remained: Our demand was bound to outstrip our supply if we did not dramatically change our habits.

The Carter administration placed a great deal of emphasis on conserving limited energy resources. In general, his administration's focus was on learning to do more with less rather than developing more resources. He announced his decision to stop research aimed at perfecting fast breeder technology and to halt construction of factories to reprocess irradiated fuels. Following President Carter's speech to Congress on the energy problem on April 20, 1977, I recorded some thoughts about the issue and reported on an interview with *Time* magazine:

From 6 to 6:30 p.m. Dianne and I heard President Carter's speech to Congress on the energy problem. He described a program depending on (1) conservation, (2) production, (3) conversion, and (4) development—a pretty good program with perhaps too much emphasis on voluntary conservation and not enough on production. He included conventional nuclear power, called for a speedup of the licensing process, and suggested a slowdown on breeder reactor production.

Immediately following the president's energy message to Congress, at 6:40 p.m. James Wilde of Time magazine (San Francisco office) called, as scheduled, to obtain my reactions. He was concerned primarily with my assessment of the moral fiber and potential response of the American people. I indicated that I thought the president had gone a long way in convincing the public that there is a serious problem and that he will get a partially positive response, but in the long run it will be necessary to include some compulsory aspects in order to implement the plan. I said that we are in trouble today because we didn't conduct the necessary scientific research and technological development in the past, starting some 20 years ago—such as providing for the gasification and liquefaction of coal, the production of more energy-efficient appliances and small automobiles, and so forth.

Wilde asked if I thought this peacetime crisis was the greatest the American people had faced since the Depression, and I said, "Yes." I said our next crisis will be in the resources field and that we must convert from a philosophy of obsolescent products to nonobsolescent products and of recycling our materials, and that we should start now to avoid being compelled to adopt such drastic solutions as we face in the energy problem.

Privately, I was more frank about my dismay that the president had taken the drastic measure of canceling the fast breeder program. I believe he was misguided in his hope that this unilateral action on the part of the United States would slow development of the technology in other countries. I understood that the urgent needs of countries such as Japan would make technological developments an imperative for them. I continued to give speeches advocating the use of nuclear power, including funding for the fast breeder reactor, as well as alternative resources.

My views on these issues are summarized in a journal entry describing my advice to Senator Ted Kennedy when he was a presidential candidate at the end of the Carter administration. When I boarded a flight from Washington, DC, to San Francisco on April 11, 1980, I discovered that the senator was seated behind me:

After breakfast was served, Ted Kennedy joined me, and we had a long discussion on energy policy. I told him I think he lost a lot of voter support, from the segment of the population where he needs it most, on the basis of his antinuclear stance. I said we need nuclear energy as a part of our energy mix, that the United States has never been in as vulnerable a position as it is today, and that this could even lead to our involvement in a war in the Mideast. In answer to his question, I said I believe that the nuclear waste disposal problem has been solved technically; what is needed is a political decision to adopt a system. I described the action of ion exchange in surrounding soil as an insurance against the migration of fission products. I described the Oklo incident in Gabon, Africa, and told him about Sweden's progress in this field. We commented on the recent favorable vote in Sweden on nuclear power.

In answer to another question, I estimated that practical power from nuclear fusion is 30–50 years away, adding that we should nevertheless develop nuclear fusion. I told him that energy from ocean thermal gradients doesn't look too hopeful. We agreed that geothermal energy will not make an appreciable contribution. In reply to his observation that the U.S. demand for electricity is going down, I cited the growing number of new families in the United States, the needs of poor people, and the worldwide need for energy and said we must cooperate in meeting this need. With respect to solar energy, I agreed that solar heating of homes can make a contribution, but practical solar electricity (as advocated by Jerry Brown) is actually many years away. I emphasized the reliance of countries such as France on nuclear power.

In answer to his questions about France's and the Soviet Union's progress on the breeder reactor, I said they have forged ahead of us. I decried President Carter's policy on the breeder reactor and explained why this is counterproductive in preventing the proliferation of nuclear weapons. I said nuclear power is relatively safe compared with other forms of energy and that coal burning releases more radioactivity to the atmosphere than what comes from nuclear plants. I said that nuclear wastes decay at a rate so that, after 500–1000 years, the amount of radioactivity is less than in uranium ores, which are widespread throughout the earth. I commended the National Academy of Sciences CONAES (Committee on Nuclear and Alternative Energy Systems) report, with which he was familiar, that was put together by hundreds of the most competent scientists in the United States. I suggested he adopt its recommendations as his energy platform.

President Carter and I did not see eye to eye on the question of nuclear power. Our views on the importance of limiting the proliferation of nuclear weapons and on achieving a comprehensive test ban (CTB) treaty were much more similar.

On June 5, 1979, scientists and others were called to the White House for a briefing on the pending Strategic Arms Limitation Talks (SALT) II Treaty and a discussion of its implications:

I walked to the White House and went to a reception room where I joined those waiting to attend the breakfast at which President Carter and others were scheduled to brief us on SALT II. Present were Bob McNamara (president, World Bank), James Van Allen (Iowa State University), Paul Doty (Harvard), Isidor Rabi (emeritus professor, physics department, Columbia University), Frank Borman (chairman of the board, American Airlines), Alvin Weinberg (director, Institute of Energy Analysis, Oak Ridge Associated Universities), Frank Long (Cornell University), Carl Sagan (director, Laboratory Planetary Studies, Cornell University), Phil Abelson (editor, Science magazine), Lee DuBridge (president emeritus, Cal

Tech), Lewis Branscomb (chief scientist, IBM), E.L. (Ed) Goldwasser (vice chancellor, University of Illinois), Norman Hackerman (president, Rice University, and member of the National Science Board), Lyman Fink (industrial executive), Herman Postma (director, Oak Ridge National Laboratory), and 10–15 others, along with Frank Press (science adviser to the president), Ann Keatley (assistant to Frank Press), Spurgeon Keeny (deputy director, Arms Control and Disarmament Agency), and Ann Wexler (assistant to the president).

At a little after 8 a.m. we went into the family dining room, where four tables were set up for breakfast, with a speaker's podium at the far end. I sat with McNamara, Van Allen, Doty, Rabi, and others. During breakfast I had a good opportunity to talk with McNamara. He mentioned that he still gets a thrill every time he is in the White House, and I told him that I feel the same way. We reminisced about our work together in the Kennedy and Johnson administrations. He recalled with especial pleasure his visit to Berkeley in spring 1962, when President John F. Kennedy gave his exciting speech at the Charter Day exercises in a completely full Memorial Stadium. At that time I also attended the dinner of the Alumni Association to accept, on his behalf, McNamara's Alumnus of the Year Award. We discussed John Kennedy's excellent leadership in successfully obtaining the Limited Test Ban Treaty, and I told him that I am using my notes to write about this period and this accomplishment; he said that he thinks this is a good idea, and he hopes that I have kept some notes (apparently, he has not).

I asked him what he is doing these days, and he indicated that, as president of the World Bank, he has made a breakthrough in aiding poor people in developing countries; the route he has developed is to aid the poorest segment directly rather than use the old theory that aid to the government of a country will filter down to the poor and hence benefit them because the country as a whole has increased its productivity.

We discussed the energy problem, and he vehemently agreed with me that our country is making a terrible mistake to continue to rely on the tremendous outflow of U.S. dollars in order to buy oil. He, too, thinks there should be

"President Carter and I did not see eye to eye on the question of nuclear power. Our views on the importance of limiting the proliferation of nuclear weapons and on achieving a comprehensive test ban (CTB) treaty were much more similar."

A Chemist in the White House: From the Manhattan Project to the End of the Cold War

more emphasis on the development of an energy supply and says we should have started at least five years ago with a program of liquefying coal. I told him I believe that we should have in the past or at least should now develop an "energy budget" and then proceed to try to produce the energy needed to meet that budget. I asked him if he would nationalize the oil industry and he said definitely not; he would use the marketplace to encourage production, although he conceded that it probably would be necessary for the government to guarantee the price of oil produced by the liquefaction of coal or other means.

After we finished breakfast, at about 8:40 a.m. or so, Press went to the podium and told us that the reason for the meeting was to brief the high-level members of the scientific community on SALT II. The briefing would be done by President Carter, followed by Defense Secretary Harold Brown and Deputy Defense Secretary for Research and Engineering William Perry. Press emphasized the importance of the scientific community in supporting SALT II and the need for scientists to testify before Congress in support of SALT II. Such testimony was important for the ratification by the Senate of SALT II. He also mentioned the historical role of scientists, beginning with the famous Albert Einstein letter that initiated the atomic bomb project, their role in raising the DNA issue, and so on.

President Carter entered the room at about 8:45 a.m. and went immediately to the podium. He thanked us for coming and said this is one of a series of meetings with key people for briefings on SALT II. He said that SALT II enhances our security, contributes to world peace, and contributes to the whole process of arms limitations started by President Eisenhower. He said the SALT Treaty is fair and well balanced, will increase the confidence of our allies in our intentions, will increase the stability of our relations with the Soviets, and will help prevent the proliferation of nuclear weapons. He said that many nations are now in a position to produce nuclear weapons within a few months, mentioning, for example, Brazil, Argentina, Israel, Pakistan, South Africa, and South Korea. The president said that if SALT II is rejected by the Senate, it would be ridiculous to ask these nations to desist

from building nuclear weapons. He said the treaty is verifiable and is not based on trust of the Soviets, and without SALT II our abilities for verification would be greatly curtailed.

He then described some of the provisions of SALT II, such as the limitations it places on the number and new types of reentry vehicles. Without SALT II we would be in great doubt about the Soviet capability. Without the restraints imposed by SALT II, the Soviets' momentum in building up still more weapons of destruction would have to be met by us in an escalating race.

The president said that we have been tough negotiators in arriving at the final text of SALT II, but so have the Soviets. Negotiations have proceeded during seven uninterrupted years and have involved the president as well as some 30 members of Congress. Throughout the process the American public has been kept informed. The specificity of the treaty far exceeds that of any previous treaties. For example, its constraints require the Soviets to dismantle about 10% of its missiles.

The president said that, from the standpoint of his presidency, his involvement with SALT II and the decisions he has made in this connection are the most important. He emphasized that the issue is still in doubt; he said he could not face the world as the president of the United States if SALT II is rejected. Under these circumstances, the countries of the world would have to look eastward, the effectiveness of the North Atlantic Treaty Organization (NATO) would be jeopardized, and the Soviets could then project themselves as a strong peace-loving nation in contrast to the non-peace-loving nature of the United States.

The president then asked for questions. Doty asked why there wasn't a common option on heavy missiles on both sides, and the president indicated that this is coming—the Soviets have agreed to a limitation of a total of 10 SSATs (Surveillance Satellites), and we would then have 10 Mxs (mobile intercontinental ballistic missiles, never deployed). Rabi asked whether SALT II is part of a general momentum toward further progress, that is, toward more significant steps to follow, and the president replied that the Soviets look to SALT as a continuance of peaceful coexistence or détente, and that SALT II should be followed by SALT III, the CTB, and antisatellite agreements. Long asked whether we could be optimistic about a CTB, and the president answered "Yes," saying that we have made progress; the Soviets have finally agreed to drop peaceful uses of nuclear explosives and are now ready to accept detection devices on their own soil. However, if there are 10 devices in the Soviet Union (the number now under discussion), they want 10 devices in the United States and 10 in the United Kingdom. In present discussions a three-year ban on all nuclear tests, to be followed by some proof testing of existing nuclear weapons, has been suggested.

DuBridge asked what kinds of amendments can be made to satisfy members of Congress so that they will change their opposition to a vote for the treaty, and the president replied, "practically none." He said that the Soviets would reject any amendment that would be to their detriment, and obviously the United States would not suggest amendments that would be to its detriment.

Sagan said there has been much criticism that SALT II has the form but not the content of a serious treaty and asked, therefore, if the president was satisfied. The president replied, "No," because actually there will be, even with SALT II, an increase in missiles on both sides. However, this is the first time a limit has been put on existing missiles, on the "MIRVing" (multiple independently targetable reentry vehicle) of missiles, and so on. SALT III will be discussed between Carter and Brezhnev at their Vienna meeting, the primary goal of which is to reduce numerical numbers and obtain better verifiability.

Fink recognized the difficulty of going to a total elimination of nuclear weapons, and the president replied, "This is my goal." He has discussed this with Deng Xiaoping (vice premier of the People's Republic of China), James Callaghan (prime minister of the United Kingdom), and Valery Giscard d'Estaing (president of France); and they all agree that this is an ultimate goal.

Another question concerned the importance of the European Theatre, and the president replied that SALT III will provide for consultation with our allies on theatre nuclear weapons.

The president concluded by saying that we have had remarkable uniformity of support for SALT II throughout the government, although some individuals in the Joint Chiefs of Staff would prefer some different aspects in the treaty—this seemed to imply that they might not support it in its present form.

The president left the meeting at about 9:30 a.m.

Brown then went to the podium and described the content of SALT II, stating that it will not solve all our strategic problems, because beyond the military aspects are broader aspects such as nonproliferation, relations with our allies, and other concerns. He said our relationship with the USSR will remain competitive even with SALT II, but SALT II will help ameliorate the competitiveness. He concluded by picturing the horrible effects of nuclear explosions, with which a number of those present are familiar, and said it is the prevention of such explosions that is at stake.

DuBridge asked Brown whether there is a difference of opinion in the military concerning the support of SALT II. Brown answered this obliquely by saying that some military people know what they can do in a military way but are uncomfortable with what SALT II can do. He said this is not the attitude of the Joint Chiefs of Staff but then said that not all the Joint Chiefs of Staff will necessarily support SALT II. Weinberg raised the question of whether the United States should have a civil defense program, and Brown replied that this is not a very effective tack; when it was tried in the early 1960s it was demonstrated that it doesn't come easily in the United States. After another question or two, Brown brought his part to an end and called on Perry to make his presentation.

I left the meeting at this time (about 9:50 a.m.).

The SALT II Treaty was hotly debated in the Senate in summer 1979, then put on the shelf as part of the adverse reaction to the Soviet invasion of Afghanistan at the end of the year. However, the provisions of SALT II were nevertheless honored until they were repudiated early in the Reagan administration.

I recall another occasion, on Oct. 20, 1979. I went to Boston for the dedication of the John Fitzgerald Kennedy Library, where I saw President Carter play a central role in the dedication ceremonies:

I had breakfast in the Howard Johnson 57 Park Plaza Hotel with Sol and Toni Linowitz (who I happened to meet as we were standing in line). Sol told me that he has again talked to President Carter about my having one of the four part-time positions on the Energy Mobilization Board. I talked to California ex-Governor Pat and Bernice Brown; he said he is stumping the country for his son Jerry.

After breakfast, we took one of the buses to the site of the Kennedy Library at Columbus Point on Dorchester Bay. Because Sol and Toni had tickets for the reserved area, I followed along with them and we managed to find seats in the second row on the right-hand side, just in front of the speaker's rostrum.

At a little after 10:30 a.m. those to be seated on the platform entered, and shortly after that President and Rosalynn Carter entered. The president kissed Joan Kennedy (who reacted cordially) and Jacqueline Kennedy Onassis (who

jumped back to try to avoid the kiss by Carter) as he passed down the line of those seated in the front row. In the front row from right to left (from the standpoint of those seated in the audience) were (1) the woman who served as translator of the proceedings, using sign language for the deaf; (2) Sargent Shriver; (3) Mrs. Sargent Shriver (Eunice Kennedy); (4) Patricia Kennedy Lawford; (5) Mrs. Steve Smith (Jean Kennedy); (6) Steve Smith; (7) Mrs. Robert (Ethel) Kennedy; (8) Joseph P. Kennedy, II; (9) Sally Willamen (the winner of an essay contest concerning the John F. Kennedy Library and its function); (10) Reverend Herbert Meza, Church of the Pilgrim, Washington, DC; (11) Archbishop Humberto Cardinal Medeiros, Archbishop of Boston; (12) Rosalynn Carter; (13) President Carter; (14) Joan (Mrs. Edward) Kennedy; (15) Edward (Teddy) Kennedy; (16) Lady Bird Johnson; (17) an unidentified man; (18) Jacqueline Onassis; (19) John F. Kennedy, Jr.; and (20) Caroline Kennedy.

The dedication proceeded according to the program. During the opening musical selection by the Boston Pops Esplanade Orchestra, conducted by Harry Ellis Dickson, President Carter and Joan Kennedy talked animatedly, and the president held Joan's hand part of the time.

Stephen E. Smith, president of the John Fitzgerald Kennedy Library Corporation, opened the speaker's program with some introductory remarks. He acknowledged the help of those who had worked to make the Kennedy Library possible, notably Mrs. Patricia Lawford.

Caroline Kennedy then made a short welcoming statement, speaking extemporaneously; she did very well. She then introduced her brother, John F. Kennedy, Jr., who read a poem by Steven Spender. When he returned to his place next to Caroline, she gave him a kiss expressing approbation for his creditable performance. The invocation was then given by Archbishop Humberto Cardinal Medeiros, followed by the playing and singing of the National Anthem.

Joseph P. Kennedy, II (the eldest son of Robert and Ethel Kennedy), then gave his talk, "The Unfinished Business of Robert F. Kennedy." He talked in rather strident tones, emphasizing the

President Jimmy Carter and First Lady Rosalynn Carter at the dedication of the John Fitzgerald Kennedy Presidential Library. Columbia Point, MA. Oct. 20, 1979. The other people are identified in text. Courtesy of the John Fitzgerald Kennedy Library, Columbia Point, MA.

need for more attention for the poor. He criticized the recent action of the chairman of the Federal Reserve Bank, looking at President Carter as he did so, which was directed toward creating a recession in order to fight inflation.

Next, John F. Kennedy, Jr., and President Carter came to the podium and John, Jr., without making any remarks, handed the keys of the library building to President Carter, an aspect that symbolized the gift of the library building to the U.S. government.

This action served as the introduction for President Carter's remarks. He used the late president's own words to jokingly advise JFK's brother, Senator Edward Kennedy, not to seek the presidency. He recalled that President Kennedy had been reminded at a March 1962 press conference that younger brother Ted had observed the ravages of the presidency on his elder brother and commented that he might never want to be president. "Would you run again for the presidency?" and "Would you recommend it to anyone else?" a reporter asked him at that conference. Carter recalled the response: "I do not recommend it to others—at least for awhile." Then he added: "President Kennedy's wit, and also his wisdom, is certainly as relevant today as it was then." (In President Carter's references to legislation passed after the death of Kennedy, he failed to mention Lyndon B. Johnson's name.)

President Carter's address was followed by an address by Edward M. Kennedy. He began without making a salutation to President Carter, but he did turn toward him and make references to him in his talk, which was mainly on the accomplishments of his brother. He told the audience his brother had taught him to ride a bicycle, toss a football, and sail against the wind. He said that his brother was the person he looked to for help when he was a boy and that John had returned to Boston as his home when he needed spiritual help.

Stephen Smith, as master of ceremonies, introduced Jacqueline Kennedy Onassis, Lady Bird Johnson, and Ethel Kennedy, who each stood up.

At this point recordings were played of excerpts from a number of John Kennedy's speeches, and hearing them was a very moving experience. The great similarity between the voices of John and Edward was apparent to those present.

The program was brought to an end by Reverend Herbert Meza giving the benediction and the audience singing "America the Beautiful." Those on stage and members of the various "families Kennedy" sitting in the first two rows center then filed out and went to visit the library (President and Rosalynn Carter had visited the library earlier and soon left).

During the ceremony and afterward I met and talked to (or greeted) the following people: Robert and Margaret McNamara; McGeorge and Mary Bundy; Willard and Mary Jane Wirtz; Tazwell and Julia Ann Shepard (naval aide to President Kennedy); Averell and Pamela Harriman (in response to my question, he told me he had not written an article on the test ban treaty and said the best source of information is Butch Fisher, who accompanied him to Moscow for the final negotiations); John Macy; Anthony Celebreeze; Stewart and Ermalee Udall; Orville and Jane Freeman; Douglas Dillon; Newton and Josephine Minow; James and Estelle Ramey; Joseph Swidler; The Jim Faheys (he is the author of Pacific War Diary and will send me a copy; our picture was taken together); James Carr (he asked for a copy of my statistics used in my "Energy" talk); Walter and Elspeth Rostow (she is dean of the Lyndon B. Johnson School of Public Affairs at the University of Texas at Austin); George McGhee; Dean Rusk (in distance); Ed Day (in distance); Jody Powell (in distance); Stu Eizenstadt (in distance); Peter Duchin (son of orchestra leader Eddie Duchin); and the Kennedy children nearby.

In December 1981 I received a gracious letter from "Jimmy" in response to my sending him a copy of my book, *Kennedy, Khrushchev, and the Test Ban.*

I have had many encounters with Carter's vice president, Walter F. Mondale, beginning in 1965 when he was a U.S. senator from Minnesota. Helen and I also became acquainted with his wife Joan. I recall, especially, a fascinat-

JIMMY CARTER

December 4, 1981

To Glenn Seaborg

Thank you for sending me Kennedy, Khrushchev, and the Test Ban. Yours is an impressive study, and I am honored to receive your thoughtfully inscribed book.

With best wishes,

Sincerely,

Jimmy

Mr. Glenn Seaborg
1 Cyclotron Road
Berkeley, California 94720

Former President Jimmy Carter's letter thanking Seaborg for sending him a copy of Seaborg's book, *Kennedy, Khrushchev, and the Test Ban.* Dec. 4, 1981.

ing talk on the importance of the arts that Joan Mondale gave at the Conference on Arts, Education, and Americans at the National Academy of Sciences in Washington, DC, on May 24, 1977.

An interesting encounter with Mondale occurred at the Hilton Hotel in San Francisco on Monday, Aug. 30, 1976, during his successful campaign for the vice presidency of the United States as Carter's running mate:

I went down to the Imperial Ballroom to hear Senator Mondale's talk at the Commonwealth Club. I sat at a Commonwealth Club Board of Governors table with Mr. and Mrs. Howard Vesper, Vernon Goodin (attorney and prominent Cal alumnus), and others. Mondale opened his talk, "A Major Foreign Policy Address," with some clever ad libs, then followed his text with some cuts. He handled the question-and-answer period quite well. This and the talk included a good deal of scientific-technical material on energy sources as well as nuclear arms buildup and arms control.

Afterward, I talked to him, complimented him on his handling of the technical material (he expressed some

Minnesota Senator and vice-presidential candidate Walter F. Mondale speaking before the Commonwealth Club at the Hilton Hotel, San Francisco. Aug. 30, 1976. Courtesy of David Perlman, Science Editor, *San Francisco Chronicle*, San Francisco.

apprehension about it), and then told him about my request to see Carter to discuss (as a representative of the American Chemical Society, of which I am president this year) energy policy and science policy; he said he is seeing Carter next week and will bring my previous written request to his attention. Mondale asked me to brief his assistant, Jim Johnson, who was standing nearby, on this matter, which I did.

I was then interviewed by Linda Schacht of Channel 9 on Mondale's speech. I said he handled the technical aspects quite well; that I probably will vote for him, depending on campaign postures; and that I knew him quite well as a friend during my tenure in Washington.

I had another encounter with Carter at the Salute to Excellence dinner of the American Academy of Achievement in Minneapolis, MN, on July 7, 1984, which Helen also attended:

At the reception Helen and I talked to many interesting people. Carter approached us and said I am one of his heroes, and he has been a great admirer

of me for years. I mentioned my interest in a CTB, and he asked Ted Turner (chairman, Turner Broadcasting Systems, Inc., and owner of the Atlanta Braves baseball team) to join us, and we described to Ted the values of a CTB. Carter said he failed to get a CTB in part because the British insisted on their quota of on-site inspections. I gave Ted a copy of my Chemical & Engineering News editorial on the CTB. An academy photographer took pictures of me with Carter and Turner. Helen and I also met and talked with Rosalynn Carter.

Although, as I noted earlier, I do not agree with Carter's negative views on civilian nuclear power, in view of his background as a nuclear engineer I have felt especially close to him. I also have tremendous admiration for his devotion to human rights and to the pursuit of peace.

As president, Carter gave unprecedented priority to human rights around the world, making observance of human rights a primary criterion in determining U.S.

Former President Jimmy Carter, Ted Turner (back to camera), and Seaborg at the American Academy of Achievement's Banquet of the Golden Plate "Salute to Excellence." Minneapolis, MN. July 7, 1984. Courtesy of the American Academy of Achievement, Malibu, CA.

foreign policy toward individual nations. He earnestly sought to curb the nuclear arms race by reaching arms control agreements with the Soviet Union. He and Soviet Premier Brezhnev signed the SALT II agreement in June 1979. He also made greater progress toward agreement on a compre-

hensive test ban than any other president before Clinton.

Carter has played an impressive role as an international statesman during the years following his presidency. He has personally monitored or observed elections in several Latin American countries and made his judgments public,

Former First Lady Rosalynn Carter at the American Academy of Achievement's "Salute to Excellence." Jimmy Carter is seated left. Minneapolis, MN. July 7, 1984. Courtesy of the American Academy of Achievement, Malibu, CA.

denouncing, for example, the election in Panama that kept Manuel Noriega in power and extolling the 1990 election in Nicaragua. He tried to mediate between the Chinese government and discontented Tibetan subjects; he convened negotiations between the Ethiopian government and the Eritrean rebels; and he has made repeated trips to the Middle East to speak to leaders of the major countries in the interests of peace. In 1994 he visited North Korea and opened the way for continued talks between the United States and North Korea about the use of nuclear materials. He also made steps toward bringing the leaders of North and South Korea together in the first summit between the two countries since they were divided in 1945. It can be fairly claimed that in his tireless pursuit of peaceful initiatives, Carter has been the most productive ex-president in the country's history.

I am pleased that Carter also served as co-chairman for the dinner honoring me on my 80th birthday at the Palace Hotel in San Francisco on Nov. 9, 1992.

A Chemist in the White House: From the Manhattan Project to the End of the Cold War

chapter 9 RONALD WILSON REAGAN

1981–1989

AN AMIABLE FELLOW

The high points of my contacts with the Reagan administration

were my participation with the National Commission on Excellence in Education (NCEE) on the preparation of the

famous report "A Nation at Risk: The Imperative for Educational Reform," the "cleansing" of my journals, and my unsuccessful attempts to obtain U.S. support for a comprehensive test ban (CTB).

I first encountered Ronald Wilson Reagan when he was governor of California, during which time I met him on a number of occasions (including the breakfast at the Bohemian Grove on July 23, 1967, as described in Chapter 6).

Soon after Reagan became president, in the summer of 1981, I was appointed to NCEE. Secretary of Education Terrel H. Bell phoned me on Aug. 17, 1981, to describe the proposed commission, its goals, meeting agenda, and 18-month time scale for presenting to President Reagan its report and recommendations for the renovation of precollege education. After I turned him down (because I thought I just didn't have the time), I received another phone call on Aug. 19, 1981, from the designated commission chairman David Pierpont Gardner (at that time president of the University of Utah; president of the University of California in 1983). He persuaded me to reverse my decision and to serve on the 18-member commission. Yvonne W. Larsen, immediate past president of the San Diego City School Board, served as vice chairman, and Milton Goldberg served as executive director of the commission.

In 1957 the American public was shocked by the announcement that the Soviet Union had launched an unmanned satellite into space. Sputnik heightened the nation's concern about and attention to public education. A renewed emphasis was placed on improving cognitive learning. New national testing, particularly assessment of student performances, was developed. Creative new methods of teaching and learning were explored, including the debut of the Public Broadcasting System's preschool program "Sesame Street" in 1969. I served as chairman of the steering committee of the Chemical Education Material Study (CHEM Study) in the early 1960s. At present, 24 audiovisual productions in the CHEM Study series are available as films and as VHS and PAL videos. These were last revised in 1989. Millions of copies of the textbook and lab manual have been sold around the world. The books have been translated into 17 languages and the films into 8 languages.

However, after Neil Armstrong's dramatic moon landing on July 20, 1969, a sense of "mission accomplished" infused the country; and the nation began to turn its attention to other issues: the Vietnam War, the domestic social unrest and student protests, Watergate, and the oil crisis and economic downturn of the 1970s. Public education was pushed off center stage of the national agenda. The

years of neglect exacted a heavy toll.

In 1981, when NCEE was convened, American public education was practically in the throes of a crisis. Average achievement scores of high school students on standardized tests were lower than when Sputnik was launched. Science achievement scores of 17-year-olds in the United States, as measured in national assessments of science in 1969, 1973, and 1977, showed a steady decline; mean mathematics scores on the Scholastic Aptitude Test (SAT) dropped from 502 in 1963 to 466 in 1980. More than 20 million Americans were functionally illiterate, and about 13% of all American 17-year-olds could be considered functionally illiterate; and the figure was perhaps as high as 40% among minority youth. The facts and statistics were startling:

- A study conducted by the International Association for the Evaluation of Education Achievement in 1970–71 showed that, although the United States had by far the largest percentage of students still in school at age 18, American 18-year-olds scored the lowest in achievement and achievement gain among the students of the 15 developed and 4 developing countries in the survey. U.S. students had the lowest achievement scores in 7 subjects of the 19 academic tests.
- Enrollment in remedial mathematics courses at four-year public colleges increased 72% between 1975 and 1980. In contrast, the total student enrollment increase during that period was 7%. Remedial mathematics courses constituted 25% of all mathematics courses taught at those institutions.
- At community colleges, 42% of classes covered only sec-

"However, after Neil Armstrong's dramatic moon landing on July 20, 1969 . . . Public education was pushed off center stage of the national agenda."

ondary-level material.

- From 1971 through 1980, the number of secondary school science and mathematics teachers being trained dropped by 65% and 75%, respectively.
- A 1981 survey cited a shortage of secondary school chemistry teachers in 38 states. In nine of these states, the shortage was classified as ''critical.'' The same study also cited 42 states with a shortage in physics teachers, classifying 27 states as critical, and 43 states with a shortage in mathematics teachers, 18 of them critical.
- In 1982 the total number of science teachers graduated from all California schools of education was only 191.
- The Department of the Navy reported that 25% of its recent recruits could not read at the ninth-grade level, the minimum level necessary to understand simple written safety instructions.

The commission, which became operative on Aug. 16, 1981, met eight times, generally in Washington, DC. The first meeting was Oct. 9–10, 1981, the decisive penultimate meeting was Jan. 21–22, 1983, and the final meeting was April 26, 1983. In addition, there were about a dozen public hearings, panel discussions, and symposia, including testimony from some 250 experts, and about five dozen commissioned papers by educational experts to help us formulate our conclusions and recommendations.

On Oct. 9, 1981, the commission met with President Reagan. I think the following journal entry gives a good feeling of the president's interest in our assignment:

Friday, Oct. 9, 1981—Washington, DC

We then adjourned to go via cars and minibuses to the White House for our appointment with President Reagan. We entered via the southwest gate, waited in the Roosevelt Room, and then went into the Cabinet Room to take our assigned places at the table. Undersecretary William C. Clohan (Department of Education) stood behind me.

I sat in the vice president's chair, directly across from the president's chair, and there were large jars of jelly beans at my place and the president's place. There were large pictures of Coolidge and Eisenhower on the west wall (i.e., the wall neighboring the Oval Office) and a smaller picture of Lincoln on the east wall over the fireplace.

President Reagan entered the room a little before 11 a.m. and was announced by Secretary Bell, who then took him around the table to introduce to him each commission member individually. The president was very cordial and friendly as he recognized me while we shook hands. After the president took his place at the table (with Secretary Bell on his right and Gardner on his left), he began his remarks. He apologized for using notes (5 in. × 7 in. cards with typed notes interspersed with his handwritten insertions), saying he didn't want to make too many mistakes, especially in talking to the prestigious members of this commission.

He said the happiness and anticipation of this occasion are tempered by the sorrow and remorse following Egyptian President Anwar Sadat's death. He emphasized the worldwide reaction of sadness.

President Reagan then went on to emphasize the importance of schools and colleges to the general well-being of our country. As an aside, he said he is sure that Gardner will forgive him for any trouble he gave the University of California during his governorship (when Bell introduced Gardner to the president, he mentioned he had been a vice president of the University of California during Reagan's term as governor; Reagan responded, "And you still will speak to me!"). President Reagan said that the problems of and challenges to the educational enterprise in our country are enormous. The decline in our quality of education has contributed very much to the present decline in our economy.

President Reagan then said that excellence demands competition. He said without a race there could be no champion. As an example of how ill-informed some people now are, he cited a letter from a correspondent urging that we send our "Jaywalks" to Saudi Arabia. He emphasized that diversity is central to our mode of life. He said we are a pluralistic society. As an aside, he recalled his daughter's response when he told her something was "morally wrong." She responded, "By whose standard?" He also recalled a meeting with a group of black parents in Los Angeles while he was governor of California. One of the mothers pleaded with him, "My one request to you is to get my son back in the classroom." (The child had played hooky for the previous 10 weeks.) In conclusion, he asked that we bring God back into the classroom. He said there have been 4 billion laws enacted since the beginning of time, with the overall result that they are no better than the Ten Commandments. Finally, he urged that we go to the people with our report.

Secretary Bell thanked President Reagan. He said the college entrance scores in the United States have declined for each of the past 10 years. He said there are, of course, some outstanding schools, and that 3 out of every 10 Americans are involved in one way or another in the education enterprise. He said both public and private schools need to be strengthened. He cited Defense Secretary Caspar Weinberger's request for better educated recruits in the armed forces. We must turn our nation around and improve the quality of education, he said, and our report can be implemented only through persuasion.

Gardner then made his remarks, also using notes (he said he surely needs to avoid mistakes if his president does). He summarized some basic objectives in our commission's program.

President Reagan then responded with his thanks for the opportunity to talk to us. He recalled an episode of "60 Minutes" in which a student in Chicago smuggled a volume of Shakespeare to read in bed. The president remarked that he probably hadn't appreciated his reading of Shakespeare as much as he should have when he was in school. He also recalled a visit to a room of fifth graders in Arizona, where he was impressed by the performance of the students in solving problems posed to them verbally. When he asked the teacher how this could be taught so well, the teacher replied, "Practice." President Reagan said he is happy to report, however, a marked increase in high school graduates in armed forces enlisted personnel—it is now up to 80%. The president left us at a little after 11:15 a.m. with words of encouragement and emphasized the importance of our mission. I think he did very well. He was a little nervous, which I found rather charming.

Our final report, "A Nation at Risk: The Imperative for Educational Reform," was handed to President Reagan, in a ceremony attended by interested members, government officials, and the press, at the White House on April 26, 1983:

Tuesday, April 26, 1983

I took a taxi to the Ramada Renaissance Hotel (1143 New Hampshire Ave.) where, in the New Hampshire 2 Room, I joined the luncheon meeting of NCEE members and staff.

Present were Gardner (chair); Larsen (vice chair); William O. Baker (chairman [retired], Bell Telephone Laboratories); Anne Campbell (former commissioner of education, Nebraska); Emeral A. Crosby (principal, Northern High School, Detroit, MI); Charles Foster, Jr. (immediate past president, Foundation for Teaching Economics, San Francisco, CA); Norman Francis (president, Xavier University of Louisiana); Shirley Gordon (president, Highline Community College, Midway, WA); Robert V. Haderlein (immediate past president, National School Boards Association, Girard, KS); Gerald Holton (Mallinckrodt Professor of Physics and professor of the history of science, Harvard University); Annette Kirk (Kirk Associates, Mecosta, MI); Margaret Marston (member, Virginia State Board of Education); Albert H. Quie (former governor, Minnesota); Jay Sommer (National Teacher of the Year, 1981–82; high school teacher, New Rochelle High School, New Rochelle, NY); Richard Wallace (high school principal, Lutheran High School East, Cleveland Heights, OH); and myself and essentially all of the staff, including executive director Milton Goldberg.

We were given copies of our NCEE report, "A Nation at Risk: The Impera-

"We must turn our nation around and improve the quality of education, he said, and our report can only be implemented by persuasion."

tive for Educational Reform." (There is no indication of an Open Letter to the American People on the cover, only a small lettered mention in the inside title page.) There are 36 pages in the main (essay) section, which includes the recommendations. We were also given a flyer with the same title describing the report and including a report order form (cost $4.50).

At the luncheon table, after we finished eating, Gardner and Bell made some remarks of appreciation and Bell presented each commission member with a certificate of appreciation, individually and with photographs taken. After the luncheon we went out into the foyer and asked our fellow commissioners to autograph our copies of the report. We re-assembled in the New Hampshire 2 Room at 2 p.m. for our final meeting. We formally adopted the report by a unanimous vote and then discussed methods of distributing it. (Each commissioner was given 5 copies and will be sent 100 additional copies.) Copies will be sent to every member of Congress, all state governors, state legislatures, the head of AFL–CIO, the National Education Association (Mary Futrell, secretary-treasurer), the American Federation of Teachers (Albert Shanker, president), the American Association of School Administrators (Paul Salman, executive director), the National Association of Secondary School Principals (Scott Thompson, executive director), the United States Chamber of Commerce, the Business Roundtable, and many other organizations. Goldberg will send commission members a list of organizations that received the report and a batch of newspaper clippings of coverage of the report.

At the conclusion of the meeting, at about 3 p.m., I called Pat Rupp (my secretary) at my Lawrence Berkeley Laboratory office, took care of a few business items, and then rode to the White House with the commission members in a van. Staff and friends went in automobiles. First we went to the Red Room before entering the State Dining Room, where the ceremony was to be held before an audience of some 200–300 invited people and numerous representatives of the press (newspapers, radio, TV, etc.).

Beginning at about 3:45 p.m., there were speeches by Gardner (who, along with Larsen and Bell, sat on the stage back of the lectern) and then Bell, who spoke until President Reagan arrived. I sat next to Gary Jones (undersecretary of education). President Reagan spoke from a prepared text and in large measure embraced and endorsed our report. However, and unfortunately (in my opinion), he interpreted our report as a "call for an end to federal intrusion." He said it is "consistent with our task of redefining the federal role in education" and added, "We'll continue to work in the months ahead for passage of tuition tax credits, vouchers, educational savings accounts, voluntary school prayer, and abolishing the Department of Education." He included the charming comment that when Eureka College recently awarded him an honorary degree, he didn't really deserve it because when he graduated he considered the degree they awarded him then as an honorary one. Also when he misspoke, saying "compromise" instead of "comprise," he said it must be a Freudian slip because he is preoccupied by attempts to arrive at compromises with Congress.

Upon completion of his talk, which lasted between 5 and 10 minutes, he stepped down and went to the front row to shake hands with some members of Congress and with some NCEE members, including me. Photographers took pictures (including a picture of me shaking hands with President Reagan, which appeared in the New York Times the following day, Wednesday, April 27).

After the ceremony I was approached by reporters who asked me to identify the places in our report where we advocated such concepts as "vouchers" and "prayer in schools." I replied (diplomatically, I believe) that our report made no references to any such recommendations. This part of the president's remarks was, fortunately, ignored by the press.

The members of our commission had gone through a period of contention before we could agree on these dramatic opening paragraphs:

> Our Nation is at risk. Our once unchallenged preeminence in

To Glenn Seaborg
With best wishes, *Ronald Reagan*

commerce, industry, science, and technological innovation is being overtaken by competitors throughout the world. This report is concerned with only one of the many causes and dimensions of the problem, but it is the one that undergirds American prosperity, security, and civility. We report to the American people that while we can take justifiable pride in what our schools and colleges have historically accomplished and contributed to the United States and the well-being of its people, the educational foundations of our society are presently being eroded by a rising tide of mediocrity that threatens our very future as a Nation and a people. What was unimaginable a generation ago has begun to occur—others are matching and surpassing our educational attainments.

If an unfriendly foreign power had attempted to impose on America the mediocre educational performance that exists today, we might well have viewed it as an act of war. As it stands, we have allowed this to happen to ourselves. We have even squandered the gains in student achievement made in the wake of

Seaborg shakes hands with President Ronald W. Reagan in the White House State Dining Room at the presentation of the National Commission on Excellence in Education report "A Nation At Risk." Washington, D.C. April 26, 1983. Courtesy of the White House.

the Sputnik challenge. Moreover, we have dismantled essential support systems which helped make those gains possible. We have, in effect, been committing an act of unthinking, unilateral educational disarmament.

We succeeded in drawing almost unprecedented attention from educators, parents, the public, and the press.

After "A Nation at Risk" was issued, our NCEE members, and even President Reagan, were involved in a number of meetings throughout the country to encourage the implementation of our recommendations. One such occasion, in which President Reagan participated, occurred in Whittier, CA, on June 30, 1983:

I had breakfast in the coffee shop and afterward, NCEE members, panel members, and others rode in several cars to Pioneer High School in Whittier. Here we went to the principal's office and then the gymnasium, where about 1500 people were assembled, all of whom had gone through a security check. The forum was opened by Gardner, seated at a table in the center of the gym. He introduced Mrs. Eve Burnett (president of the Whittier Union High School District Board of Trustees), who welcomed the audience. Senator William Campbell (Senate minority leader, California) spoke briefly on behalf of Governor George Deukmejian.

Gardner then introduced today's panel members and the five NCEE members present: Charles Foster, Norman Francis, Margaret Marston, Richard Wallace, and me. Directly after the introductions, at 9:30 a.m., President Reagan entered with Secretary Bell. The president began to speak immediately. He praised Pioneer High School and said that everyone is proud of it. He praised our report and said the nation should respond as it did in the 1950s. He said that we federal officials should identify with the national interest in education but that local leadership also is important. He praised Governor Deukmejian's education program. He emphasized that home and family are the foundation for the support of educa-

tion and called for educational renewal and reform. Education must be the highest priority in America today. Pioneer High School was selected as the first to be honored for excellence in education among the 144 exemplary high schools chosen by the government.

The president, flanked by Bell and Gardner and seated at the central table, then rose and presented to Superintendent Norman Eisin and Principal Robert Eicholz a huge flag with the letters "Pioneer High School, Excellence in Education, 1982–83." Superintendent Eisin then responded. He said the basic criterion for participation in athletics will be attendance, not academic accomplishment. Bell then introduced Jaime Escalante (famous math teacher, Garfield High School, Los Angeles). He asked President Reagan to sign his teacher's certificate. He decried the "Mickey Mouse" classes and told about the increased math requirements at his high school.

Bell then called on Eicholz, who spoke about the emphasis on basic skills (reading, writing, math, etc.) at Pioneer High School. However, he also emphasized the importance of athletics, fine arts, business, and other subjects. He placed emphasis on attendance and said he was getting help from parents in this and other regards. Bell then introduced Ralph Figueroa (student, California High School, Whittier), who praised the job Eicholz does at Pioneer High School. Bell said the other 143 winners will be announced at noon today.

Again the president spoke, emphasizing the importance of extracurricular activities and describing his contact with his high school principal in Dixon (who emphasized that he cared only what Reagan would think of him 15 years from now, and not now). President Reagan also described how some of his fellow students were turned on by science classes.

The president started to leave at 10:15 a.m., recognized me as he came by, and stopped to shake hands with me and other NCEE members as well as other people. I mentioned to him that my son Eric (president of the American Hiking Society) had been present when he signed the Amendment to the National Trails Act. He left the gym at about 10:20 a.m.

To promote our report, I published an editorial, "A Call for Educational Reform," in the July 15, 1983, issue of *Science* magazine, summarizing our most important recommendations:

A Call for Educational Reform

"If an unfriendly foreign power had attempted to impose on America the mediocre educational performance that exists today, we might well have viewed it as an act of war. As it stands, we have allowed this to happen to ourselves. We have even squandered the gains in student achievement made in the wake of the Sputnik challenge. Moreover, we have dismantled essential support systems which helped make those gains possible. We have, in effect, been committing an act of unthinking, unilateral educational disarmament."

This warning sets the tone for the clarion call for educational reform made by NCEE. Our report, handed to President Reagan on 26 April, has attracted national attention, and there are indications that it will spark widespread remedial action.

There has been an alarming deterioration of our precollege educational system during the past 15 to 20 years. This adversely affects the capacity of individuals to adapt to the changing demands of our complex age and the ability of our nation to compete in today's world of high technology. The deficiency in the quality and quantity of teaching of science and mathematics—subjects that are emphasized in a number of countries that are our competitors—is undoubtedly a factor in our country's economic decline. Lack of scientific literacy threatens the efficient, or even adequate, functioning of our democracy in this scientific age.

More than 20 million American adults cannot read, write, or comprehend the English language. Two-thirds of our high schools now require only one year of science and one year of mathematics for graduation. Low salaries for science and mathematics teachers have driven large numbers of them to better-paid positions in industry and have discouraged college students from entering the teaching profession. There is a critical shortage of mathematics and science teachers in some 40 states, and half of those newly employed are not qualified to teach these subjects.

The Commission's report includes some 40 implementing recommendations in five major categories. Following are some of the more radical, but also some of the more important, of these. Requirements for high school graduation should include four years of English, three years of mathematics, three years of science, three years of social studies, and one-half year of computer science. Salaries for the teaching profession need to be increased and should be professionally competitive, market-sensitive (this means differential pay), and based on performance (not merely years of service). Salary, promotion, tenure, and retention decisions should be tied to an effective evaluation system that includes peer review so that superior teachers can be rewarded, average ones encouraged, and poor ones improved or terminated. Recent graduates with mathematics and science degrees, graduate students, and industrial and retired scientists should, with appropriate preparation, be allowed to teach immediately in these fields. The capabilities of science centers should be used for educating and retraining teachers. University scientists, scholars, and members of professional societies, in collaboration with master teachers, should participate in the development of more effective curricula, as they did with success in the post-Sputnik period. The federal government should provide national leadership in the field of education and assume primary responsibility for the support of curriculum improvement; for research on teaching, learning, and management of schools; and for teacher training in areas of crit-

ical shortage of key national needs. It should also provide financial assistance for college students, research, and graduate training.

The Commission calls on all Americans to insist on excellence in education and to assist in bringing about the educational reforms proposed in its report.
—Glenn T. Seaborg, Lawrence Berkeley Laboratory, University of California, Berkeley 94720, and member of NCEE

Members of NCEE (after Gardner had assumed the presidency of the University of California) again met with President Reagan in the Cabinet Room of the White House on May 11, 1984:

After the breakfast I rode in a van with the other 10 NCEE members to the northwest gate of the White House. We went through the gate as our identifications were checked and then, after waiting in an anteroom, went to the Cabinet Room and took our places around the table. I sat on the chair labeled "Secretary of Energy." President Reagan entered the room at about 11 a.m. and took his regular seat with Gardner on his left and Bell on his right. The president opened his remarks with the humorous comment that our national commission has been so successful that he should have had it working on the national budget deficit problem. He indicated that the impact of our report has been widespread. He referred to the deterioration of education in the United States and expressed the wish that the universities and colleges would eliminate the "bonehead" courses that have become rampant. Secretary Bell presented President Reagan with a copy of "The Nation Responds," a 220-page report summarizing the activities across the country directed toward implementation of the recommendations in "A Nation at Risk." The president responded that this would give him reading material for the weekend. When the length of the report was called to his attention, the president quipped that this additional paper has been made available because of the widespread cancellation and curtailment of government publications that

Meeting of the National Commission on Excellence in Education, White House Cabinet Room. Washington, D.C. May 11, 1984. Clockwise around table: Yvonne Larsen (back to camera) (President, San Diego City, CA, School Board, 1982), Emeral Crosby (Principal, Northern High School, Detroit), William O. Baker (Chairman of the Board [retired], Bell Telephone Laboratories), unidentified individual), Seaborg, Jay Sommer (National Teacher of the Year, 1981-82) (seated at head of table), Richard Wallace (Principal, Lutheran High School East, Cleveland Heights, OH), Terrel Bell (U.S. Secretary of Education), President Ronald Reagan, and David P. Gardner (President, University of California, and Commission Chairman). Cut off from picture, right of Gardner: Milton Goldberg (Executive Director, NCEE). Courtesy of the White House.

have taken place under his administration.

Secretary Bell then called on Gardner for his comments. Gardner began by noting that the president was wearing Golden Bear cufflinks similar to his own, and the president responded by acknowledging this and calling attention to his Golden Bear tie clip. Gardner said that the reason our report made a difference was that it made sense. He alluded to the fact that "A Nation at Risk" made no mention of dollar amounts needed to reform education, saying the issue is, instead, whether we have made recommendations. It is in this manner that our society works. Gardner emphasized that the report was addressed not to the government but to the American people. If they think that it makes sense, then things will happen, for this is the way our country works best. He said that he had explained this to French President François Mitterrand when he talked to him at the luncheon we hosted for him in Berkeley. Gardner

then added that our commission is grateful to President Reagan for moving the cause of education ahead. The reference to President Mitterrand led President Reagan to make the remark that another Frenchman, de Tocqueville, made a similar comment when he visited the United States in 1830, that, "no bureaucracy was involved in moving our education forward."

President Reagan then told the story about the basketball star who, as a kid, was already six feet five inches tall in junior high school but who couldn't read or write by the time he reached the 11th grade. He wanted a diploma but wasn't able to get the required courses, which included "History of Baseball." He went back to school in Chicago in a fourth-grade class and then planned to attend one of the University of Illinois campuses.

President Reagan then presented Gardner with a large, glass-covered citation with a copy of the front page of "A Nation at Risk." The inscription, "To David P. Gardner, in appreciation of your service to our Nation as a member of NCEE," was signed by Secretary Bell and President Reagan. The president stated that our commission produced results in an almost magic manner: "We threw a bomb, created a miracle, parted the waters."

Following the presentation, the president presented to each NCEE member similar large, glass-covered certificates. A photographer took a picture of each commission member with the president during each presentation. I complimented President Reagan on his meeting with John Lawrence and Alex Hollaender yesterday.

We all went to the south lawn for the presentation by the president of the Academic Fitness Awards. In addition to the 60 students were some 130 guests of the students, some 80 educators, and a large contingent from the news media. The president emerged from the White House, strode to the lectern, looking unusually fit and in excellent physical condition, and then gave an outstanding talk on the educational situation and on how our report has led to the needed education reform in the United States. He quoted the sentence, "If an unfriendly foreign power had attempted to impose on America the mediocre educational performance that exists today, we might have viewed

To Glenn Seaborg
With appreciation and best wishes,

Ronald Reagan

President Ronald Reagan presents a plaque to Seaborg after a National Commission on Excellence in Education meeting at the White House. Washington, D.C. May 11, 1984. Courtesy of the White House.

it as an act of war." (Last night during a conversation, NCEE staff member Susan Tremaine said that NCEE staff refer to this sentence as the "Seaborg Sentence.")

"A Nation at Risk" ignited a movement. [Readers interested in learning more about NCEE recommendations are referred to the report, "A Nation at Risk: The Imperative for Educational Reform," April 1983, available from the U.S. Government Printing Office, #065-0000-00177-2.] We succeeded in drawing almost unprecedented attention from educators, parents, the public, and the press. The precollege educational crisis and the urgent need for educational reform became broadly perceived as a top priority. Organizations and agencies that had been working for years to improve precollege education received a boost in publicity. Since the issuance of "A Nation at Risk," more than 100 reports by a wide spectrum of American organizations have emphasized the deplorable state of precollege education in science and mathematics in the United States today. Specifically, I want to cite some of the following statistics and trends, as of 1990:

• Thirty-seven states now require 4 years of English, 28

require 3 or more years of social studies, and 10 require 3 years of mathematics. However, only 4 states had upgraded their requirements to 3 years of science for graduation from high school. Although many states had raised their high school graduation requirements, their requirements still fell far short of the recommendations in our report.

- Longer school days were reported in 40% of high schools, 30% of middle schools, and 34% of elementary schools. A longer school year had been established in 17% of high schools, 16% of middle schools, and 18% of elementary schools. More homework had been required in 27% of high schools, 30% of middle schools, and 32% of elementary schools.
- The number of science centers in the United States had nearly doubled since the appearance of our report, to a total of several hundred, and these centers serve as an integral part of the educational community. They served 10 million students annually—6 million at the centers and 4 million through outreach programs. Some 100 centers conducted teacher training workshops, serving more than 50,000 teachers annually.

Unfortunately, although some progress has been made, the goal of implementing the report's recommendations "over the next several years" was never realized in a substantial manner. (In the next chapter, on the Bush administration, I mention an action conference that was co-chaired by Department of Energy (DOE) Secretary James D. Watkins and me at the Lawrence Hall of Science in Berke-

ley, CA, in 1989.) Needless to say, the state of precollege education in our country is still at critical levels and the challenge of creating a public educational system for the 21st century remains a daunting task.

In Chapter 4, I alluded to the "cleansing" of my journals during the Reagan administration. I shall describe this unfortunate episode in some detail. The core of the journal was a diary, much of which I wrote at home each evening. The diary was supplemented by copies of correspondence, announcements, minutes of meetings, and other relevant documents that crossed my desk each day. Both in the diary and in the addition of the supporting documents, rigorous attention was given to excluding any subject matter that could be considered classified information under the standards of the day. My purpose was to provide for historians and other scholars a record that might not be available elsewhere of what occurred at high levels of government regarding the Atomic Energy Commission's (AEC) important areas of activity.

Illustrative of the general recognition that my journal was unclassified was the fact that in 1965 the AEC historian microfilmed for public access in the John F. Kennedy and Lyndon B. Johnson libraries portions that correspond to those presidencies. To assure myself further that the journal contained no classified material, I had it checked by the AEC Division of Classification during summer and fall 1971, just before my departure from AEC. It was cleared, virtually without deletions. (Unfortunately, I received no written confirmation of this action, which is perhaps understandable, given obvious unclassified origin of the material.) A copy, which I will refer to as copy #1, was then transmitted by AEC to my office at the University of California, Berkeley. Also, at about this time, AEC transferred another copy of the journal, referred to hereinafter as copy #2, first to my Berkeley office, then to the Livermore laboratory and, soon thereafter, to my home in Lafayette, CA. It was known that neither at my Berkeley office nor my home was there any provision for the protection of classified material, and the fact that AEC saw fit to ship the journal to those places is a clear indication that the commission regarded the journal as an unclassified document.

The office and home copies of the journal remained accessible to scholars for the ensuing 12 years. Then the problems began. In July 1983 DOE's chief historian asked to borrow a copy for use in the next phase of the History Division's long-term project, the writing of *A History of the United States Atomic Energy Commission*. Volume IV of the history was to be devoted largely to the years of my chairmanship. The historian promised to return the journal within three weeks, as soon as copies had been made. I sent him copy #1, the one in my Berkeley office.

Despite my repeated entreaties, the historian's office did not return the journal in three weeks, nor in three months, nor in a year and a half. Nor was any explanation ever offered to me for the delay. Finally, I was informed in February 1985 that the journal had indeed been found to contain classified information. Accordingly, DOE ordered its San Francisco Area Office to pick up copy #2, the one that I kept at home, so that it also could be subjected to a classification review. At first I said I would not allow this. But then I was told that, legally, the journal could be seized and that I could be subject to arrest if I resisted. Faced with this disagreeable prospect, I acceded to a compromise plan (the best of several unsatisfactory alternatives) whereby DOE provided me with a locked storage safe, complete with burglar alarm, so that I could continue to have access to the journal, which at that time I was preparing for publication. It was no longer, however, to be available for use by scholars.

Then in May 1985 I was contacted by DOE's San Francisco area manager. He had been instructed by DOE headquarters to institute a classification review of copy #2 at my home. He added that if I did not agree to this, the FBI would seize the papers under court order. The weakness of my case, if I chose to resist, was that there was no record of the journal ever having been declassified by AEC. Thus, I could be accused of having illegally removed classified material when I left the commission. He noted that if legal proceedings were instituted, I could, of course, hire a lawyer to defend myself, but he knew of no case like this that the government, with all its resources, had lost.

Under this ultimatum, I agreed to the classification review with the understanding that it would be completed within 10 days. The reviewer started work in my home on May 9, 1985, kept at it for several weeks (not the promised 10 days), and came up with 162 deletions of words, phrases, sentences, or paragraphs, affecting 137 documents.

Then in May 1986 I learned that copy #1, the one borrowed by the DOE historian, was also undergoing a classification review. This review was completed in October 1986 and led to deletions from 327 documents. In addition, 530 documents were removed from the journal entirely pending further review by DOE or other government agencies.

At the same time that reviews of my complete journal were being undertaken in DOE and in my home, a further review was taking place in the Bethesda, MD, home of Benjamin S. Loeb, who was then collaborating with me on my book, *Stemming the Tide—Arms Control in the Johnson Years,* which was to be published in 1987. Copies had been sent to Dr. Loeb of just those portions of the journal that related to arms control. Beginning July 10, 1986, as many as six DOE Division of Classification staff members sat around his dining room table for a few days, selecting a large number of documents that they then took back to DOE headquarters in Germantown, MD. In due course, most of these were returned with deletions; however, a number of documents that required review by U.S. government agencies other than DOE, or by the United Kingdom, were not returned until August 1990.

But there was more. In October 1986 I was informed that the DOE classification people wanted to perform another review of copy #2, the one in my home, in order to "sanitize" it, a euphemism for a further classification review of the already reviewed journal. I was informed that the sanitization procedure would take place at Livermore, that it would last three to six weeks, and that it would involve from eight to twelve people. Copy #2 was duly picked up at my home and delivered to Livermore on Oct. 22, 1986. When the "sanitized" version was returned almost two months later, it had been subjected, including the prior review, to about 1000 classification actions. These included the entire removal of about 500 documents for review by other U.S. agencies or, in a few cases, by the British. Over my objection, an unsightly declassification stamp was placed on every surviving document.

Finally, DOE sent to the Lawrence Berkeley Laboratory a team of about 12 people to begin a "catalog," or itemized listing, of all the personal correspondence I had brought from AEC as well as the contents of my journal and files for the 25 years of my working life prior to becoming AEC chairman.

Beginning on April 29, 1987, the team spent about two weeks at this task. In March 1988 another DOE group visited me for about a month to complete the catalog. The motives of DOE in undertaking this task were not clear. They may well have intended to be helpful to me. Before they finished, however, the two groups uncovered some additional "secret" material.

My grammar, high school, and university student papers, stored in another part of my home, were overlooked by the DOE classification teams and thus far have escaped a security review.

My journal was finally reproduced in January 1989 in 25 volumes, averaging about 700 pages each, many of them defaced with classification markings and containing large gaps where deletions had been made. In June 1992 a 26th volume was added. It contained a batch of documents initially taken away for classification review and subsequently returned to me, with many deletions, after production of the other 25 volumes in January 1989. (Many other removed documents have still not been returned.) All 26 volumes are now publicly available in the expurgated form in the Manuscript Division of the Library of Congress.

This, then, is a summary narrative of the rocky voyage of my daily journal amid the shoals of multiple classification reviews. Those interested in a more detailed account can find it among the daily entries in my journal for the period after I left AEC. This is available in the Manuscript Division of the Library of Congress and, fortunately, has not yet been subjected to classification review.

What can be concluded about this sorry tale? One conclusion I have reached is that the security classification of information became in the 1980s an arbitrary, capricious, and frivolous process, almost devoid of objective criteria.

> "One conclusion I have reached is that security classification of information became in the 1980s an arbitrary, capricious, and frivolous process, almost devoid of objective criteria."

> "I had in mind that there might come a day when a more rational approach to secrecy might prevail and permit wider access, especially to historians, of the complete record."

Witness the fact that the successive reviews of my journal at different places and by different people resulted in widely varying results in the types and number of deletions made or documents removed. Furthermore, some of the individual classification actions seem utterly ludicrous. These include my description of one of the occasions when I accompanied my children on a "trick or treat" outing on a Halloween evening and an account of my wife Helen's visit to the Lake Country in England. One would have to ask how publication of these bits of family lore would adversely affect the security of the United States. A particular specialty of the reviewers was to delete from the journal many items that were already part of the public record. These included material published in my 1981 book (with Benjamin S. Loeb), *Kennedy, Khrushchev, and the Test Ban*. Another example concerned the code names of previously conducted nuclear weapons tests. These were deleted almost everywhere they appeared, regardless of the fact that in January 1985 DOE had issued a report (NVO-209, Revision 5) listing, with their code names, all "Announced United States Nuclear Tests, July 1945 through December 1984." A third category of deletions concerned entries that might have been politically or personally embarrassing to individuals or groups but whose publication would not in any way threaten U.S. national security. In fact, I would go so far as to contend that hardly any of the approximately 1000 classification actions (removals of documents or deletions within document), taken so randomly by the various reviewers, could be justified on legitimate national security grounds.

Consistent with this belief, I have requested repeatedly throughout this difficult time that a copy of my journal as originally prepared (before all the classification reviews) be kept on file somewhere. I had in mind that there might come a day when a more rational approach to secrecy might prevail and permit wider access, especially to historians, of the complete record. There are indications that, especially with the end of the cold war, such an era may be at hand or rapidly approaching. Although DOE has made no commitment to honor my request, I am informed that DOE's History Division does maintain an unexpurgated copy for its own use. Perforce, it is handled as a classified document.

I would like to emphasize that I received fine and sympathetic treatment from many DOE individuals who made it clear to me that they disagreed with the treatment accorded me and my journal during the process recounted above. In fact, more than one person in DOE has told me informally that evidence does indeed exist to verify that my journal had received a clearance before my departure from AEC in 1971.

The problems posed by classification and declassification of sensitive materials are major ones and require wise people who must make sophisticated decisions. It requires a range of individuals who, on the one hand, have some vision with regard to the whole range of scientific and national security policies and, on the other, have time to read pages of detailed descriptions in a wide range of areas. Sometimes this complex goal gets derailed by those who see the trees and not the forest. Those in charge of classification should have an appreciation of the need, in our open society, to publish all scientific and political information that has no adverse effect on national security (realistically defined).

Although I have in general received sympathetic treatment, I cannot help but note that this treatment has produced quite different conclusions at different periods in the country's history. Actually, AEC, from its beginning in 1947, initiated and executed an excellent progressive program of declassification with an enlightened regard for the need of such information in an open, increasingly scientific society. By the 1960s, this program was serving our country very well. Unfortunately, during the 1980s, the program had retrogressed to the extent of reversing many earlier declassification actions. Fortunately, the present situation is very much improved so we can look forward to the future with considerable optimism.

Helen and I attended a dinner in the White House in honor of Prime Minister Ingvar Carlsson of Sweden on Sept. 9, 1987:

After awhile we changed into our formal clothes and took a taxi to the east gate of the White House, where we arrived at about 7:20 p.m. to attend a dinner honoring the Prime Minister of Sweden, Ingvar Carlsson, and Mrs. Ingrid Carlsson. We went through the metal detector and walked toward the west end of the White House. Our arrival was announced to the press corps, which consisted of a bevy of reporters and photographers. We went through, past this line, where our picture was taken by numerous photographers. The reporter from the Washington Times, Lisa McCormack, asked me what I planned to take up with the president. I told her that I did not have any fixed agenda; it was a social occasion. She asked what I would ask if I were going to take advantage of the opportunity and press some point with the president. I said I would urge him to support basic research in the United States so that we as a nation can compete in the high-technology world. Reporters asked Helen whether this was our first visit to the White House, and she said it was for the Reagan administration, but not with respect to earlier administrations.

As we emerged from the press line, we ran into Curtis and Arleen Carlson (friends with the Swedish Council of America) and exchanged greetings with them. We then went upstairs to the East

Room (also known as the Gold Room), where again our entrance was announced. Here we talked to a number of dinner guests. I first talked to Howard Baker. I recalled our pleasant and instructive relationship when he was a member of the Joint Committee on Atomic Energy, and he agreed. He said he missed those days. I urged him to run for president, but he said he has had enough of the White House during his six months there; he doesn't want to be president. He invited me to have lunch with him the next time I come to Washington and said that he would include the presidential science adviser, William Graham, if I would call him to make the appointment.

Helen and I had a rather long conversation with Supreme Court Chief Justice Bill Rehnquist and his wife Ann. We compared our origins, he in Wisconsin and I in northern Michigan. He has a Swedish parent. Later, Curtis Carlson and Helen talked to him about the possibility of his accepting a Great Swedish Heritage Award. They had only partial success, because the indications are that he would not find time to travel very far to receive this honor. Among others we talked to were Swedish Ambassador Wilhelm Wachtmeister, Professor Burt Boland (science adviser to the Swedish prime minister), Professor Bengt Samuelsson of the Karolinska Institute, Daniel J. Boorstin (ex-librarian of Congress) and his wife Ruth (I promised to send him a copy of Stemming the Tide, and he promised in return to send me one of his books), Holly Coors (special representative to the 1987 National Year of the Americas), Utah Senator Orrin G. Hatch, M. Carl Holman (president, National Urban Coalition) and his wife Mary Ella, Florida Representative Bill McCollum, and James H. Webb, Jr. (secretary of the navy) and his wife Joanne.

We then went through the reception line (which included President and Nancy Reagan). As I passed President Reagan, I greeted him as a fellow Californian, and when I passed Prime Minister Carlsson, he made a favorable comment about our correspondence and said that we should discuss it later. As Helen passed President Reagan, she commented that he seemed to be in good shape, and he responded by leaning over to her and saying that you must

Seaborg and First Lady Nancy Reagan at a dinner in honor of Ingvar Carlsson, Prime Minister of Sweden. Ingrid (Mrs. Ingvar) Carlsson is at the left, and Helen Seaborg at the right. East Room, White House. Washington, D.C. Sept. 9, 1987. Courtesy of the White House.

not believe what you read in the news-papers.

We proceeded to the State Dining Room, where I talked to Bjørn Borg, the Swedish tennis star. He told me that other Swedish tennis stars, Mats Wilander and Stefan Edberg, had both won in the U.S. Open today and perhaps would meet in the semifinals. He agreed that a good night's sleep and top mental attitude are essential for winning a close match.

In the State Dining Room I sat at a table with Mrs. Walker Percy (her husband is a famous author), Mrs. Martin (Carol) Seretean (her husband is a carpet company executive), Mrs. Rehnquist, Frank Neville Ikard (Texas business executive), Ambassador Wachtmeister, Kenneth Walsh (White House correspondent, U.S. News and World Report), and Mary Evans (her husband is a former chairman of Union Pacific Corporation). I sat between Mrs. Percy and Mrs. Seretean. Helen sat at a table that included Chief Justice Bill Rehnquist (she sat next to him), Curtis Carlson, and Carl Holman (he sat next to her). My table was on one side of the table that included President Reagan, and Helen's table was on the other side, so we were quite close to him. Candice Bergen, as I learned later, was at the president's table.

After dinner we were serenaded by about a dozen people playing violin music. President Reagan then made some welcoming remarks, saying that our countries have been friends for as long as the United States has been a country. He also remarked that the Carlssons will celebrate their 30th wedding anniversary next week and will visit Northwestern University, where Prime Minister Carlsson was a student in 1961.

Prime Minister Carlsson, who was at the table across the room from the president's table, responded with remarks that included reference to the historic breakthrough in the endeavor the United States is now engaged in to start dismantling nuclear weapons. He also commented on his nostalgic return visit to Northwestern University.

After dinner I talked to Ulla Wachtmeister as well as to Pehr Gyllenhammar (chairman and CEO of the Volvo Company) and his wife Christina.

Because most people had gathered for coffee in the Blue Room (there were very few people in the neighboring Green Room and Red Room), Helen and I went in there. Here I had a long conversation with Prime Minister Carlsson. He told me that he has circulated our correspondence to his circle of advisers and that they will come to a conclusion about whether to support the amendment approach to a nuclear test ban. Then, at a meeting of the leaders of the Six Nation Initiative to be held in Stockholm in January, this issue might be taken up. I told him that most other leaders of the Six Nation Initiative support the amendment approach, and he agreed. He said that in his meeting with President Reagan today he offered the services of the Six Nation Initiative to the verification of a CTB, and President Reagan seemed to be more amenable than he had been to a CTB. He recalled that he had met me when he was a boy on one of my visits with Swedish Prime Minister Tage Erlander. I told him that Sweden was a very important nation for showing leadership in the test ban amendment approach, and he agreed.

Helen and I then went back to the East Room for the entertainment, which consisted of six songs and an encore by the singer Marilyn Horne. After dancing awhile in the foyer entrance area, Helen and I walked back to the University Club. We retired a little before midnight.

Prime Minister Carlsson's statement that "he has circulated our correspondence to his circle of advisers" refers to my letter to him urging that he support the amendment approach to a test ban. The Partial Test Ban Treaty could be

Helen Seaborg shakes hands with President Ronald Reagan at the dinner for Swedish Prime Minister Ingvar Carlsson, East Room, White House. Washington, D.C. Sept. 9, 1987. Courtesy of the White House.

To Helen Seaborg
With best wishes,

Ronald Reagan

converted to a comprehensive nuclear test ban through an amendment procedure, which would, in some respects, be easier to do than adopting a separate CTB treaty. An international organization, "Parliamentarians for World Order," took the lead in trying to implement the amendment approach.

I spoke to several members of the Reagan administration, including Secretary of State George P. Shultz, Secretary of the Treasury James A. Baker, and Secretary of Defense Caspar W. Weinberger to urge resumption of negotiations toward a CTB treaty, which had been suspended near the end of the Carter administration, but I was not successful. To bolster my argument, I had an editorial, "Seaborg Proposal: Support a Comprehensive Test Ban," published in the June 13, 1983, issue of *Chemical & Engineering News*:

Seaborg Proposal: Support a Comprehensive Test Ban

Sir: I urge all who wish early progress in nuclear arms control to throw their support behind a comprehensive test ban (CTB) as the simplest and quickest way of moving forward.

A CTB would have great benefits to the United States in slowing and reversing the nuclear arms race, in strengthening international efforts to prevent further proliferation of nuclear weapons, and in providing a new momentum in arms control negotiations.

A CTB would halt that aspect of the arms race that is most threatening, the qualitative improvements in nuclear weapons. Such improvements in offensive weapons continue to make them ever more dangerous. Improvements in defensive weapons might tempt either side to launch a first strike on the assumption that this can be done with relative impunity or needs to be done before the other side achieves an effective defense.

A CTB is absolutely essential if the tenuous nonproliferation regime now in effect is not to unravel. At the insistence of nonnuclear weapon states, the superpowers pledged in both the Limited Test Ban Treaty of 1963 and the Non Proliferation Treaty (NPT) of 1970 that they would move earnestly and quickly to the negotiation of a treaty to end all nuclear testing. The fact that they have not followed through on these pledges is repeatedly called to their attention, as, for example, when they try to enlist new adherents to the NPT. More frightening is the rising tide of revolt on this issue among those who *have* signed. At the last (1980) NPT Review Conference, dissension was so rife that it proved impossible to issue the customary communiqué at the end of the meeting. A clear warning was issued that if, by the time of the next Review Conference in 1985, there has not been very substantial progress toward a CTB, withdrawals from the NPT could be expected.

A great virtue of a CTB would be its simplicity. The two sides would simply agree to stop testing! Negotiators for the United States, U.K., and the USSR reported to the UN Committee on Disarmament in July 1980 that they had agreed on the main outlines of a CTB. It would rely on automatic seismic detection stations on the territories of the three powers, supplemented by a system of voluntary onsite inspections designed to ensure that any claim of a possible violation, or any rejection of such a claim, was based on serious information. Ultimate recourse to the UN Security Council was provided for in case disagreement persisted. What was lacking in 1980 was the political will to approve what the negotiators had agreed to. It is still lacking, and therein lies a task for those who would influence events.

As I have described in my book, *Kennedy, Khrushchev, and the Test Ban*, negotiations for a CTB in the early 1960s broke down over U.S insistence that obligatory onsite inspection was needed in order to be sure that the Soviets would not cheat. I think now, in retrospect, that even at that time we were wrong. The likelihood that the Soviets would have risked international censure by trying to cheat was very small when they could not have been sure of what we would detect. What they could have gained from the very small tests they might have sneaked by could not have affected the military balance. With each passing year since then, the ability to monitor compliance with a CTB has become more assured. At this time, tests above one kiloton have a high probability of being detected and identified, whether through seismic or satellite means, or through intelligence sources (see, e.g., Lynn R. Sykes and Jack F. Evernden, ''The Verification of a Comprehensive Test Ban,'' *Scientific American*, October 1982, pages 47–55). Even more than in the 1960s, it seems unlikely today that the Soviets would take large political risks for the chance of making the insignificant military gains they could achieve through clandestine tests under a CTB.

By contrast with the simplicity of a CTB, modalities being negotiated for actual arms reduction are immensely complex. Even if an agreement were possible, it is likely to require years of negotiation—years during which new and more dangerous weapons could be fashioned to increase the planet's peril. The issues involved—whether to count missiles or warheads; whether to count British and French weapons on our side and weapons deployed in Asia on the other side; definitions of which weapons are to be considered strategic, intermediate, or tactical; how to deal with bombers and submarines; how to count the various categories and how to define equality; how to define a new, versus an improved, weapon—are brainbusting, approaching the metaphysical, requiring extended argument just to define the questions. Obviously, these obstacles must ultimately be overcome and nuclear arms reduction must be achieved if the preservation of human life is to be assured. In the meantime, let's proceed with the easily understood and quickly attainable goal of CTB. It can prevent the current situation from becoming much worse.

The concept of a mutual nuclear freeze seems simple enough, but the requirement that there be verification procedures inside the other country poses great difficulties. Like arms reduction, a comprehensive freeze would require extensive and elaborate negotiations involving several teams of negotiators over a period of years. It seems unlikely that it can be negotiated if one side considers itself inferior and, in the crazy mathematics of the arms race, it is highly likely that one side or the other will consider itself inferior.

As an objection to the CTB, it has been argued that ''prooftesting'' of weapons in the stockpile is needed from time to time to assure that they remain operable. Those who offer objections of this sort ignore the fact that any inconvenience we suffer under a CTB would be visited also on our adversaries. In contradiction to this objection, moreover, Norris Bradbury, former director of the Los Alamos Laboratory has stated: ''The assurance of continued operability of stock-piled nuclear weapons has in the past been achieved almost exclusively by nonnuclear testing—by meticulous inspection and disassembly of the components.'' One should be quite clear about this: It is *improvements* in nuclear weapons that require testing and the aim of a CTB is specifically to prevent or impede such improvements.

The benefits of a CTB far outweigh the risks. If we had been able to negotiate a CTB with the USSR in 1963, when it was necessary to settle for a Limited Test Ban, we and the rest of the world would be much better off. We are negotiating today at a higher and more dangerous level. It would be to our great advantage to achieve a CTB now before we proceed to an even higher and still more dangerous level.

Glenn T. Seaborg
Lawrence Berkeley Laboratory
University of California, Berkeley

Some progress in arms limitation was made during the Reagan years. The Intermediate-range Nuclear Forces (INF) Treaty, eliminating American and Soviet land-based intermediate missiles (300–3300 miles in range) was signed at a Gorbachev–Reagan summit in Washington at the end of 1987. The Strategic Arms Reduction Talks (START I), placing limits on deployed ballistic missiles (ICBMs and SLBMs) and heavy bombers as well as long-range, sea-launched cruise missiles (SLCMs), with the subsequent cooperation of the Bush and Clinton administrations, finally went into force at the end of 1994. START II, specifying substantial reductions in such missiles and the prohibition of multiple warheads (MIRVs) on land-based missiles, has not yet (1997) been

"The concept of a mutual nuclear freeze seems simple enough, but the requirement that there be verification procedures inside the other country poses great difficulties."

approved. Despite this, because of the importance I place on attainment of a CTB treaty, I am convinced that the Reagan administration's policies on issues such as nuclear arms control and the institution of silly secrecy rules were not in the best interests of our national security. (And, I still find it difficult to understand how the American people were nearly sold on the Strategic Defense Initiative, which was never more than a science fiction fantasy on which much money was wasted.) Nonetheless, I believe his administration should be given deserved credit for recognizing that the declining standards of education in our country were a quite genuine threat to the future security of our nation.

On June 28, 1990, I heard President Reagan give an inspirational but somewhat rambling talk in Chicago to the attendees at the Salute to Excellence weekend of the American Academy of Achievement:

He emphasized that the people should tell the government what to do, and he exhorted the kids in the audience to go out with the view that we, the people, are the boss; and he urged everyone to vote along these lines. In the question-and-answer period he was asked what he thought about Gorbachev, and he gave a long answer, essentially praising him for the new Russian attitude toward nuclear war (he said nuclear war cannot be won), crediting the recognition of the horrors of nuclear war as coming from the Chernobyl accident. He was also asked whether he took credit for the fall of communism, and he said he did. In response to another question, he blamed the American family for the present attitudes of young people. In a question about Bush's change of mind about a tax increase (Bush is now supporting it), he made a long statement to explain that Bush was forced to do this in order to compromise with the Democrats about balancing the budget. As a closing statement, he said he was repeating what he advocates in his speeches nowadays: presidential line-item veto power, a constitutional amendment forbidding the running of a national budget deficit, and removal of the two-term limit for the presidency.

As a side note, I might mention that during the Reagan administration I became active in a campaign for a single six-year term for the presidency. The arguments for this constitutional change are, I think, persuasive. Perhaps the most dramatic of these is the budgetary process. The first budget for an incoming president is that of his predecessor; then he has his own budget during the next two years, after which the last budget goes to his successor for automatic revision, whether it is needed or not. In our present highly competitive, high-technology society, we cannot operate efficiently under the old system of the four-year presidential terms when such a large fraction of the president's time is taken up with involving himself on the political front in order to ensure his reelection. We need a system in which the president can devote full attention to the problems that our country needs to solve in today's complicated society.

I found President Reagan very personable and articulate. He was quick in repartee, which was not always clearly relevant. The Reagan Presidential Library and Museum in Simi Valley, CA, instituted in 1994 a quarterly newsletter entitled *Tomorrow.* I am pleased that Reagan also served as co-chairman for the dinner honoring me on my 80th birthday at the Palace Hotel in San Francisco on Nov. 9, 1992.

The high point of my contacts with President George Herbert

Walker Bush came when I was called to Washington to brief him on "cold fusion." Other contacts included the several visits he made to talk to the 40 finalists in Westinghouse Electric Corporation's Science Talent Search (STS). I made an unsuccessful attempt to persuade him to support and advocate a comprehensive test ban (CTB).

I first got to know George Bush during the Nixon administration, when he was serving as U.S. ambassador to the United Nations. I recall that on Sunday, Jan. 31, 1971, he was among the passengers on a chartered Pan American plane on which Helen and I flew to Cape Kennedy to witness the launching of Apollo 14. I met him on numerous occasions after I left Washington. For example, on Saturday, July 24, 1976, while he was serving as director of the Central Intelligence Agency, I had an interesting visit with him during the Summer Encampment of the Bohemian Club at the Bohemian Grove in northern California.

Science Talent Search finalists with Vice President George H. W. Bush and Seaborg, Indian Treaty Room, Executive Office Building. Washington, D.C. March 1, 1982. Courtesy of Science Service, Inc., Washington, D.C. Copyright, Westinghouse Photographic Unit.

On March 1, 1982, I walked from the Mayflower Hotel to the Executive Office Building with the 40 STS finalists. We went to the Treaty Room, where Vice President Bush greeted us. He made a fine talk about the importance of science and young people participating in it. He greeted me during his talk. Afterward, pictures were taken. I stood next to Vice President Bush in one of them and mentioned my acquaintance with Joanne Hall, who was his sister's roommate in college.

Helen and I met Vice President Bush and his wife Barbara at a reception hosted by Swedish Prime Minister Ingvar and Ingrid Carlsson at the Swedish embassy on Sept. 10, 1987:

We went by the reception line— Swedish Ambassador Wilhelm and Ulla Wachtmeister, Prime Minister Ingvar and Ingrid Carlsson, and U.S. Vice President George and Barbara Bush. Carlsson told me he had had successful meetings today with the House and Senate Foreign Relations Committees. He discussed the CTB with them and found a friendly reception.

After supper there was a welcoming talk by Ambassador Wachtmeister (he told us that George Bush and Bjørn Borg had defeated him and Vitas Gerulaitis at tennis this morning). Prime Min-

ister Carlsson spoke next (he said he has had a very productive visit to the USA, and he and Ingrid look forward to visiting Northwestern University, their old school, tomorrow). We were then entertained by Birgitta Svenden, an opera singer from the Royal Theater of Stockholm. Vice President Bush spoke, expressing satisfaction with the fine relations between the United States and Sweden and recalling with pleasure his and Barbara's fine visit to Sweden in 1983. Ambassador Wachtmeister then brought the program to an end.

I talked further with Prime Minister Carlsson. He again told me about his productive meetings with the House and Senate Foreign Relations Committees today. I told him (as I learned today from Aaron Tovish of Parliamentarians for World Order) that the European Parliament has endorsed the amendment approach to a CTB; this was news to him. I mentioned that the West German Social Democratic party has also endorsed this approach, and he said he knew about this. I also mentioned the favorable attitude of the central party in Denmark, and he told me the elections in Denmark were held just the other day. He said he will keep in touch and let me know about the conclusions that he and his advisers reach on the amendment approach.

Early in his administration as president, Bush came to talk to the STS finalists at the National Academy of Sciences (NAS) in Washington, D.C., on Mar. 3, 1989:

President Bush arrived promptly on schedule at 2:09 p.m. at the secondary C Street entrance. John Marous (chairman, Westinghouse) and I, followed by Paul Lego (president, Westinghouse), Frank Press (president, NAS), and Eileen Massaro (director of corporate relations, Westinghouse), led President Bush into the Great Hall, where we conducted a tour of the exhibits of the STS finalists. He took notes on several of the exhibits, including that of Scott Schiamberg. Marous described most of the exhibits and I added comments on a number of them, including the one on "Chaos from Simplicity" by Michael Stern. I told him about the importance of chaos in recent years and tried to explain to him what this means. On one occasion I said (on the basis of a predetermination that I would do so) that the project was so complicated that I didn't understand it and that we would probably need the student to explain it to us.

Following our tour, we went into the auditorium where the finalists were seated on the stage. The auditorium was nearly full. Lego, Press, Marous, and I entered in that order and stood in our designated places on the stage. Then President Bush entered and stood in the center between Press and Marous.

The program began with about two minutes of comments by Press, in which he made references to NAS, the STS finalists, and the importance of the work. (He didn't mention Science Service.) This was followed by about two minutes of comments by Marous, who mentioned the importance of the STS, the role of Westinghouse, and so forth. (He didn't mention Science Service, either.) Marous then introduced the president, who spoke for about 5–10 minutes, using text typed on cards. He talked about the importance of work on stopping the depletion of the ozone layer and mentioned the plan to proceed with construction of the Superconducting Super Collider. He began with references to the tour he had of the exhibits and, using the notes he had made during the tour, commented on several of the exhibits, saying that he and his wife Barbara would discuss this at bedtime in order to pursue the matter further. At the end of his talk he was thanked by Marous. President Bush turned to the STS finalists and was told that Scott Schiamberg was the person who had prepared the exhibit on the "Preparation and Behavior of Inhibitors of Sialidases: Research on a Potential

Swedish Embassy reception. Washington, D.C. Sept. 10, 1987, From left to right: Bush, Ingrid Carlsson, Ulla Wachtmeister, Prime Minister Ingvar Carlsson, Barbara Bush, and Swedish Ambassador to the United States Wilhelm Wachtmeister. Courtesy of the National Archives and Records Administration, Washington, D.C.

To Dr. Glenn Seaborg
With best wishes,

Geo Bush

48th annual Science Talent Search held at the National Academy of Sciences. Washington, D.C. March 3, 1989. From left to right: President George H. W. Bush, Frank Press (President, National Academy of Sciences), Paul Lego (President, Westinghouse Corporation), and Seaborg. Courtesy of Science Service, Inc. Washington, D.C. Copyright, Westinghouse Photographic Unit.

Service is a nonprofit corporation, of which I was chairman (1966–95), that administers the STS and the International Science and Engineering Fair and publishes *Science News* (a weekly news magazine of science).

In April 1989 I was called back to Washington to brief President Bush on cold fusion, the totally unexpected phenomenon discovered by University of Utah scientists. A couple of days earlier, the purported codiscoverer of cold fusion, University of Utah electrochemist Stanley Pons, had spoken to an enthusiastic standing-room-only audience of chemists at the semiannual meeting of the American Chemical Society (ACS) in Dallas, TX. His talk had attracted so much attention that, apparently, the news had reached the White House. After briefing White House Chief of Staff John Sununu, I went into the Oval Office to brief President Bush on April 14, 1989:

Cancer Therapy" to which he had referred. President Bush thereupon gave Schiamberg the first page of his speaking cards after he had signed it with "Good wishes, George Bush."

As he was leaving, President Bush shook hands with and spoke to each of us on stage. He addressed me as "Glenn" and expressed pleasure at having the chance to talk to me again. He left at about 2:35 p.m., exactly on schedule, and we greeted him again as he left through the C Street entrance.

My reference to Science Service deserves an explanation. Science

Department of Energy (DOE) Secretary James Watkins, DOE Undersecretary John Tuck, and DOE Director of Office of Energy Research Robert Hunter and I then rode in a DOE car to the White House for an 11:15 a.m. appointment with John Sununu, White House Chief of Staff, and a number of his aides—Andy Card (deputy to Sununu), Ed Rogers (one of Sununu's aides), and David Bates (secretary to President Bush's cabinet).

We met for about 20 minutes in Sununu's office, during which I briefed him and his aides on the situation with respect to the announcement of the cold fusion discovered by the University of Utah scientists. I said that some scientists, including Steven E. Jones (physicist, Brigham Young University), have observed a low level of neutron emission, during the electrolysis of deuterium oxide—something at the level of only one neutron per second—and that about 10^{10} as much corresponding heat energy had been observed by the University of Utah scientists (led by Pons), who have some rather exotic explanations in terms of aneutronic (no emis-

A Chemist in the White House: From the Manhattan Project to the End of the Cold War

sion of neutrons) reactions. I indicated that there is a good deal of doubt about the validity of both observations. I said that there had been a tremendous reception in favor of this work at the ACS meeting in Dallas earlier this week. (The meeting was attended by about 7000 chemists, including Clayton Callis, ACS president.) I said that, on the other hand, a number of nuclear physicists take a very dim view of this development and, in effect, do not believe it. I indicated that my own view is one of some skepticism, but I believe it has to be investigated, and I have recommended the creation of a panel to look into it. Watkins indicated that the appointment of such a panel is under way.

Sununu was very alert and, because of his engineering background, seemed to understand what I was saying. He asked me a number of penetrating questions, which I believe handled fairly well. Most of the conversation was between Sununu and me.

I mentioned to Sununu that I was the codiscoverer of the radioactive iodine that was used a couple of days ago to treat Barbara Bush, which interested him very much. He said that he would see whether the president was available to see us and left us to go down the hall to check on this. He came back very soon and said that the president would be glad to see us, and we started off down the hall to the Oval Office to meet the president. We noticed that Department of Energy Secretary Watkins wasn't accompanying us (apparently he felt modestly that he shouldn't be included), but Sununu went back to get him.

We then went into the Oval Office, where President Bush greeted me cordially, addressing me again as "Glenn," and pictures were taken of Bush and me and of Bush, Watkins and me. We all sat down to start our discussion, which lasted about 10 minutes. I told him about my role in the discovery of the radioactive iodine that had been used to treat Barbara and said that a similar treatment with radioactive iodine had effected a miraculous cure for my mother, who was suffering from the same condition as Barbara. The president facetiously said that Barbara is now radioactive and is not allowed to kiss their dog as long as this condition prevails, but he implied that it didn't

seem that this prohibition included himself. (In May 1991, President Bush benefited from treatment with the same radioactive iodine [iodine-131].) I then described briefly the situation with respect to cold fusion. I indicated that this is something that has to be viewed dispassionately; there are some indications that it is not a valid observation (i.e., not due to nuclear fusion) but, on

Seaborg briefs President Bush on the "cold fusion" phenomena in White House Oval Office. Washington, D.C. April 14, 1989. Courtesy of the Bush Presidential Library.

the other hand, it must be looked into, and I mentioned the creation of a panel that will do so.

The president seemed very interested and convinced by my assessment, and he encouraged us to go ahead with an investigation of this intriguing situation. We also touched on the problem of precollege science and math education, and he indicated that this is something that also must receive his attention and that of the country.

I might add that the panel I recommended to study the purported cold fusion process was created and, about six months later, published a report disputing the validity of the observation, pretty much in line with the view I had adopted in my briefing with the president.

As cochairmen of "Fifty Years with Nuclear Fission"—a conference held in the Washington, DC, area, April 25–28, 1989, Emilio Segré and I wrote to President Bush on March 14, 1989, to invite him to be a speaker.

Unfortunately, President Bush did not find it possible to accept our invitation.

I tried, but failed, to convince Vice President Bush to support negotiations for a CTB treaty, as illustrated by my exchange of correspondence with him on this subject:

June 6, 1988
The Honorable George Bush
Vice President of the United States
The Capitol
Washington, D.C. 20510
Sir:

I am writing to you at the suggestion of Donald P. Gregg, with whom I met on April 13th. Mr. Gregg urged me to share with you the reasons why I strongly support a comprehensive test ban treaty.

The supreme tasks of statesmanship for the next president are to prevent nuclear war from occurring during his term and to create conditions that will make nuclear war less likely thereafter. A comprehensive test ban can be a most effective way of bringing about these results.

Seaborg's and Emilio Segrè's invitation to President Bush to attend the Fifty Years with Nuclear Fission" conference. March 14, 1989.

Nuclear war by deliberate action of either superpower is not now an imminent danger. This is because we have now, as we have had for nearly forty years, a situation in which neither side can strike first without fear of unacceptable retaliation. What can threaten this situation, known as strategic stability, is the introduction of new weapons that increase the danger to, or threaten to impair the effectiveness of, either side's deterrent force. The side that feels so threatened might be tempted to let fly in the irrational atmosphere that can prevail in a crisis. Conversely, the side that feels it may have the advantage might do so.

The preeminent value of a CTB is that it would prevent the development

April 27, 1989
The National Institute of Standards and Technology
Plenary: 50 Years of Fission, Science and Technology,
at the NIST headquarters in Gaithersburg, Maryland.
[This would also be an historically appropriate forum
since President Franklin D. Roosevelt delegated
responsibility in 1939 for an investigation into
possible uses of atomic energy released from nuclear
fission to the "Uranium Committee," which was chaired
by Dr. Lyman J. Briggs, director of the National
Bureau of Standards. As you know, in 1988, the NBS
was renamed the National Institute of Standards and
Technology and was given the responsibility of helping
to lead U.S. industry toward a strong competitive
position in the world marketplace. An ideal time on
April 27th would be at 9 a.m. at the opening of the
conference there.]

April 27, 1989
At the "Pioneer's Banquet" at the Gaithersburg
Marriott Hotel at 7 p.m. DOE Secretary Admiral James
Watkins and NRC Chairman Lando Zech have also been
asked to attend and make a few remarks.

We sincerely hope that you will honor us with your
presence at this golden anniversary celebration.

Respectfully yours,

Glenn T. Seaborg

Emilio Segrè

P.S. Thank you for your support
of the Science Talent Search
#38

of weapons sufficiently innovative to upset strategic stability. Even though some of these weapons have emerged onto scientists' drawing boards, their full development and deployment can yet be prevented if nuclear testing is terminated.

It so happens that we are at a time in the history of the arms race when some particularly destabilizing new weapons are looming on the horizon. One example is the class of depressed trajectory weapons that can reach their targets with much less warning time. This development would place a premium on rapid response, perhaps leading to a launch-on-warning strategy that could increase the possibility of nuclear war occurring by accident or through misinterpretation of information. Another new class of weapons that can be particularly destabilizing are those that depend not on explosive impact but on the propagation from space of electromagnetic energy that can cause severe effects on earth. Knowledge that the opponent is acquiring such weapons can greatly reduce the confidence of either side in its deterrent and lead to unpredictable behavior under crisis conditions.

Military planners argue that by continued testing they can enhance the effectiveness of our deterrent—e.g., make our weapons more accurate, more destructive, more rapid, etc. Soviet tests can do the same for their deterrent. When the sum of our efforts and theirs are weighed in the balance the effect will be that the heightened deterrent effects achieved by both will be outweighed by the increased mutual fears of preemptive nuclear strikes. One cannot fault military planners on either side for advancing such arguments. They are exercising their responsibility for making the respective deterrents as effective as possible. It is the responsibility of political leaders, however, to view the issue in its larger and more essential terms. In short, political leaders must judge whether, as military planners assert, the security of the United States requires that we continue to develop new types of nuclear weapons or whether such development lessens the security of the United States by making nuclear war more likely. If by making destruction of Soviet targets more certain and complete our weapons enhance the likelihood of nuclear war, we will not have served our country's interests or those of civilization— quite the reverse. Even if the U.S. were first to achieve a new military capability, our security would be impaired, on balance, if the Soviets then acquired the same capability. This has been the story of the last forty years. We have achieved gain after gain in power—the H-bomb and MIRVed missiles are perhaps the two most gigantic advances— yet our security has diminished because the Soviets have followed relatively quickly with the same increments in power. The result, as Averell Harriman said, is that each time we negotiate it is at a higher level of danger. The hope apparently nurtured by some that we could gain a decisive advantage by dint

"The lesson that our political leaders must learn is that it is not necessary to improve our weapons in order to field a survivable and effective deterrent."

of superior scientific and technical prowess has again and again proved illusory.

The lesson that our political leaders must learn is that it is not necessary to improve our weapons in order to field a survivable and effective deterrent. The fact is that we already have such a deterrent and it will remain so unless by massive amounts of testing of new weapons the Soviets find a way to overwhelm or neutralize it. A prime benefit to the United States from a comprehensive test ban is that it would prevent such Soviet testing.

A CTB can also act to prevent nuclear war from another source, namely, the proliferation of nuclear weapons capability to additional countries. Such proliferation is held in check now by a somewhat shaky nonproliferation regime, of which the 1968 Nonproliferation Treaty is the central pillar. In this treaty the non-nuclear weapon parties insisted, as a price for their participation, on a pledge by the superpowers (Article VI) "to pursue negotiations in good faith on effective measures relating to cessation of the nuclear arms race at an early date." The non-nuclears also insisted on a means of redress if progress on Article VI was not forthcoming or was too slow. This was the origin of the provision limiting the life of the NPT to twenty-five years—the superpowers wanted it to be of unlimited duration—and of the provision for review conferences every five years "to assure that the purposes and provisions of the Treaty are being realized." Both before and since the signing of the NPT, the non-nuclear parties have made it abundantly clear that what they had in mind in Article VI was a comprehensive test ban. They have not in the least been placated by SALT, START, or other negotiations held since 1968. Discord over the absence of CTB negotiations has been the central feature of each of the NPT Review Conferences held thus far. Many consider that the NPT's survival would be threatened if a CTB is not in force by 1995, when a vote on the treaty's renewal must be taken.

The Nonproliferation Treaty is not the only arena in which the international community is expressing its demand for a CTB. Last fall the UN General Assembly adopted, 128 to 3, a resolution calling on the non-nuclear

weapon parties to the Partial Test Ban Treaty to activate the amendment procedure of the Treaty "with a view to convening a conference at the earliest possible date to consider amendments to the treaty that would convert it into a comprehensive nuclear test ban treaty." If an amendment proposal gains support by one-third of the parties, the treaty requires that an amendment conference must be held. It is true that any amendment voted by such a conference must be approved by all three "original parties"—U.S., U.K., USSR. It remains to be seen whether a future U.S. administration would wish to stand alone in vetoing the expressed determination of the international community. (The Soviet Union has indicated it would support the amendment.) We can anticipate that strong public pressures against such an isolated position would be released in this country, especially in the Congress.

In its continued insistence on a CTB, the international community has conspicuously turned a cold shoulder toward the timid, step-by-step process agreed to by the superpowers last year. It is abundantly clear that that process cannot result in significant testing limitations for many years, during which time significant, perhaps lethal damage can be done by the development of dangerous new weapons. As the Six Nation Initiative leaders said in their 1987 Stockholm Declaration, "Any agreement that leaves room for further testing would not be acceptable."

The Reagan administration has consistently opposed a CTB except as an "ultimate objective," offering several reasons why the United States must continue testing. The most basic of these purposes is to "[continue] our weapons modernization program." I have dealt with the need for weapons modernization above and will only repeat here my belief that the development of new nuclear weapons reduces rather than enhances U.S. security.

A second reason offered as to why we must continue testing is "to detect deterioration or other problems that may occur with stockpiled weapons." There is no question that so-called "proof testing" can serve this purpose. There is intense controversy in the scientific community, however, as to whether such testing is the only means to accomplish this objective, a respectable body of opinion maintaining that disassembly, remanufacture, and non-nuclear testing of components are sufficient. I can testify that during the ten years when I was chairman of the Atomic Energy Commission (1961–71) no proof tests of the type in question were performed. It is also clear that any comprehensive test ban that prevents such proof testing on our side will also prevent it on the Soviet side, and that any doubts they may incur as to the reliability of their strategic forces may be all to the good.

Without engaging in protracted argument, I would contend that other purposes the administration has advanced as reasons why it is necessary to continue testing, namely, to improve the safety and security features of nuclear weapons, or to improve the survivability of command and control equipment against nuclear attack, can either be achieved without nuclear testing or are not of an order of priority to justify such testing.

There is some controversy among those desiring an end to all testing as to whether this should be accomplished all in one gulp or whether tests should continue to be allowed below the threshold—say one kiloton—where verification is believed to be uncertain. My own strong preference is for a fully comprehensive treaty at the outset on the following grounds:

1. It is far simpler to verify whether a test has been conducted at all than whether it exceeds a particular threshold.
2. The military advantages of clandestine testing at very low levels are so minor compared to the political risks of being discovered in violation of a treaty that, particularly, in its new mood, the Soviet Union is not likely to undertake such clandestine tests. If it were to do so, the consequences to U.S. security would be minor in relation to the benefits of a CTB.

In essence, the issue boils down to a single question: would the United States (and, incidentally, the survival of civilization as we have known it) be more secure with a comprehensive test ban than without one? The evidence, overwhelmingly, is for an affirmative answer.

Sincerely yours,
Glenn T. Seaborg

President Bush responded. I then wrote him again:

August 25, 1988
The Honorable George Bush
Vice President of the United States
The Capitol
Washington, D.C. 20510
Sir:

I appreciate your taking time during a busy season to send your reply of July 27th to my letter advocating a comprehensive test ban. I am taking the liberty of writing you again on this matter because I regard it as being of crucial importance.

Let me begin by acknowledging that there would be risks in having a CTB. I do not agree that these are as serious as the administration has portrayed them to be, but I do not wish to belabor those aspects. The essential point I wish to make is that there would also be risks in *not* having a CTB. This is something that the administration has never acknowledged. Let me just list some of these:

1. *An escalating arms race.* If both sides can test, each will develop new and more dangerous weapons, requiring the other side to react with countermeasures, requiring measures to counteract the countermeasures, etc. This is the costly cycle we have been pursuing since the nuclear age began—more and more weapons, ruinous costs, and, despite all the expenditures and weaponry on both sides, less security for both.
2. *Nuclear war by inadvertence.* If both sides can test, each will develop weapons that threaten the survivability or effectiveness of the other's deterrent. Under such circumstances one or both may adopt a launch-on-warning strategy during times of crisis. Launch-on-warning carries with it the grave risk of false warning.
3. *New nuclear weapons countries.* Non-nuclear nations have long regarded negotiation of a CTB by the superpowers as a key test of

THE VICE PRESIDENT
WASHINGTON

July 27, 1988

Dr. Glenn T. Seaborg
Lawrence Berkeley Laboratory
1 Cyclotron Road
Berkeley, CA 94720

Dear Doctor Seaborg:

Thank you for your thoughtful letter concerning a Comprehensive Test Ban (CTB) treaty. I wholeheartedly agree with you that the major tasks for the next President are to prevent nuclear war and continue to foster conditions that will make nuclear war less likely. The arms control agenda, including issues of nuclear test limitations pursued by the Reagan-Bush administration, has been aimed at achieving precisely those ends.

Careful consideration has been given to all major areas of arms control as basic elements of a comprehensive foreign policy and national security framework. In this context, the full range of possible approaches to nuclear test limitations were examined. We concluded that as long as we must rely on nuclear weapons for deterrence, we must continue to test them as we have each year they have been a part of our deterrent. A CTB can only be achieved in the context of a very significantly reduced U.S. dependence on nuclear deterrence to ensure western security and international stability.

The Administration concluded after exhaustive analysis that we should work toward such nuclear test limitations in a deliberate step-by-step process. As a first step the highest priority has been given to improved verification procedures for the Threshold Test Ban Treaty and the Peaceful Nuclear Explosions Treaty. Admittedly this approach has not yielded dramatic results in the short term, but it has been fruitful. The two sides have gained a better understanding of their respective views and by agreement to conduct measurements of tests at each other's test sites -- a development unimaginable a few years ago. These steps should lead to an agreement on adequate verification methods and put us in a better position to address further test limitations.

I do not believe a CTB would in itself help to reduce the number of nuclear weapons. It could be argued that a CTB could make reductions in nuclear weapons even more difficult. Without testing there could be a growing loss of confidence in the reliability of existing nuclear weapons leading to pressure to add additional weapons to the stockpiles.

Vice President George Bush's response to Seaborg regarding his lack of support for a comprehensive test ban treaty. July 27, 1988.

whether the latter mean to end their nuclear arms race. The absence of such negotiations can threaten the Nonproliferation Treaty when it comes up for review in 1995. It can also make nuclear-capable nations, several of whom have not signed the NPT, more pessimistic about their own security, and induce them to take the few remaining steps across the nuclear weapons threshold. What we come down to, then, is a matter of relative risks. I submit to you, urgently, that *the risks of not having a CTB are far greater than the risks of having one.*

It behooves every statesman to contemplate, prayerfully, not just one side, but both sides of this equation.

I can well understand that it might not be possible for you to do an about-face on this issue under the pressures of a presidential campaign. But should you become president I fervently trust that you will give renewed and earnest thought to ending the qualitative race in nuclear arms—a race far more deadly than the quantitative race—by agreeing to a simultaneous comprehensive nuclear test ban.

One area that causes me great concern is the problem of verification. In your letter you imply that clandestine testing might occur under a CTB and that the political risks of discovery would be so high that the Soviets would be unlikely to cheat. Certainly risk of detection has not deterred the Soviet Union from committing serious violations of several nuclear arms control agreements. Nations do what is in their interest and will often test the limits of agreement and verification. Although the costs might be high, the benefits from clandestine testing, even at low levels could be significant. The benefits from one or two of these tests, especially at very low yields, might well be worth the cost to the Soviets. The consequences to our security could be great indeed.

I believe the long-term risk associated with reaching a CTB <u>now</u>, rather than at a time when all of the necessary groundwork for its success and the protection of our security has been laid, would more than offset any near-term gain. In sum, I must disagree with your conclusion that the United States should move now to negotiate a CTB. While we hope that a CTB will be an appropriate measure in the future and continues to be an ultimate goal, the time has not yet come. Nuclear war has been avoided for over forty years with a strong nuclear deterrent backing up our conventional capability and without a CTB. I am yet to be convinced that free world security and international stability would be enhanced by such a measure at this time.

Sincerely,

George Bush

Let me compliment you an your excellent acceptance speech in New Orleans.

Cordially yours,
Glenn T. Seaborg

I didn't receive a reply to my second letter to Bush.

However, President Bush was active in some other arms control measures. In July 1991 he and Soviet President Mikhail Gorbachev signed START I in Moscow. The START I Treaty reduces the nuclear arsenals of each country to 6000 accountable warheads within the seven-year implementation period. A delay in ratification occurred in December 1991, when the Soviet Union ceased to exist as a geopolitical entity and the Commonwealth of Independent States (CIS), consisting of 12 new independent states (including four with nuclear weapons on their territories) was created. Ratification finally went into force at the end of 1994. In June 1992 President Bush and Russian President Boris Yeltsin signed a Joint Understanding estab-

Seaborg gives a guided tour to Vice President J. Danforth Quayle and Marilyn Quayle (center) at the annual Science Talent Search held at the National Academy of Sciences. Washington, D.C. March 3, 1990. Courtesy of Science Service, Inc. Washington, D.C. Copyright, Westinghouse Photographic Unit.

lishing the framework for START II. Overall, the concomitant end to the cold war led to a much more optimistic outlook for advances in arms control.

I first met Vice President J. Danforth (Dan) Quayle and his wife Marilyn on March 3, 1990, when I showed them the exhibits of the STS finalists at NAS:

I then went to the entrance into the back room area (where President Bush entered last year), and we formed a receiving line for the impending visit of Vice President Quayle. The receiving line consisted of Lego (incoming chairman, Westinghouse), Massaro (vice president, public relations, Westinghouse), Ted Sherburne (president, Science Service) and Sam Thier (head, Research Institute of Medicine). When Vice President Quayle and his wife Marilyn arrived, individual pictures were taken as they passed [through] the receiving line, and then Lego and I escorted them into the Great Hall to view the STS exhibits. We made a tour. Marilyn Quayle was particularly interested in Exhibit 24, which had to do with breast cancer; she was scheduled to participate in the making of a movie on this subject immediately following their visit. During the tour pictures were

taken by Westinghouse photographers and many other photographers. I tried to engage Vice President Quayle in conversation but without much success. I did, however, talk a good deal with Marilyn. I told her something about the STS, the plans for a 50th anniversary celebration next year, when the 2000 STS finalists will return, the geographic diversity of the present participants, and so forth.

Following the tour Vice President Quayle walked up to a microphone on a little stage. Lego was supposed to introduce him, but the confusion made this impossible. Vice President Quayle gave a short talk, describing from notes the exhibits that he had seen and extolling the virtues of the STS finalists. He and Mrs. Quayle then left through the same area of the building where they had entered.

I had the impression that Marilyn had a much better understanding of the science exhibits than did Vice President Quayle.

As early as his first State of the Union address in 1989, President Bush announced his National Education Goals (primarily for precollege education). One result of this action was the president's famous Education Summit, held in September 1989 with the state governors in Charlottesville, VA. Then, on Oct. 8–10, 1989, Watkins and I served as cochairmen of the Math/Science Education Action Conference at the Lawrence Hall of Science in Berkeley, CA. The conference, sponsored by DOE, the Lawrence Hall of Science, and the Lawrence Berkeley Laboratory, was attended by nearly 250 scientists, educators, business executives, and government leaders. The objective was to develop a concrete plan of action for restructuring and revitalizing precollege mathematics and science education. Proposals, ideas, and recommendations from the conference have resulted in many constructive actions, especially by DOE. But much more needs to be done!

President Bush addressed the STS finalists and a large audience of returning STS finalists and their friends at the 50th anniversary celebration at the Washington Hilton Hotel on March 4, 1991:

At a little before 5 p.m. I went back to our room, where Helen and I donned our formal clothes for the evening's banquet. We then went down to attend the VIP reception in the Crystal Ballroom. At about 6 p.m. the STS finalists, along with Lego, Richard Gott (Princeton physicist, chairman of judges for the STS), Sherburne, Carol Luczsz (director of youth programs, Science Service), Massaro, and other Westinghouse officials went to a special room to wait for the arrival of President Bush. He arrived at 7 p.m. and went by a receiving line, consisting of Mr. and Mrs. Lego, me, Sherburne, and Gott. After greeting a number of the STS finalists, the president was brought by to visit with the finalists standing by their exhibits—Judson Berkey (Thomas Jefferson High School for Science and Technology, Alexandria, VA), Susan Criss (Fox Chapel Area High School, Pittsburgh, PA), Tara Bahna-James (La Guardia High School of Music and the Arts, NY), and Yves Jeanty (Stuyvesant High School, NY). The first visit was with the Berkey exhibit, "The Optimal Launch

Angle of a Baseball," and I took the occasion to comment to the president on his appearance in the college world series in 1947 as a member of the Yale team, when our Berkeley team beat them for the collegiate championship. He said that, unfortunately, he recalls this all too vividly! President Bush and his entourage left first and went to an area backstage in the International Ballroom. Then we all went to the ballroom, and the finalists took their places on the stage. Helen and I sat at a table with Mr. and Mrs. Lego, Mr. and Mrs. Sherburne, Gott, and Barbara Franklin (Westinghouse executive).

President Bush entered with Lego to the playing of "Hail to the Chief" and was introduced by Lego. In his opening salutation the president referred to me as "my old friend, Dr. Glenn Seaborg." He paid tribute to the contribution of the STS in identifying future scientists and engineers—so important in our society—and also made reference to the success of our armed forces in the war in the Persian Gulf. He left to a standing ovation. Lego then joined us at our table.

Left to right: Seaborg, Admiral James Watkins (Secretary of Energy), Vice Admiral Richard H. Truly (U.S. Navy, retired, and Administrator, National Aeronautics and Space Administration), University of California President David P. Gardner at Watkins/Seaborg Math/Science Education Action Conference held at the Lawrence Hall of Science, University of California–Berkeley. Oct. 9, 1989. Courtesy of the Lawrence Berkeley National Laboratory, Berkeley, CA.

President George Bush and Seaborg at the 50th anniversary of Science Talent Search at the Washington Hilton Hotel's International Ballroom. Washington, D.C. March 4, 1991. Courtesy of Science Service, Inc. Washington, D.C. Copyright, Westinghouse Photographic Unit.

President George Bush awards the National Medal of Science to Seaborg in a ceremony at the White House Rose Garden. Washington, D.C. Sept. 16, 1991. Courtesy of the Bush Presidential Library.

To Glenn Seaborg
With best wishes,

George Bush

The last time I saw President Bush in his role as president was when he presented me with the National Medal of Science in a Rose Garden ceremony at the White House on Sept. 16, 1991:

Helen and I left our hotel about 9:30 a.m. and walked to the east entrance of the White House grounds where we joined (our son) Eric (Pete, our oldest son, came a little later) and the recipients of the National Medal of Science and the National Medal of Technology, their guests, implementing staff of the National Science Foundation (NSF) and the White House, and others. From here we went to the Rose Garden, where we talked to a number of the National Medal of Science recipients and their guests, including Elvin Kabat (Columbia University), Ronald Breslow (Columbia University), Guy Stever (former director, NSF; science adviser to President Gerald Ford), Arthur Schawlow (Stanford University), Steven Weinberg (University of Texas), and Robert Kates (Brown University). The national medal winners then went to their assigned seats and the others took seats in the designated area in the Rose Garden.

President Bush entered at precisely 10:30 a.m. and introduced Robert A. Mosbacher (secretary of commerce), D. Allan Bromley (assistant to the president for science and technology), Walter E. Massey (director, NSF), and others. In his introductory remarks (about 10 minutes long), President Bush emphasized the importance of science and technology in today's society and made special mention of several medal winners. He joked that, as a result of his efforts to learn to operate computers during the past six months, he has cut down by a factor of 5 the time he takes to make mistakes.

Massey read the citations as each recipient came up to receive the National Medal of Science from the president, who made a few remarks individually to each person, and then pictures were taken. Bromley also shook hands with the recipients. When I received the medal, the president remarked that he was surprised I hadn't received all the honors of this type years ago. I mentioned that he had met my son Eric in the Oval Office in June

at the time of the hike across the United States, and he professed to recall the occasion.

I am pleased that President Bush wrote me a gracious letter of congratulations for the occasion of the dinner celebrating my 80th birthday at the Palace Hotel in San Francisco on Nov. 9, 1992.

I saw Bush most recently on July 24, 1993, at the Bohemian Grove in Northern California:

I then had a chance to talk with Bush. He immediately recalled that he

President Bush's letter congratulating Seaborg on his 80th birthday. Oct. 13, 1992.

THE WHITE HOUSE
WASHINGTON

October 13, 1992

Dear Dr. Seaborg:

I am delighted to send congratulations as your friends and admirers gather to celebrate your 80th birthday. This event not only marks an important milestone in your own life, it also honors your devotion to education, science, and the betterment of life for your fellow citizens and for people throughout the world.

Your accomplishments -- including the Nobel Prize in Chemistry, Chairmanship of the Atomic Energy Commission, Chancellorship of the University of California at Berkeley, and, most recently, Chairmanship of the Lawrence Hall of Science -- comprise a unique record of dedication to scientific excellence and service to America. These achievements, however, represent just some of the highlights of your distinguished career. As a Professor of Chemistry, you have offered guidance and leadership to over two generations of scientists. This is indeed one of the finest legacies that any individual can leave.

While many would be content to rest on such laurels, you continue to be involved in activities relating to science education. As you work to develop and adapt new technologies to benefit our society, you remain an important force in the effort to build a strong and economically healthy America, for today and for future generations, as well.

Barbara joins me in sending best wishes for a memorable celebration and for continuing success in all of your endeavors.

Sincerely,

G. Bush

Dr. Glenn T. Seaborg
Lawrence Berkeley Laboratory
Berkeley, California 94720

had been the speaker at events for the STS. When I reminded him that I had briefed him on cold fusion, he said he recalled this, but he said, jokingly, "I didn't understand a word of what you said." He went on to say, "You're the greatest."

I believe my many contacts with George Bush, extending over more than 20 years during his service as UN ambassador, vice president, and president, and during his post-presidential years, have been amicable and pleasant. I regret that I was unable to convince him to support a CTB treaty.

chapter 11 WILLIAM JEFFERSON CLINTON

1993–1997

RENEWED HOPE FOR A COMPREHENSIVE TEST BAN

Soon after his election as president, I sent William Jefferson Clinton

whom I had not yet met, a copy of my monograph, "National Service with Ten Presidents of the United States." I was very pleased to receive his gracious acknowledgment.

In my role as chairman of Science Service, I joined my colleagues in inviting President Clinton to meet with the 40 finalists in the 1993 Science Talent Search (STS). As a result, he invited us to come to the White House to meet with him on Thursday, March 4, 1993:

Thursday, March 4, 1993— Washington, DC

At about 3:15 p.m., I rode in a bus to the White House with the STS finalists as well as Judge and Chairperson Richard Gott, Science Service President Fred McLaren, and John Yasinsky (president, Westinghouse Power Systems and Environment), who had flown down this afternoon from Pittsburgh through the rainstorm. We went to a room in the East Wing, where the group was lined up on a sort of steps to have a picture taken with President Clinton. I was in the center of the back row at the top of

**Office of the President-elect
and Vice President-elect**

January 9, 1993

Glenn Seaborg
Chairman
Science Service
1719 N Street, NW
Washington, DC 20036

Dear Glenn:

Thank you so much for the monograph "National Service with Ten Presidents of the United States." Your thoughtfulness and your encouragement mean a lot to me.

Sincerely,

Bill Clinton

Bill Clinton

BC/dis

1120 Vermont Avenue, N.W., Washington, DC 20270-0001   202-973-2600

President-elect William J. Clinton's letter thanking Seaborg for sending him a copy of Seaborg's booklet, ''National Service with Ten Presidents of the United States.'' Jan. 9, 1993.

the center of the back row at the top of the picture. When President Clinton came in and took his place, he spoke to the press and TV cameras, describing the occasion as the Westinghouse Science Talent Search and saying that the future of our country depends on people like these STS finalists. John (Jack) Gibbons (Presidential Science Adviser) called out that Glenn Seaborg was present, who had worked with and had had his picture taken with the previous nine presidents. President Clinton turned around to shake hands with me and said he was honored to meet me. Pictures, both still and videotape, were taken by a large number of newsmen and White House photographers.

Then the president went over to the side of the room so that each STS finalist and the rest of us could come by and shake his hand while individual pictures were taken. All of the others just went by to have their picture taken in quick order. When I came by, he talked to me for several minutes. I mentioned that I had worked with each of the presidents, beginning with Franklin Roosevelt, and he again said he was very honored to meet me. I also said I had heard that at a news conference he was asked about cold fusion. He said this was the case and that he gave a noncommittal response because he wasn't too sure of his facts. I told him he should be very skeptical of the claims for cold fusion. A number of newsmen and some of the STS finalists took pictures of me with Clinton.

I have known Vice President Albert Gore's father, Albert Arnold Gore, since early 1961, when he began service on the Joint Committee on Atomic Energy (JCAE) and I started as chairman of the Atomic Energy Commission. (Young Albert had just reached his 13th birthday.)

Helen and I met Albert Arnold and Pauline Gore on numerous occasions, notably the luncheon they gave for President-elect Guillermo Leon Valencia of Colombia at Marwood (the country estate of Col. Grady Gore) in Potomac, MD, on June 24, 1962, at which I believe young Albert was present.

I recall meeting Albert and Tipper Gore at the ''Salute to Excellence'' weekend of the American Academy of Achievement:

Friday, July 1, 1988—Nashville, TN
We then rode by bus to the Cumberland River, where we boarded the

President Bill Clinton greets Seaborg on the occasion of the Westinghouse Science Talent Search finalists' visit to the White House, March 4, 1993. Courtesy of Ernst H. ''Kass'' Kastning III, Radford, VA.

Congratulations and best wishes, Bill Clinton

Seaborg and President Bill Clinton with the 1993 Science Talent Search finalists in the White House. March 4, 1993. Courtesy of Sharon Farmer, the White House.

General Jackson showboat for a boat ride during dinner. During the evening Helen and I talked with Tennessee Senator Albert Gore and his wife, Tipper.

After dinner there was entertainment by a country music group, which included imitations of such famous country music singers as Jimmy Rogers, Eddy Arnold, and Tennessee Ernie Ford. This was followed by the evening program, for which Gen. Richard F. Abel served as master of ceremonies. I was among those introduced by Abel. The speakers were Senator Albert Gore, Jr., Johnny Cash (country music singer), and Herschel Walker (professional football star). After this program we rode back to the hotel in a bus, then walked to our room with Sol and Toni Linowitz and Mr. and Mrs. David Jones (former chairman, Joint Chiefs of Staff, 1978–82). I met Walker in the hall near our room and talked with him for awhile.

I also met Vice President Albert Gore on my visit to the White House on March 4, 1993:

I talked to a number of White House assistants, including Ann McCoy and Beth Pritchard. When McCoy learned that I would like to see Al Gore, she got in touch with his office and

Gore invited me to come over immediately. I was escorted to the West Wing of the White House to the vice president's working office and immediately brought in to see him, along with a photographer. I told him I had met the previous 10 or so vice presidents and the previous 9 presidents of the United States and had been a friend of his father when his father was a member of JCAE. He said that he would pass on my greetings to his father, who he said is in fine shape now at the age of 86. A number of pictures of Vice President Gore and me were taken by a White House photographer.

During 1994–1995 I served as Honorary Chairman and active member of an international *Special Panel on the Protection and Management of Plutonium,* operating under the sponsorship of the American Nuclear Society. Our Panel issued a report *Protection and Management of Plutonium* recommending the reactor irradiation option for disposition of surplus U.S. and Russian weapons plutonium.

On January 24, 1996, I wrote to President Clinton to express my deep concern over the disposition of excess weapons-grade plutonium throughout the world. Since the Soviet Union dissolved, I have considered this issue to be critical for ensuring global security—a view not always shared by government leaders. I enclosed the special panel report of the American Nuclear Society on this subject, *Protection and Management of Plutonium*. My letter to the President follows:

January 24, 1996
President Bill Clinton
The White House
1600 Pennsylvania Avenue, NW
Washington, DC 20500
Dear Mr. President:

I write as a lifelong Democrat, and the co-discoverer of plutonium for which I was awarded the Nobel Prize in 1951. I am perturbed that the United States does not have a proper policy for handling this useful, but dangerous, substance. This is in spite of two committees of distinguished people who have during the last two years made recommendations about it.

The first was a committee of the National Academy of Sciences on "Management and Disposition of Excess Weapons Plutonium." They concluded that the existence of excess "weapons grade" material in stockpile, particularly in Russia, poses a "Clear and Present Danger" to the United States because it can all too easily be diverted to weapons uses.

I had the honor to chair the second committee, convened by the American Nuclear Society, which included distinguished people who had served in both Democratic and Republican administrations and members from overseas. The report had contributions from all members and was unanimous. The committee had a broader mandate to consider future civilian uses also.

The most urgent issue is the disposition of weapons plutonium. The rest of the world expects the United States and Russia to show good faith not only in reducing the stockpile of weapons pointed at each other (which threat has, as you pointed out in the State of the Union address, disappeared—at least for awhile) but actually to render the plutonium unusable. The National

Academy Committee considered two options: the first one, mixing the plutonium with radioactive waste to render it difficult to use for the next hundred years, and also to bury it. The second one is to burn it in one of the many electricity-producing reactors in our country or elsewhere.

Our ANS committee much preferred the second option because it is irreversible; much of the plutonium disposed of in this way is destroyed and the remainder is degraded to a point that makes it permanently unattractive for weapons use. This is also the proper approach to follow for the much larger quantities of plutonium from civil power reactors now being accumulated in many countries, in both purified and unpurified form.

In view of the urgency in actual threat to the United States and in the public perception of third world countries, I urge you to look carefully at these reports and to initiate action as soon as possible. I enclose a copy of the ANS report. I note in the Preface that production of plutonium was shown to be feasible at the end of 1942 and only two years later we had plutonium being produced in usable quantities. We

Seaborg with Vice President Al Gore in the vice president's office in the White House during a visit of the 1993 Science Talent Search finalists to the White House. March 4, 1993. Courtesy of Callie Shell, The White House.

should be able to burn our plutonium on a reasonable time scale if we set our minds to it.

I finish with congratulations on your fine State Of The Union speech last night.

Yours sincerely,
Glenn T. Seaborg
Nobel Laureate, 1951
Chairman, Atomic Energy Commission, 1961–1971
enc: ANS special panel report: *Protection and Management of Plutonium*

On April 12, 1996, President Clinton replied to the issues I raised in January. Meanwhile, I had participated with others who share my concerns in further urging the President to act boldly during his summit with President Yeltsin to discuss this subject on April 19–20, 1996. I served as a member (1995–1997) of the Senior Technical Review Group of the Amarillo National Resource Center for Plutonium, which advised the DOE (and the president) on the safe disposition of weapons–grade plutonium. We also strongly urged that the major part of the country's weapons plutonium be burned in electricity–producing reactors. When this option was adopted as a path for the disposition of excess weapons plutonium, I wrote then–DOE Secretary Hazel O'Leary on January 8, 1997 to applaud her decision, which has since become U.S. policy.

In the Spring of 1996 I was asked to endorse a petition urging President Clinton and Congressional leaders to continue the federal government's support of fundamental scientific research. Funding for basic research had been seriously threatened by the

President Clinton's response to Seaborg indicating that the subject of handling and disposal of weapons-grade plutonium would be discussed at the upcoming Nuclear Safety Summit in Moscow. April 12, 1996.

THE WHITE HOUSE
WASHINGTON

April 12, 1996

ЯЗЯ 4/17/96

Dr. Glenn T. Seaborg
Associates Director-at-Large
Ernest Orlando Berkeley
    National Laboratory
Building 70A, Room 3307
One Cyclotron Road
Berkeley, California 94720

Dear Glenn:

Thank you very much for writing and for sending the booklet regarding the management and disposition of excess weapons-grade plutonium. I appreciate having your perspective on this important matter, and I agree that it is vitally important for the United States and Russia to ensure that fissile material removed from dismantled nuclear weapons is disposed of safely, permanently, and in a fully transparent manner.

The safety, security, and ultimate disposition of surplus nuclear weapons material stockpiles in the United States and former Soviet Union have been among my Administration's top priorities, and President Yeltsin and I have declared our nations' joint resolve to actively cooperate to address the dangers posed by such stockpiles. At the Nuclear Safety Summit in Moscow, we will be taking action to move this agenda forward. Your efforts and those of the National Academy of Sciences have contributed greatly to the public debate and aided my Administration's work on this issue.

Although many challenges remain, I am pleased with our and Russia's joint commitment to improving the safety and security of nuclear materials and to assessing the technology options for the disposition of surplus weapons material. In addition to these ongoing measures, my Administration is reviewing the environmental and economic aspects of plutonium disposition to determine which method, or combination of methods, best serves the interests of the United States.

Be assured that I remain committed to the safe management and disposal of surplus nuclear resources to eliminate the possibility of their use in the production of weapons. I am confident that your continued involvement and the capabilities of our scientific, technical, and industrial sectors will enable us to meet this vital challenge.

Sincerely,

Bill Clinton

cost-cutting zeal of the Congress. Joining me in signing the petition were 59 Nobel Laureates from across the United States. The American Chemical Society (ACS) sponsored the effort, which was part of a larger endeavor to encourage the government to continue its commitment to basic research. The June, 1996 letter was sent to the President, Vice-President Gore and every member of Congress:

19 June 1996
The Honorable William J. Clinton
President of the United States
Washington, D.C. 20500
Dear Mr. President:

As men and women who have helped to shape the modern scientific age and who care deeply about the future of our nation, we urge you to reaffirm the fundamental role of the federal government in supporting basic scientific research.

Americans have been awarded more than one-half of all Nobel Prizes in physics, chemistry and medicine since 1945. This impressive success is no accident, but the result of a firm and consistent commitment by the federal government to basic science research at our universities. Our nation's policymakers and public have been prudent investors because their support has paid off in tremendous ways.

America's investment in research over the last fifty years has been a vital source of our economic and political strength around the world, as well as the quality of life Americans enjoy at home. The polio vaccine, computers, jet propulsion and disease resistant grains and vegetables are some of the thousands of advances pioneered at our universities that have had dramatic benefits for our health, economy, security and quality of life.

New and equally breathtaking advances may be just around the corner. Genetic research, for example, gives promise of better treatments for Alzheimer's, cancer and other diseases. Lighter and stronger composite materials may be developed with important applications in transportation, medicine and the military. Continuing support for university-based research will not only pave the way for these important breakthroughs, but will also train the next generation of pioneers and Nobelists.

The engine of scientific innovation and discovery cannot fuel itself. Our own achievements and the benefits they have brought would not have been possible without the government's 'patient' capital. Discoveries are rarely made instantaneously, but result from years of painstaking work by scientists in a variety of fields. With competition forcing industry to focus research investments on returns over the shorter term, the government is left with the crucial role of making the longer term investment in discovery.

America's future prosperity will depend on a continued commitment to producing new ideas and knowledge, and the people educated to apply them successfully. They will be central to our economic opportunity in the face of intense global competition, to our protection against renewed threats to our security and environment, and to ensuring the health of Americans. Federal funding for university-based research is an investment in our future that should be maintained.

Sincerely,
[Signers of the petition]

President Clinton's reply of September 13 was a stirring affirmation of our principles, and fully supported our position that basic research is the cornerstone of America's future scientific and technological—and therefore, economic—leadership in the world.

As I have indicated, START I went into force in December 1994. This was the result of complicated, successful conferences and negotiations by the Clinton administration with Russian President Boris Yeltsin and the other nuclear weapons members of the Commonwealth of Independent States. The success of START I enables the United States and Russia to advance the START II ratification process. The START II Treaty, signed in January 1993, would reduce the strategic nuclear

THE WHITE HOUSE
WASHINGTON

September 13, 1996

Glenn T. Seaborg, Ph.D.
Associate Director-at-Large
Lawrence Berkeley Laboratory
Nuclear Science Division, MS80A-3307
University of California
Berkeley, California 94720

Dear Dr. Seaborg:

Thank you very much for your letter regarding our national commitment to basic scientific research. You and your fellow American Nobel Prize winners, relying on talent, hard work, and our public investment in research and education, have made enormous contributions to this nation's scientific and technological leadership. I pledge to continue the research and education investments needed to preserve that leadership and build upon it for the benefit of future generations.

You have eloquently stressed the importance of the government's "patient" capital for providing the scientific and technological foundation of our continuing prosperity, security, health, and quality of life. My Administration remains dedicated to nurturing science in America. This requires both maintaining federal investments in research and education and creating the economic climate that fuels private-sector investment, as well as private-public partnerships, in science and technology. Our commitment is and has been to sustain priority investments in education, the environment, research, and technology while making the difficult choices needed to continue our success in deficit reduction and to reach a balanced budget.

Your letter emphasized the special importance of maintaining federal funding for university-based research. I share that view. America's research colleges and universities are the bedrock of American leadership in science and technology. As the Nobel Prize awards themselves demonstrate, our universities are without peer in producing new knowledge, stimulating innovation, and training the scientists and engineers who will do so much to shape our future. Research at America's colleges, universities, and medical schools is and will remain a priority for the Administration.

President Clinton's reply to Seaborg affirming the government's support of fundamental science and university-based research programs. September 13, 1996.

arsenals of each country to 3000–3500 warheads by the year 2003 (if not sooner). President Clinton and Boris Yeltsin met at a summit in Helsinki, Finland in March, 1997 during which they agreed on guidelines for negotiations of a START III treaty to limit the number of nuclear warheads kept by Russia and the United States at 2,000 to 2,500 weapons each; the total reductions to be completed by the year 2007.

Because of pressure from Congress, President Bush acceded in 1992 to the initiation of a morato-rium on nuclear weapons testing. Russia and Great Britain, and later France and China agreed to participate in the moratorium. President Clinton encouraged the nuclear weapons testing moratorium while actively pursuing the attainment of a comprehensive test ban treaty. On September 11, 1996 the Comprehensive Test Ban Treaty (CTBT) was endorsed by a majority of countries comprising the United Nations General Assembly. Thirteen days later, first President Clinton, then representatives of the four other world nuclear powers,

As you know, there have been calls for sharp cutbacks in the investments necessary to sustain America's capacity for leadership in the twenty-first century. We need a national debate about the commitments we must make for our future. Responsible voices must be heard. Given your deep concern about the need to maintain U.S. leadership in science and technology and higher education, and your unique stature, I urge you to examine and help clarify the public policy alternatives facing our nation. The American people deserve the benefit of your analysis and perspective.

We look to you and your colleagues to help promote opportunity for all Americans in a new age of scientific discovery and advancement.

Sincerely,

Bill Clinton

---

signed the CTBT, thereby fulfilling a momentous aspiration that began 33 years ago with President Kennedy and the signers of the Limited Test Ban Treaty. Although significant obstacles remain to ratification (notably India's refusal to endorse the treaty), it is likely the CTBT will be reviewed in three years and, if the remaining countries possessing nuclear reactors have not yet signed, other means may be found to bring the treaty into force.

After signing the treaty, President Clinton addressed the U. N. General Assembly. In a passionate speech emphasizing his enduring commitment to global security, he called for the control of all weapons capable of massive destruction, whether nuclear, chemical, or biological. Only seven months later, he presided over the U.S. ratification of the United National Chemical Weapons Convention, which came into force on April 29, 1997.

Throughout his presidency, Bill Clinton has shown admirable leadership in the effort to curtail the developement and use of all catastrophic weapons.

CONCLUSION

I have had the privilege of serving the last 10 presidents of the United States—Franklin D. Roosevelt, Harry S. Truman,

Dwight D. Eisenhower, John F. Kennedy, Lyndon B. Johnson, Richard M. Nixon, Gerald R. Ford, Jimmy Carter, Ronald W. Reagan, and George H.W. Bush. An equal number (five) represented each political party; for the Democratic party—Roosevelt, Truman, Kennedy, Johnson, and Carter; for the Republican party—Eisenhower, Nixon, Ford, Reagan, and Bush. I have also met President Bill Clinton (and Herbert Hoover) and thus nearly one-third of all presidents of the United States.

I knew the last 13 men who served as vice president—Henry A. Wallace, Harry S. Truman, Alben W. Barkley, Richard M. Nixon, Lyndon B. Johnson, Hubert H. Humphrey, Spiro T. Agnew, Gerald R. Ford, Nelson A. Rockefeller, Walter F. Mondale, George H.W. Bush, J. Danforth Quayle, and Albert Gore. Five of these men went on to serve as president.

I have also met all of the first ladies at times before, during, and/or after the presidential terms of their husbands—Eleanor Roosevelt, Bess Truman, Mamie Eisenhower, Jacqueline Kennedy, Lady Bird Johnson, Patricia Nixon, Betty Ford, Rosalynn Carter, Nancy Reagan, and Barbara Bush. Each brought a unique and inimitable charm to the White House, and Helen and I often recall many of the pleasant social occasions we enjoyed with the first ladies and their husbands.

I am pleased that my knowledge and experience as a scientist (and, perhaps, as an administrator) placed me in a position to provide national service to these presidents of the United States. I am also pleased that I have been able, much of this time, to be an active researcher, teacher, or university administrator.

I am proud of my contributions

as an ambassador of goodwill in my visits to some 60 countries, most of them during the time I served as chairman of the Atomic Energy Commission (AEC).

The enthusiasm engendered by the U.S. Atoms for Peace Program led in 1955 to the convening in Geneva of the huge UN Conference on the Peaceful Uses of Atomic Energy. The success of this conference led to a second one held in 1958, a third in 1964, and a fourth in 1971. At the first two Geneva conferences I was a member; at the third, chairman of the U.S. delegation. I had the honor of being elected president of the fourth conference. Another repeated occasion for travel abroad was the International Atomic Energy Agency's (IAEA) General Conference. During my 10 and one half years as AEC chairman, I, along with one or more of my fellow commissioners, attended this annual event 11 times. The conference was held in Vienna each year except in 1965, when it was held in Tokyo.

It became my practice to visit other countries before and after the various conferences I attended. Thus, in 1965, when the IAEA General Conference was held in Tokyo, I visited nine countries in a trip around the world. A presidential plane was placed at my disposal for three of my trips: in January 1967, when I circled the globe in visiting five countries; in January 1970, for a trip to six African countries, Spain, and Germany; and in July 1967, when I visited six South American countries. One highlight of my travels abroad occurred in September 1964. Leaving the third Geneva conference for a weekend, I served as host to high-ranking officials of 15 national nuclear energy organizations aboard the NS *Savannah*, the world's first nuclear-powered cargo–passenger ship. The *Savannah*, which had started operation in August 1962, was complet-

ing a tour of the Scandinavian countries and was at anchor in Halsingborg, Sweden. My guests and I spent the night aboard ship, then cruised the Baltic the next day.

Throughout the 1960s fruitful cooperation on peaceful uses of the atom was enjoyed with the USSR. This was accomplished pursuant to several bilateral Memoranda on Cooperation in the Field of Utilization of Atomic Energy for Peaceful Purposes negotiated between the U.S. AEC and the Soviet State Committee for the Utilization of Atomic Energy. The first of these was signed in 1959 by AEC Chairman John A. McCone and his Soviet counterpart, Professor Vasil Emelyanov. I and my counterpart, Andronik M. Petrosyants, signed succeeding memoranda in May 1963, July 1968, and early 1970.

One of the fruits of the Memoranda of Cooperation was exchanges of visits by American and Soviet scientists to laboratories and facilities in one another's countries. A notable exchange of visits occurred in 1963. In May I led an American delegation on a tour of Soviet nuclear energy facilities. Everywhere we went, we were treated with the warmest hospitality. Our hosts accepted unhesitatingly the itinerary we had proposed and even included some additional sites they thought would interest us. Our journey achieved a number of "firsts." We were the first foreign group to visit the Soviet reactor testing station at Ulyanovsk and the site of the high-energy accelerator at Serpukhov, the first Western visitors since World War II to visit the Radium Institute in Leningrad, and the first foreign group to see certain industrial reactors and other scientific equipment. Overall, I believe this visit contributed to the improved relations that made possible the negotiations, some two months later, of the Limited Test Ban Treaty.

The first Soviet–American experiment in the nuclear sciences began in 1970. Pursuant to the fourth Memorandum on Cooperation, six U.S. physicists were assigned for six months to the High Energy Physics Institute at Serpukhov, working with Soviet scientists at the 70-BeV (billion electron volts) accelerator. In return, Soviet scientists were to be assigned to the 200-BeV accelerator at Weston, IL, when it would be completed.

Another exchange of scientist visits led by Chairman Petrosyants and me took place in 1971. The Soviet group visited nuclear facilities throughout the United States from April 15 to April 28. Our return tour took place between August 4 and August 20. Following visits to laboratories in the Moscow area, an extensive 10-day tour by our party utilized a specialized Aeroflot plane used by Premier Kosygin on some of his trips. Traveling a distance of 12,110 kilometers, we visited nuclear facilities in and around eight cities: Minsk, Leningrad, Ulyanovsk, Novosibirsk, Tashkent, Erevan, Tbilisi, and Schevchenko, with a stop at Samarkand. I also attended meetings and visited research laboratories in Moscow after our tour.

On entering the Soviet Union at this time, I had the newly acquired and rarely bestowed status of foreign member of the USSR Academy of Sciences. This honor had been conferred on me during the academy's general assembly in March.

These trips involved extended separations from my family, disruptions of normal eating and sleeping habits, exhausting schedules at nearly every stop, intensive in-flight "homework" to prepare for the next visit, a host of minor frustrations and inconveniences, and (on return) a mountain of accumulated work. But the rewards were great. I am convinced that my personal discussions with scientists and statesmen of other nations, and visits to their scientific facilities, contributed significantly to the constructive use of the peaceful atom and nuclear safeguards as well as better international relations. It was gratifying to know that President Johnson, for one, in repeatedly urging me to take such trips, felt the same way.

During my travels I met a rather large number of heads of state or high government officials: President Juan Carlos Ongania (Argentina); Chancellors Josef Klaus and Alfors Gorbach and State Secretary Karl Gruber (Austria); Foreign Minister José de Magahaes Pinto (Brazil); Foreign Minister Mitchell Sharp (Canada); Premier Zhou Enlai (China); Emperor Haile Selassie and Crown Prince Asfa-Wossen Haile Selassie (Ethiopia); President Urho Kekkonen (Finland); Prime Minister Kofi A. Busia (Ghana); Prime Minister Harold Macmillan (Great Britain); Prime Minister Indira Gandhi (India); President Suharto (Indonesia); Prime Minister Amir Abbas Hoveyda (Iran); President Eamon De Valera (Ireland); Prime Minister Levi Eshkol (Israel); Vice President Daniel arap Moi (Kenya); Foreign Minister Antonio Carrillo Flores (Mexico); Foreign Minister Mohamed Syilnassi (Morocco); Prime Minister Petrus J. S. deJong (The Netherlands); President Ayub Khan (Pakistan); President Nicolae Ceausescu (Rumania); President Park Chung Hee (South Korea); Chairman Nikita S. Khrushchev, President Leonid I. Brezhnev, Foreign Minister Andrei A. Gromyko, and V.M. Molotov (Soviet Union); Foreign Minister Gregorio Lopez Bravo and Prince Juan Carlos and Princess Sofia (Spain); Prime Ministers Tage Erlander, Olof Palme, Ingvar Carlsson, and Carl Bildt, and Kings Gustaf VI and Carl XVI Gustaf (Sweden); Presidents Chiang Kai-shek and Teng-Hui Lee and Premiers C.K. Yen and Huan

Lee (Taiwan); Prime Minister Kittikachorn Thanan (Thailand); Foreign Minister Habib Bourguiba (Tunis); Vice President Aleksandar Rankovic (Yugoslavia); and UN Secretary Generals Trygve Lie, Dag Hammarskjold, U. Thant, Kurt Waldheim, and Javier Perez de Cuellar.

The trips were not without some personal "spin-off"—the Danube at Budapest on a clear September day, Roman paving-stones on the Appian Way, the Bibi Khanym Mosque in Samarkand, Inca ruins in Peru, the Great Buddha at Kamakura, the Temple of Bacchus at Baalbek, the Acropolis in Athens, the ruins of Carthage, the house where Beethoven composed "Fidelio," the mighty Congo 2000 feet below me winding through green jungle toward a dam construction site, canals in Venice, the charm of exotic animals in Australia, sunset over Scotland's downs, the Great Wall of China—kaleidoscopic contacts with nature and the history of man.

I would like to conclude with some comments in a more personal vein. There were a number of instances in which diplomacy was required to avoid disaster in our congressional hearings. I recall a humorous incident that occurred in April 1970, when I testified in defense of our FY 1971 budget before the Subcommittee on Public Works of the Senate Appropriations Committee. During a discussion that involved plutonium, Chairman Allen J. Ellender (senator from Louisiana) asked me derisively, "What do you know about plutonium?" In order not to embarrass him and to fend off his threatened large cuts in the AEC budget, I answered in a way that camouflaged my knowledge of plutonium and made no reference to my role in its discovery. As a follow-up we persuaded Senator

Ellender to give a talk to an enthusiasic crowd in the packed auditorium of our Germantown headquarters. His final action on our budget was reasonable.

Sometimes I had to represent the Executive Branch of our government under trying circumstances when a show of humor helped. I am recently reminded of a story that occurred during my tenure as AEC chairman. Cuts to the AEC budget had forced the agency to begin laying off some machinists at the Oak Ridge complex in Tennessee. I was called by the Joint Committee on Atomic Energy to testify before Congress to explain these cuts. I recall clearly that committee member Senator Al Gore (D-Tennessee) asked me, "What do you have against machinists?" I responded by saying, "I don't have anything against machinists. In fact, my father was a machinist. My grandfather was a machinist. And my great-grandfather was a machinist." After a brief pause, I continued and said, "And if I had any talent for it, I would've been a machinist." At this remark, everyone, including Senator Gore, broke into laughter and the hearing essentially ended.

I have found my opportunities for national service to be very interesting and instructive. The office of the president of the United States holds many pressures, and I feel privileged to have personally known so many of the men who have held that office and to have observed and learned from the characteristics and values that each brought to his position. My relations with each president were good and, in some cases, very close. Helen and I will not forget the merit and dedication of some of these men, whose courage and conviction sustained them in the performance of their immense responsibilities.

ACRONYMS

AAAS	American Association for the Advancement of Science
ABM	Antiballistic Missile
ACDA	Arms Control and Disarmament Agency
ACS	American Chemical Society
AEC	Atomic Energy Commission
AICBM	Anti Intercontinental Ballistic Missile
AID	Agency for International Development
AIF	Atomic Industrial Forum
AUA	Argonne Universities Association
BOB	Bureau of Budget
BOQ	Bachelor Officers' Quarters
CASE	Committee on Academic Science and Engineering
CCR	Commission on Civil Rights
CERN	European Organization for Nuclear Research
CHEM	Study Chemical Education Material Study
CIA	Central Intelligence Agency
CIRUS	Canadian-Indian Reactor Uranium System
CIS	Commonwealth of Independent States
CONAES	Committee on Nuclear and Alternative Energy Systems
COSPUP	Committee on Science and Public Policy
CP-1	Chicago Pike-1
CTB	Comprehensive Test Ban
CTBT	Comprehensive Test Ban Treaty
DOD	Department of Defense
DOE	Department of Energy
EBR 1	Experimental Breeder Reactor No. 1
ENDC	Eighteen Nation Disarmament Conference
ERC	Emergency Relocation Center
ERDA	Energy Research and Development Agency
EURATOM	European Atomic Energy Community
EXCOM	Executive Committee of the National Security Council
FCC	Federal Communications Commission
FCST	Federal Council on Science and Technology
FSWP	Armed Forces Special Weapons Project
GAC	General Advisory Committee
GSA	General Services Administration
HEW	Department of Health, Education, and Welfare
HFIR	High-Flux Isotope Reactor
HILAC	High-Ion Linear Accelerator
HUD	Housing and Urban Development
IAEA	International Atomic Energy Agency
ICBM	Intercontinental Ballistic Missile
ICO	Inter-Agency Committee on Oceanography
INF	Intermediate-range Nuclear Forces
JCAE	Joint Committee on Atomic Energy
KAPL	Knowls Atomic Power Laboratory
KBI	Kawecki Berylco Industries

LAMPF	Los Alamos Meson Physics Facility
LMFBR	Limited Metal Fast Breeder Reactor
LRL	Lawrence Radiation Laboratory
LTBT	Limited Test Ban Treaty
MAD	Maintenance and Disassembly Building
MATS	Military Air Transportation Service
MIRV	Multiple Independent Reentry Vehicle
MLC	Military Liason Committee
MLF	Multilateral Force
MSFC	Manned Spacecraft Flight Center
MTR	Materials Testing Reactor
MURA	Midwestern University Research Association
NAD	National Academy of Science
NASA	National Aeronautics and Space Administration
NASC	National Aeronautics and Space Council
NATO	North Atlantic Treaty Organization
NBS	National Bureau of Standards
NCAA	National Collegiate Athletic Administration
NCEE	National Commission on Excellence in Education
NDEA	National Defense Education Act
NERVA	Nuclear Engine for Rocket Vehicle Application
NET	National Educational Television
NIH	National Institutes of Health
NOAA	National Oceanic and Atmospheric Agency
NPT	Nonproliferation Treaty
NRC	Nuclear Regulatory Commission
NDRC	National Defense Research Committee
NRDS	Nuclear Rocket Development Site
NRTS	National Reactor Testing Station
NSAM	National Security Action Memoranda
NSDM	National Security Decision Memoranda
NSF	National Science Foundation
NSSM	National Security Study Memoranda
NUMEC	Nuclear Materials and Equipment Corporation
OAS	Organization of American States
OEP	Office of Emergency Planning
OMB	Office of Management and Budget
ORNL	Oak Ridge National Laboratory
OSRD	Office of Scientific Research and Development
PLUTO	Nuclear Ramjet Missile Propulsion System
PSAC	President's Science Advisory Committee
RTA	Reciprocal Trade Act
Rover	Nuclear Rocket Program
SALT	Strategic Arms Limitation Treaty
SAMOS	Satellite And Missile Observation System
SAT	Scholastic Aptitude Test
SCAE	State Committee on Atomic Energy
SLBM	Sea-Launched Ballistic Missile
SLCM	Sea-Launched Cruise Missile

SNAP	Space Nuclear Auxiliary Power
SST	Supersonic Transport
START	Strategic Arms Reduction Talks
STG	Space Task Group
STS	Science Talent Search
TLA	Tariff-Lowering Act
UCLA	University of California, Los Angeles
UN	United Nations
USIA	U.S. Information Agency
USN	United States Navy
XSSS	Experimental Space Station

SUBJECT INDEX

NOTE: PAGE NUMBERS FOR PHOTOS ARE IN **BOLDFACE**

Business Roundtable, 275

C

Cabinet meetings
 Johnson, 146-**147**
 Kennedy, 90-**91**
Canadian-Indian Reactor Uranium System (CIRUS), 151
Carrier compounds, 8, 9
Carriers, nuclear, 159
Central Intelligence Agency (CIA), 128, 211, 291
Cerium flouride, 4
Charlottesville, Virginia, Bush's Education Summit in, 303
Chattanooga, Tennessee, U.S. Junior Chamber of Commerce banquet in, 200-201, 207
Chemical Education Material Study (CHEM Study), 66, 272
Chemists
 German, 1
 ultramicro-chemists, 10
Chernobyl nuclear accident, 118
Chile, NPT negotiations with, 152
China. *See* People's Republic of China
Christmas Island, U.S. atmospheric testing on, 122-126, 127
Civil Rights Bill, 1964, 188
Civil rights, fair housing and, 188-189
Civil Service Commission, 204
Civilian Nuclear Power Program, 158, 174
Clinton Engineer Works/Clinton Laboratories (Site X), 9, 11
Clinton, Tennessee, 9
Cold fusion, 291, 294-296, 306, 309
Cold war, 23, 53
Columbia River, Washington, 10, 11
Commission on Civil Rights (CCR), 188
Committee on Academic Science and Engineering (CASE), 216
Committee on Nuclear and Alternative Energy Systems (CONAES), 262
Committee of Principals, 82, 120, 165-166, 223
Committee on Science and Public Policy (COSPUP), 216-217
Committee of Scientific Society Presidents, 253, 254
Commonwealth of Independent States (CIS), 301, 313
Communication Satellite Policy, 89-90
Comprehensive Test Ban (CTB), 131, 138
 Bush administration, 291, 292, 296-301, 306
 Carter administration, 262, 264, 268, 269
 Clinton administration, 314-315

Nixon administration, 211, 212, 213
 Reagan administration, 271, 287, 288, 290
Copper, 4
Coprecipitation phenomena, 8
Cuba, NPT negotiations with, 152
Cuban Missile Crisis, 128-129, 141
Cyclotron
 bombardments, 3, 6, 8, 10
 invention of, 2
Czechoslovakia, Soviet invasion of, 144, 153

D

Decontamination process, plutonium, 8, 9
Democratic National Convention, 1944, **18**, **19**
DuPont Company (E.I. DuPont de Nemours), 9, 42, 208

E

Education
 Bush's Education Summit, 302-303
 graduate, 53-54, 64-66, 67-70, 72
 National Commission on Excellence in Education (NCEE), 271- 281
Egypt, NPT negotiations with, 151
Eighteen Nation Disarmament Committee (ENDC)
 Johnson administration, 144-145, 148, 152
 Kennedy administration, 124
 Nixon administration, 211, 212, 213, 223
Eisenhower World Affairs Institute, 75
Emergency Relocation Center (ERC), 128, 129
Energy agency article, Seaborg's, 238-239
Energy budget policy, 249, 253
Energy Crisis of 1973-74, 251, 253
Energy Independence Act, 253
Energy Independence Authority, 253
Energy Mobilization Board, 265
Energy Policy and Conservation Act, 253
Energy Reorganization Act of 1974, 251
Energy Research and Development Administration (ERDA), 238, 251, 260
Enrico Fermi Award, 75, 87, **104**-105, 185-**186**, **187**
Equal Employment Opportunity Commision (EEOC), 188
European Atomic Energy Community (EURATOM), IAEA and EURATOM safeguards, 149-150
Executive Committee of the National Security Council (EXCOM), 128

Experimental Breeder Reactor No. 1 (EBR-1), 182

F

Federal Council on Science and Technology (FCST), 60
 committees, 216
 Inter-Agency Committee on Oceanography (ICO), 164, 165
 Kennedy administration, 80, 82, 87-88
 Nixon administration, 215-217
Federal Energy Administration, 260
Federal Power Commission, 102, 103, 260
Federal Radiation Council, 82
Federal support for basic scientific research, 312-313
Fission
 discovery of nuclear fission, 1-2, 54
 "Fifty Years with Nuclear Fission" conference, 296
 fissionable isotopes of uranium, 2, 7
 Johnson administration cutbacks in fissionable materials, 144
 neutron-induced fission of plutonium isotope, 6
 separating fission products, 8
Foreign Aspects of United States Security conference, 56-57
France
 breeder reactor program, 240, 261-262
 EURATOM program in, 161
 LTBT, 130
 moratorium on nuclear weapons testing, 314
 NATO and MLF, 146
 NPT and, 149
 nuclear power program, 138

G

Garfield High School, Los Angeles, 277
General Advisory Committee (GAC). *See* Atomic Energy Commission (AEC)
General Electric Company, 23, 24, 92, 233
Geneva, Switzerland, atomic energy conference in, 54
George Washington University, 73
German Atomic Forum, 149
Germantown, Maryland
 AEC headquarters in, 81, 96, **97**
 new AEC headquarters, Emergency Relocation Center (ERC), 128, 129
Germany
 Berlin Crisis, 86-87
 discovery of nuclear fission in, 1-2, 54

N

Nagasaki bomb, 2, 13, 21
National Academy of Sciences (NAS),
    267
    Committee on Nuclear and Alternative Energy Systems (CONAES),
        262
    Committee on Science and Public
        Policy (COSPUP), 216-217
    Franklin A. Long affair, 222-223
    Kennedy administration, 83, 102,
        103, 112
    Science Talent Search, 292-294, 302
    200 BeV accelerator and, 173
National Accelerator Laboratory, 188-189
National Aeronautics and Space
    Administration (NASA), 162, 216
    Apollo 11, 233
    Carter administration, 260
    Kennedy administration, 98-99, 100
    Manned Spacecraft Flight Center
        (MSFC), 233
    Nixon administration, 216, 221
    See also Space
National Aeronautics and Space Council (NASC)
    Johnson administration, 161-**163**
    Kennedy administration, **82**, 88-90,
        97-101
    Nixon administration, **221**-222
    See also Space
National Association of Secondary
    School Principals, 275
National Bureau of Standards (NBS), 2
National Collegiate Athletic Association
    (NCAA), 201, 250
National Commission on Excellence in
    Education (NCEE), 271-281
National Council on Marine Resources
    and Engineering Development, 82,
    163-165, 210-211
National Defense Education Act
    (NDEA), 64, 65, 66, 67, 201, 202
National Defense Research Committee
    (NDRC), 2
National Education Association (NEA),
    275
National Educational Television
    (NET), 66, 70
National Environmental Policy Act, 242
National Historic Landmark, EBR-1 as
    a, **182**
National Institutes of Health (NIH),
    260
National Medal of Science, **304**-305
National Oceanic and Atmospheric
    Agency (NOAA), 211
National Press Club, 134
National Reactor Testing Station
    (NRTS), 148, 179-182

National Science Board (NSB), 75
National Science Foundation (NSF),
    78, 84, 260, 305
    AEC and, 45
    Franklin A. Long affair, 222-223
    Pell Bill and, 164-165
    PSAC and, 65
National Security Action Memoranda
    (NSAM), 223
National Security Council
    Executive Committee of the
        National Security Council
        (EXCOM), 128
    Kennedy administration, 90-91,
        99-101
    Nixon administration, 211-214, 223
    Vietnam War and Johnson presidency, 190-191, 193-194
National Security Decision Memoranda
    (NSDM), 223
National Security Study Memoranda
    (NSSM), 223
National Trails Act, 277
Nevada Test Site, 127, 160, 224
    Kennedy visit to, 107-109, **110**
New Deal, 16
New Production Reactor (NPR), 79
Nobel laureates, 19, 105-107, 176, 260,
    313
Nonproliferation Treaty (NPT), 54,
    204, 212, 224
    Johnson presidency and the, 143-198
North Atlantic Council, 150
North Atlantic Treaty Organization
    (NATO), 91, 263
    Multilateral Force (MLF), 145, 146,
        147-149
    Nixon administration, 214
N.S. *Savannah*, 159
Nuclear chain reaction
    production problem, 7, 8, 11
    25th anniversary of Fermi, 183-**184**,
        185
Nuclear energy
    AEC tour of Soviet nuclear facilities,
        110-119
    atomic bomb project, 1-16
    Booster bomb, 35-36, 37
    breeder reactors, 104, 161, 199-200,
        204, 205, 207, 209, 239- 242,
        260, 261, 262
    Civilian Nuclear Power Program,
        158, 174
    Experimental Breeder Reactor No. 1
        (EBR-1), 182
    hydrogen bomb, 18-21, 41-44
    IAEA and EURATOM safeguards,
        149-150
    Materials Testing Reactor (MTR),
        217

naval reactors, 158-159
New Production Reactor (NPR), 79
nuclear electric power device on the
    moon, 233-234
nuclear power reactors, 54, 97, 98,
    118
nuclear weapons testing, 121-122,
    126-128
Private Ownership Bill, 158, 174
role of nuclear power, 101-104
salt mine nuclear waste disposal
    problem, 225
Soviet atomic bomb test, 40, 41
Soviet nuclear weapons testing, 87,
    121, 122
super bombs, 41-44
U.S. atmospheric testing, 121-122,
    126-128
water reactor safety, 225
water-cooled reactors, 104
Nuclear Engine for Rocket Vehicle
    Application (NERVA), 109
Nuclear Materials and Equipment Corporation (NUMEC), 225-226
Nuclear Regulatory Commission
    (NRC), 138, 238, 251
Nuclear Rocket Development Site
    (NRDS), 109, **110**
Nuclear test ban treaty
    JFK presidency and, 77-78, 83, 99,
        100, 119-139
    Johnson administration, 157

O

Oak Ridge National Laboratory
    (ORNL), 55, 139, **140**
Oak Ridge, Tennessee, 9, 11
Office for Emergency Management of
    the Executive Office of the President, 2
Office of Emergency Planning (OEP),
    211
Office of Management and Budget
    (OMB), 223
Office of Scientific Research and Development (OSRD), 2, 13
Oman, NPT negotiations with, 152
Operation DOMINIC, 127
Organization of American States
    (OAS), 129, 259
Organization of Petroleum Exporting
    Countries (OPEC), 251
Oxidation states, plutonium, 3, 4, 5-6, 9

P

Pacific nuclear test sites, 127
Pakistan
    atomic energy installations in, 182,
        183
    NPT and, 147, 151, 152
Partial Test Ban Treaty, 287-288

NAME INDEX

NOTE: PAGE NUMBERS FOR PHOTOS ARE IN **BOLDFACE**

Rannes, Jack Sparks, 202, **206**
Ransom, Harry, 195
Rathjens, George W., 59
Read, Benjamin, 137
Reagan, Nancy, 285, **286**
Reagan, Neil and Bess, 232
Reagan, Ronald W., 138
    Nixon and, 202, 203, **206**, 233
    photos, **276, 279, 280, 287**
    presidency, 271-290
Redfield, Robert, 15
Reedy, George, 147, 188, 191
Rehnquist, Richard and Ann, 285, 286
Reichardt, C. H., 170
Reid, Thomas R., 200
Remini, William, 233
Renner, Rudolph, 92, **93**
Reno, Janet, 190
Revelle, Roger R. D., 58, 67
Rhyne, Charles S., 59
Rice, Don, 225
Richardson, Elliot, 211
Rickover, Hyman G., 259
    dinner tribute to, 255-**256**
    Enrico Fermi Award to, 186
    GAC and, 38-39
    Nixon administration, 240, 244
Rickover, Mrs. Hyman G., 255
Riechardt, Charles, 128
Riley, William T., 225, 226
Riordan, L. P., 140
Ripley, Dillon, 210, 233
Robb, Charles and Lynda Bird, 196
Robertson, H. P., 58, 59
Robinson, Dave, 164
Rochester, Nathaniel, 58
Rockefeller, Mr. and Mrs. John D., IV, 187
Rockefeller, Nelson, 66, 75, 202, 253-**255**
Rogers, Ed, 294
Rogers, Jimmy, 310
Rogers, Thomas, 215
Rogers, William, 195
    Nixon administration, 206, 208, 211, 212, 214, 215, 227, 234, 235
Rollefson, Ragnar, **82**
Rometsch, Rudolph, 209
Romney, George W., 228, 250, 255
Romney, Mrs. George W., 228
Romulo, Carlos P., 176
Roosevelt, Eleanor, **16**
Roosevelt, Franklin D.
    ACS and, 252
    atomic bomb and presidency of, 1-16
    photos, **14, 15**
Rosen, Milton W., 221
Roshchin, Aleksey, 138
Ross, Lucille, 26

Rossi, Bruno B., 59
Rostow, Elspeth, 266
Rostow, Walter W., 136, 137, 157, 266
    Johnson administration, 165, 166, 193
Rowan, Carl, 188, 190
Rowan, Steve, 221
Rowe, Hartley, GAC and, 22, 26, 32-34, 36-37, 38, 40, 42, 45
Rowen, Henry S., 220
Rubel, John H., 59, 71, 88
Rubin, Julius, 210, 234, 242, 243
Ruckelshaus, William, 239
Rudnev, K. N., 133, 134
Ruina, Jack, 137
Ruml, Beardsley, 15
Rumor, Mariana, 74
Rumsey, Peter, 215
Rupp, Pat, 275
Rusk, Dean, 266
    Johnson administration, 165-166, 188, 191-192
    Kennedy administration, 78, 88, 119, 120, 123, **125**, 131, 132, 133, **134**, 135
    NPT and, **147**, 148, 153
    space council meeting, 161-**163**
Russell, Anna, 56
Russell, James S., 34, 40
Russell, Richard B., 126
Ryan, Mr. and Mrs. Tom, 232

S
Sadat, Anwar, 273
Sagan, Carl, 262, 264
Salet, Gen., 127
Salinger, Pierre, 109
Salman, Paul, 275
Saltonstall, Leverett, 126, 131, 132, **134**
Salvetti, Carlo, 183
Samuelsson, Bengt, 285
Sarabhai, Vikram, **183**
Saragat, Guiseppi, 183, 184
Sato, Mr. and Mrs. Eisaku, 186-188
Saxby, Sheila, 196-197
Sayre, Rev. Francis B., Jr., 74
Schacht, Linda, 268
Schachter, Leon, 220
Schawlow, Steven, 305
Schiamberg, Scott, 293-294
Schiff, Leonard, 45
Schlatter, G. F., 40
Schlesinger, Arthur, Jr., 78
Schlesinger, James R., **221**, 222, 224, 243, 255
Schlesinger, Mrs. James R., 255
Schlossberg, Caroline Kennedy, 265
Schorr, Burt, **245**
Schreiber, Raemer, 109
Schultz, George, 239

Schultze, Charles, Johnson administration, 164, 165, 171, 172, 173, 184, 191
Schulz, Carl, 215
Schwartz, Monk, 109
Schwarzschild, Julius M., 59
Schweitzer, Glenn, 210, 211
Scoville, Herbert, 58, 59, 60, 71, 165
Seaborg, David (son), 79, **124**, 179, **180, 181**, 245, 254-255
Seaborg, Dianne (daughter), 78, **124, 179, 181**, 245, 260
Seaborg, Eric (son), **179, 181**, 234, 254-255, 277, 305
Seaborg, Glenn (son), **124, 181**
Seaborg, Glenn T.
    accepts AEC chairmanship, 78, 80
    and ACS, 252-253, 313
    AEC chairmanship reappointments, 80, 139, 236, 237
    AEC chairmanship under Johnson, 80, 143, 144, 154-161, 204, 223-224
    AEC chairmanship under Kennedy, 78-141, 223-224
    AEC chairmanship under Nixon, 80, 199, 204-218, 223-227, 236, 237-248
    and AEC delegation to Soviet Union, 110-119
    Bush administration, **292, 294, 295, 302, 303-304**
    "A Call for Educational Reform" article, 277-278
    Carter and, **269**
    "cleansing" of journals and AEC, 271, 281-285
    Clinton administration, **309-310, 311**
    80th birthday celebration, 248, 257, 270, 290, 305
    Eisenhower and, **69**
    energy budget policy, 249, 253
    Enrico Fermi Award to, 75
    "For a U.S. Energy Agency" article, 238-239
    Ford administration, **250, 255**
    as GAC member, 17-49
    Johnson administration, **147, 153, 154, 161, 163, 167, 171, 174, 175, 177, 178, 182, 183, 184, 186, 187, 195, 197, 198**
    journals, 271, 281-285
    Kennedy administration, **82, 88, 89, 91, 93, 94, 96, 97, 98, 99, 104, 105, 107, 108, 110, 111, 112, 113, 121, 122, 123, 134**
    on "Meet the Press", **121-122, 243, 245**
    National Medal of Science to, **304-305**

DEMCO